D0078663

PROFILES, PATHWAYS, AND DREAMS

Autobiographies of Eminent Chemists

Jeffrey I. Seeman, Series Editor

The Right Place at the Right Time

John D. Roberts

American Chemical Society, Washington, DC 1990

Library of Congress Cataloging-in-Publication Data

Roberts, John D., 1918–
The right place at the right time.

(Profiles, pathways, and dreams)
Includes bibliographical references.

1. Roberts, John D., 1918– . 2. Chemists—United States—Biography. 3. Chemistry, Organic—United States—History—20th century.

I. Title. II. Series.

QD22.R6A3 1990 540'.92 [B] 90–259
ISBN 0–8412–1766–1 (cloth)
ISBN 0–8412–1792–0 (pbk.) CIP

The paper used in this publication meets the minimum requirements of American National Standard for Information Sciences—Permanence of Paper for Printed Library Materials, ANSI Z39.48–1984.

∞

PRINTED IN THE UNITED STATES OF AMERICA

Profiles, Pathways, and Dreams

Jeffrey I. Seeman, *Series Editor*

M. Joan Comstock, *Head, ACS Books Department*

Foreword

In 1986, the ACS Books Department accepted for publication a collection of autobiographies of organic chemists, to be published in a single volume. However, the authors were much more prolific than the project's editor, Jeffrey I. Seeman, had anticipated, and under his guidance and encouragement, the project took on a life of its own. The original volume evolved into 22 volumes, and the first volume of *Profiles, Pathways, and Dreams: Autobiographies of Eminent Chemists* was published in 1990. Unlike the original volume, the series was structured to include chemical scientists in all specialties, not just organic chemistry. Our hope is that those who know the authors will be confirmed in their admiration for them, and that those who do not know them will find these eminent scientists a source of inspiration and encouragement, not only in any scientific endeavors, but also in life.

M. Joan Comstock
Head, Books Department
American Chemical Society

Contributors

We thank the following corporations and Herchel Smith for their generous financial support of the series Profiles, Pathways, and Dreams.

Akzo nv

Bachem Inc.

E. I. du Pont de Nemours and Company

Duphar B.V.

Eisai Co., Ltd.

Fujisawa Pharmaceutical Co., Ltd.

Hoechst Celanese Corporation

Imperial Chemical Industries PLC

Kao Corporation

Mitsui Petrochemical Industries, Ltd.

The NutraSweet Company

Organon International B.V.

Pergamon Press PLC

Pfizer Inc.

Philip Morris

Quest International

Sandoz Pharmaceuticals Corporation

Sankyo Company, Ltd.

Schering–Plough Corporation

Shionogi Research Laboratories, Shionogi & Co., Ltd.

Herchel Smith

Suntory Institute for Bioorganic Research

Takeda Chemical Industries, Ltd.

Takasago International Corporation

Unilever Research U.S., Inc.

About the Editor

JEFFREY I. SEEMAN received his B.S. with high honors in 1967 from the Stevens Institute of Technology in Hoboken, New Jersey, and his Ph.D. in organic chemistry in 1971 from the University of California, Berkeley. Following a two-year staff fellowship at the Laboratory of Chemical Physics of the National Institutes of Health in Bethesda, Maryland, he joined the Philip Morris Research Center in Richmond, Virginia, where he is currently a senior scientist and project leader. In 1983–1984, he enjoyed a sabbatical year at the Dyson Perrins Laboratory in Oxford, England, and claims to have visited more than 90% of the castles in England, Wales, and Scotland.

Seeman's 80 published papers include research in the areas of photochemistry, nicotine and tobacco alkaloid chemistry and synthesis, conformational analysis, pyrolysis chemistry, organotransition metal chemistry, the use of cyclodextrins for chiral recognition, and structure–activity relationships in olfaction. He was a plenary lecturer at the Eighth IUPAC Conference on Physical Organic Chemistry held in Tokyo in 1986 and has been an invited lecturer at numerous scientific meetings and universities. Currently, Seeman serves on the Petroleum Research Fund Advisory Board. He continues to count Nero Wolfe and Archie Goodwin among his best friends.

Contents

Photographs

Preface

"HOW DID YOU GET THE IDEA—and the good fortune—to convince 22 world-famous chemists to write their autobiographies?" This question has been asked of me, in these or similar words, frequently over the past several years. I hope to explain in this preface how the project came about, how the contributors were chosen, what the editorial ground rules were, what was the editorial context in which these scientists wrote their stories, and the answers to related issues. Furthermore, several authors specifically requested that the project's boundary conditions be known.

As I was preparing an article[1] for *Chemical Reviews* on the Curtin–Hammett principle, I became interested in the people who did the work and the human side of the scientific developments. I am a chemist, and I also have a deep appreciation of history, especially in the sense of individual accomplishments. Readers' responses to the historical section of that review encouraged me to take an active interest in the history of chemistry. The concept for Profiles, Pathways, and Dreams resulted from that interest.

My goal for Profiles was to document the development of modern organic chemistry by having individual chemists discuss their roles in this development. Authors were not chosen to represent my choice of the world's "best" organic chemists, as one might choose the "baseball all-star team of the century". Such an attempt would be foolish: Even the selection committees for the Nobel prizes do not make their decisions on such a premise.

The selection criteria were numerous. Each individual had to have made seminal contributions to organic chemistry over a multidecade career. (The average age of the authors is over 70!) Profiles would represent scientists born and professionally productive in different countries. (Chemistry in 13 countries is detailed.) Taken together, these individuals were to have conducted research in nearly all subspecialties of organic chemistry. Invitations to contribute were based on solicited advice and on recommendations of chemists from five continents, including nearly all of the contributors. The final assemblage was selected entirely and exclusively by me. Not all who were invited chose to participate, and not all who should have been invited could be asked.

A very detailed four-page document was sent to the contributors, in which they were informed that the objectives of the series were

1. to delineate the overall scientific development of organic chemistry during the past 30–40 years, a period during which this field has dramatically changed and matured;
2. to describe the development of specific areas of organic chemistry; to highlight the crucial discoveries and to examine the impact they have had on the continuing development in the field;
3. to focus attention on the research of some of the seminal contributors to organic chemistry; to indicate how their research programs progressed over a 20–40-year period; and
4. to provide a documented source for individuals interested in the hows and whys of the development of modern organic chemistry.

One noted scientist explained his refusal to contribute a volume by saying, in part, that "it is extraordinarily difficult to write in good taste about oneself. Only if one can manage a humorous and light touch does it come off well. Naturally, I would like to place my work in what I consider its true scientific perspective, but . . ."

Each autobiography reflects the author's science, his lifestyle, and the style of his research. Naturally, the volumes are not uniform, although each author attempted to follow the guidelines. "To write in good taste" was not an objective of the series. On the contrary, the authors were specifically requested not to write a review article of their field, but to detail their own research accomplishments. To the extent that this instruction was followed and the result is not "in good taste", then these are criticisms that I, as editor, must bear, not the writer.

As in any project, I have a few regrets. It is truly sad that Egbert Havinga, who wrote one volume, and David Ginsburg, who translated another, died during the development of this project. There have been many rewards, some of which are documented in my personal account of this project, entitled "Extracting the Essence: Adventures of an Editor" published in *CHEMTECH*.[2]

Acknowledgments

I join the entire chemical community in offering each author unbounded thanks. I thank their families and their secretaries for their contributions. Furthermore, I thank numerous chemists for reading and reviewing the chapters, for lending photographs, for sharing information, and for providing each of the authors and me the encouragement to proceed in a project that was far more costly in time and energy than any of us had anticipated.

I thank my employer, Philip Morris USA, and J. Charles, R. N. Ferguson, K. Houghton, and W. F. Kuhn, for without their support, Profiles, Pathways, and Dreams could not have been. I thank ACS Books, and in particular, Robin Giroux (acquisitions editor), Karen Schools Colson (production manager), Janet Dodd (senior editor), Joan Comstock (department head), and their staff for their hard work, dedication, and support. Each reader no doubt joins me in thanking 24 corporations and Herchel Smith for financial support for the project.

I thank my wife Suzanne, for she assisted Profiles in both practical and emotional ways. I thank my children Jonathan and Brooke for their patient support and understanding; remarkably, I have been working on Profiles for more than half of their lives—probably the only half that they can remember! My family hardly knows a husband or father who doesn't live the life of an editor. Finally, I again thank all those mentioned and especially my family, friends, colleagues, and the 22 authors for allowing me to share this experience with them.

JEFFREY I. SEEMAN
Philip Morris Research Center
Richmond, VA 23234

February 15, 1990

[1] Seeman, J. I. *Chem. Rev.* **1983,** *83,* 83–134.
[2] Seeman, J. I. *CHEMTECH* **1990,** *20*(2), 86–90.

Editor's Note

JACK ROBERTS growls when he answers his telephone. His growls are somewhere between a snarl, a grumble, a groan, and a bellyache. He must sound that way regardless of who's calling, for no matter how smart Roberts is, and there is no dispute on that score, he can't know who's on the other end until they utter a sound.

Now, I don't mean to imply that Roberts is unfriendly. Quite the contrary. Once, he even asked if I was sick. My interpretation of that question is that he was actually sad that he hadn't heard from me in two months. Had I disappeared, down Alice's tunnel, carrying all of Profiles with me?

When I last spoke with Roberts, following a lengthy and most enjoyable repartee that included business as well, I asked him what I might include in his introductory essay. "Tell them what it's like to deal with me," he challenged. "You are lucky to get so much of my time!" he growled some more. "If my secretary were here, I would have told her I wouldn't speak with you." One reason that his secretary wasn't there is that she had recently quit. Was Roberts ever mad! "How could she do this to me?" he simultaneously moaned and complained. It seems that, after being his secretary for decades, she decided to retire, sell her modest southern California home for a fortune, and buy an estate on the North Carolina coast, all at Roberts's psychological expense and personal inconvenience.

But I jest about a man of great purpose and singularity of integrity and intent. Roberts is easily able to overwhelm one with his special blend of humor, wisdom, inspiration, scholarship, and intellectual tenacity. It's just hard not to duel with this man. "Roberts loves to be ribbed," acknowledges his longstanding good friend Bill Johnson. "He usually gets the better [of one] in these situations." Indeed, I was eventually to find that Roberts is a blend of tough, soft, and delicate. Let me tell you a little of what it's like to work with this man.

His letter of June 23, 1986, suggested that trouble was in store. He wrote, "You are certainly detailed [regarding the Profiles guidelines], and the message of what you want is clear, but the likelihood that you can get it is, in my view, not good . . . . By insisting on the broad view—biographical, scientific and societal—you are freezing out hard particulars for bland pap, and I, for one, am not about to want to do it. My own idea is to narrate only how I came to be involved in one field, carbocations, and to tell the story of some 40 years of research on the $C_4H_7^+$ cation—as yet an unsolved experimental problem." I pressed, "People want to hear about the development of so many of your contributions: benzyne structure, field–space effects, NMR, carbon–carbon rearrangements, not to mention your more recent work."

Profiles was in deeper trouble when Roberts wrote four months later, "Through some unseeming diligence, I wrote what, when double-spaced, will probably amount to perhaps 150–160 pages, but for reasons I can't explain (other than being just plain carried away), I got, chronologically, only to September 1945. At this rate, having covered perhaps some 2–3% of my scientific output, it will take a very long time indeed to get up to date."

Roberts sent page after page, and I, in his words and with his urgings and approvals, "CUT, CUT, CUT, and CUT." Roberts sent sections, I returned dissections! As Profiles progressed from a one-volume collection to a 22-volume series, Roberts was given an opportunity to expand his previously condensed version. (Admittedly, it was never trimmed too much!) Some sections were resurrected, a large and very interesting story about *Organic Syntheses* was added, and with the page proofs, Janet Dodd (ACS Books Senior Editor) asked for more information about Roberts's family, which was added as well, as were 40 more pictures! Give Roberts an inch, he will take a yard, laughing all the way!

Each package containing more sections arrived with a fascinating cover letter that always included two phrases that read approximately as follows: "For more of your 'tough' editing." And second, "The material remains copyrighted, and I reserve all rights until a satisfactory resolution of the problem with your publisher is achieved."

As a standard practice, all publishers require that they own the copyright to material they publish. Roberts pointed out that Profiles was distinctly different from journal articles, for the chapters are autobiographies, and many of the contributors may wish to reuse some of this material at some later date. This opinion is an example of the Roberts tenacity and individuality. Due to his persistence and with the cooperation and determination of Barbara Polansky, ACS Copyright Administrator, the ACS granted immediate-use privileges to each author in the Profiles series for the author's own chapter.

Roberts never fails to provide a unique view of life. In his Priestley Medal address in 1987, he described how it felt to get the medal. "This is no last-minute surprise; indeed, it is more like the water cure. The Chairman of the ACS Board will call you up in April and ask if you will accept the medal and then, if you say yes, you will have about 360 days to worry (and I'm inclined to worry) about what you are going to say when the big day arrives. And, in the meantime, you will get a three-inch stack of letters (and very nice letters, indeed) congratulating you for having already received the Priestley Medal, long before the fact. All of this at a time when you have no certainty as to whether you will even be alive for 360 more days."

Mary L. Good, President of ACS, presenting the Priestley medal to John D. Roberts at the 193rd ACS National Meeting in Denver, Colorado, April 1987. Photo courtesy of Chemical & Engineering News.

The Right Place at the Right Time

John D. Roberts

John D. Roberts

IF A BUSY PROFESSIONAL LOOKS BACK IN TIME, the distant past (anything beyond a week or so) seems dim. When faced with the need for thinking about an autobiographical assessment, fears arise about lack of material that would go beyond a simple annotation of one's *curriculum vitae* or publication list.

However, if one looks back more intently (it seems to take a couple of days to really be able to remember the details of things that happened 40 or so years ago), then it is like opening an overstuffed, catch-all storage closet. A melange of uncorrelated, unresolved, sometimes unpleasant memories comes pouring out. Soon doubt arises as to what, if anything, has historical significance and, even more, what part of the accumulation would be interesting enough for someone else to read about. Finally, there is a nagging possibility that it might be difficult, if not impossible, to stuff the memories back into storage and resume one's forward tack.

What fame I can claim comes from research. The goal of research is almost always to solve problems, to answer questions. Seldom does one do research without some kind of a plan. However, it often turns out that unanticipated things happen so that the plan fails. Most of the time, the plan fails simply because of a lack of foresight with respect to one or more important details. Sometimes the goal is not attained because something quite unexpected happens. Then a decision has to be made to abandon the goal—at least temporarily—and find out what went wrong, or else cycle around and devise another approach to the desired goal. Here intuition plays a critical role. But it is also important to have the luxury of being able to make a decision to go off on a tangent. In an environment highly oriented to particular goals, unexpected findings will rarely be investigated unless they are clearly germane to the goal at hand. I have had the good fortune of almost always being able to go in whatever research direction seemed most appealing at the time.

Life Before College

My grandparents were disparate. On my father's side, they were descended from prerevolutionary immigrants from the British Isles. My grandfather, William Pitt Roberts, was a successful real estate developer in Peoria, Illinois. My mother's father, Paul Dombrowski, was born in Danzig as a German Pole and was said to be a direct descendant of Jan Ignace Dombrowski, a sort of George Washington of Poland. As the oldest son, Paul Dombrowski was destined by his parents to go into the priesthood. However, he wished to study medicine. He left home, obtained his medical degree, emigrated to the United States, and became an esteemed member of his community, also in Peoria. My father and mother did not know each other in Peoria, but met in California and were married in 1914.

My father, Allen A. Roberts, had a strong mechanical inclination, a lively curiosity, considerable athletic ability, a bent for sketching, and a love of farming. My mother, Flora, with a more aristocratic background, was especially fond of music and literature. When I was born, my father was engaged in dairy farming in Puente, California. About

(Left to right) William A. Roberts, Mary Roberts, Flora Roberts, and John D. Roberts, about 1922.

1922, the family (elder brother, younger sister) moved to the western edge of Los Angeles. My father began manufacturing hot-air gas furnaces for homes or offices, which was quite a successful enterprise in the rapidly growing greater Los Angeles area.

Of my early childhood, I don't remember much. I guess I did well in early grammar school; I was encouraged to skip a grade or two on the basis of the aptitude tests of the time. However, I have a distinct memory of a lack of athletic prowess in school games compared to the best of the boys, many of whom later became well-known high school

The author demonstrating an early interest in exploring the radio-frequency spectrum, about 1922.

and college athletes. A rather profound change came in 1928, when, as the result of something like scarlet fever, I suffered a sudden and substantial hearing loss. School became a bit of hell. Not much could be done medically in those days for that kind of hearing loss, but my grammar school did provide special classes in lip reading. Although the classes were a great help, nonetheless my life was never quite the same again.

My family read avidly. Each week we went to the public library and exchanged armloads of books. Sometime in 1929 I happened to pick up a book telling of the lives of scientists and inventors—Newton, Herschel, Pasteur, Ehrlich, Edison, and Luther Burbank, among others—and immediately became interested in science. My horizons were greatly expanded through books and experimentation, especially in chemistry, although I was also interested in astronomy, lepidoptera, and electricity. A kitchen cabinet shelf was converted to a storage place for chemicals and equipment, and I acquired a modest qualitative knowledge of simple inorganic chemistry.

By the early 1930s, the California Institute of Technology (Caltech) had become a real scientific and engineering power in southern California. The school helped to heighten my interest in science by holding a public Open House each year, which I attended as often as possible. The high-voltage laboratory was a special favorite, with loud, high-amperage sparks and million-volt horn-gap arcs rising to the ceiling. (I later learned to do it at home, on a small scale, with a 30,000-volt neon-light transformer.) My other favorite was chemistry. The leadership of the great Howard Lucas provided marvelous exhibits of synthetic organic chemistry, heavily redolent of benzaldehyde, where toluene was chlorinated and saccharin, luminol, and azo dyes were synthesized. Caltech came through as a Valhalla of science and engineering.

The switch to junior high school was difficult at first, but in the second year I encountered an excellent science teacher, Bess Reed Peacock. She took an interest in me, in part because I rescued her when she couldn't seal a glass tube onto the top of a carrot when she was demonstrating osmotic pressure, and in part because she was a personal friend of one of my aunts. I helped set up experiments and had the run of the tiny preparation room for her course.

In junior high and later, I found that I thrived on demanding teachers, especially in mathematics, and I began to enjoy algebra. Drafting was a special treat—I often smeared the India ink through impatience, but I learned about the various ways to render three-dimensional objects. Also during this period, I read Slosson's *Creative Chemistry*,[1] which described many of the great discoveries of the 19th century. The account of the work of W. H. Perkin was especially appealing, and I often carried around a notebook to work out simple organic structures in accord with the rules of valence.

The Great Depression, beginning in the 1930s, wreaked havoc on my father's business, but he did his best to shield his children from it. My brother and I did all manner of odd jobs—mowing lawns, deliver-

Allen A. Roberts, 1930.

ing papers, doing housework, and working for the Registrar of Voters. Eventually, the furnace company went bankrupt. We lost our home through foreclosure, and my mother had to rent out her beloved grand piano. With his brother, my father took over the running of a ranch that my grandfather bought 70 miles away, while the rest of us moved to a smaller rented house.

At Mono Lake, California, summer 1934, spent at Uncle Walter Dombrowski's fishing camp on Rush Creek.

With my father home, at most, on weekends, we were all thrown very much on our own responsibility. It was understood that, while we would have room and board, all our other expenses, including college, had to be earned. For my part, besides delivering newspapers, I got a job as a salesman for a local bakery chain. At first I worked on weekends. Later, after graduation from high school, my work time was extended to six evenings a week.

The transition to Los Angeles High School in 1933 was not so difficult. While my attainments outside mathematics and science were no more than average, chemistry was very easy, and I did well in physics also. Indeed, I became the laboratory assistant in physics, setting up demonstrations, helping students with their experiments, and even grading papers.

Male chauvinism, particularly in sports, was the custom in those days. Women's athletic facilities were separate and definitely not equal. The tennis courts at Los Angeles High School were no exception. The men's courts had clean concrete with freshly painted lines; the women played on dirty tar-patched asphalt, where new tennis balls turned brown after one bounce and the lines were almost invisible. It was small wonder that the women tennis players looked avidly on the men's courts and that the more adventuresome occasionally moved right in and played until finished. This caused some consternation among the male players. One of the habitual offenders was a tallish, good-looking brunette with excellent tennis style. Eventually, in our senior year, I got to know this particular trespasser, Edith Johnson. After some ups and downs, we were married 7 years later.

Undergraduate Years at UCLA

Caltech was never far from my consciousness, and my mother wanted me to go there for undergraduate study. To my mother's disappointment, I never applied. The reasons were compelling. I only had slightly over a B average. My high school geometry teacher had been wholly undemanding, and I had responded so poorly that I lost confidence in mathematics. Also, the Caltech tuition was a whopping $200 per year, and there were no women there. The University of California at Los Angeles (UCLA)—at $54 for a year's fees, with coeducational classes, and with no reputation for requiring mathematical excellence—seemed a much better choice for me. Indeed, that turned out to be so.

I was uncertain at first about being admitted to UCLA, but once that hurdle was surmounted, the 6-month wait from winter graduation to fall matriculation provided a welcome chance to set some money aside and have a respite from school. UCLA was a very interesting place in 1936. Its forerunner was a small campus in northwest-central Los Angeles that, when the Westwood campus began in 1930, became Los Angeles City College. The Westwood campus had a large tract of very desirable land. In 1936 there were only a few academic buildings, but these had a distinctive Romanesque style.

The faculty was made up of a small, older, and generally not very distinguished pre-1930 group and a sizable coterie of excellent young professors drawn by future prospects. The powers at the University of California at Berkeley had decreed that the school would have no doctoral program, but a thriving master's program was in place. The chemistry department had already turned out two superb graduates, Glenn Seaborg and Saul Winstein, who were to contribute much to American chemistry.

I still had to work nights and Saturdays at the bakery store to make ends meet and was greatly dismayed to find that I had been scheduled for Saturday morning freshman chemistry laboratory. I called on the chairman of the department, William Conger Morgan, to request a change. Morgan was in his sixties, bald as an egg, with a countenance that could curdle milk. His answer was "No." But I pointed out that I was a chemistry major and I needed the work. "I have 10 too many chemistry majors and would be glad to get rid of one. Goodbye." I attended the Saturday morning lab classes.

Morgan was a frightful teacher, had not done any research for many years, admitted that he did not understand activity coefficients, and told the most and the worst off-color jokes I have ever heard from a professor in a mixed class. Fortunately, the quiz section and laboratory were taught by a master's candidate, Jerome Vinograd, later to be a professor at Caltech and a pathbreaker in DNA research. Vinograd worked very hard to counteract the Morgan deficit. With the benefit of my earlier experience in chemistry, the first semester was easy, even though I worked six nights a week until midnight in the bakery stores.

The next semester was another story. Morgan still gave the principal lectures, which were nearly a total loss, but this was largely counteracted again by a superb quiz-section instructor, Francis E. Blacet, a distinguished photochemist. My problem this time was in the laboratory, where we were doing inorganic qualitative analysis, based centrally on sulfide precipitation. The precipitations were carried out on a fume porch, where hydrogen sulfide was dispersed from gas jets, and the place reeked. No one seemed to know then (or care) that hydrogen sulfide was comparable to hydrogen cyanide in toxicity.

I didn't do too well with inorganic qualitative analysis. First, I was overconfident. Second, I didn't really try to understand the mass-action principles involved until it was too late. The B I got in this course I deserved. It was not a very good year scholastically, and I needed more time to study, so I began to borrow money and worked fewer nights.

The hurdle in the sophomore year was quantitative analysis. The course had an awesome reputation under William R. Crowell, who was regarded as fair but tough and somewhat distant. I had the flu and missed the first week of lab work. Then I rather badly botched the first gravimetric determination of the composition of a sodium carbonate–bicarbonate mixture by loss of weight on heating. Next was a simple acid–base titration to a methyl-orange end point. After I made quite a few tries with disparate results, Crowell came by, looked over my shoulder, and passed judgment: "I could do better with a graduated cylinder than you are doing with that burette." Crowell was kind and helpful and soon got me back on the track, but it seemed that quantitative analysis was not my forte.

However, near the end of the year we had to do a special project, and I chose one that involved an analysis from which oxygen was best excluded. Laboratory nitrogen at that time was not oxygen-free. Researchers usually passed the gas through some kind of a solution that would remove the oxygen impurity. Chromous chloride was the favorite solution at UCLA, and a special spiral bubbler was required for good gas-to-solution contact. I asked Crowell if he had such a bubbler, and he said all of his were in use. On being pressed, he said, "Well, you could make one." This was a shock. Such bubblers had a long internal spiral glass tube and a ring seal, and I had never done anything more complicated than bend glass tubing. How could I make such a thing? Crowell told me there was a glass-blowing shop in the basement that I could use, and unlocked the door so I could go to work.

The shop was a sea of glass shards, with racks of various sizes of soft-glass tubing, a couple of burners, and a thin book on laboratory glass blowing. Crowell left me there, and I began to read. I must have spent every spare minute in that room for the next 2 or 3 weeks. I learned to make spirals and ring seals before I even tried simple end seals. The soft glass was hellish to work with because it required very careful annealing before a piece of work could be allowed to cool down. The problem of how carefully it had to be heated for reworking was just as bad. The floor and bench became littered with my unsuccessful attempts to make ring seals. Finally I emerged with a crude, but workable, spiral bubbler. Unfortunately, it was too late to finish the analysis project for which it had been intended.

Looking back over my career, I think it is fair to say that this was a critical turning point for me. Completion of the spiral bubbler brought me into research. It somehow showed that, whether or not I did very well in the courses, I could persevere on my own in a new project and get results. One wonders how many schools today would turn a sophomore loose in a research glass shop and let him teach himself to blow glass the way I did.

Crowell was sufficiently impressed to ask me if I would like to do research during the coming summer. He could pay me perhaps $10 per month, which came out of his own pocket. This was a fantastic morale booster, because I was the only one chosen from the current class. To facilitate my work, I was issued a master key to all of the chemistry department doors by Professor Morgan himself.

Crowell wanted to look at the potentiometric titration of chloride–bromide mixtures with silver nitrate. There was some evidence from work by the leading analytical chemist of the time, I. M. Kolthoff, that inclusion of a peptizing agent would sharpen the end point. Crowell did not tell me why he was interested in this analysis, but I suspect it was because one of his former students worked in the Paramount Pictures film processing laboratory. Such an analysis could have been important to his work.

The junior year for chemistry majors included an introduction to organic chemistry. To my dismay, it was taught that year by Professor Morgan. The text for the course was the brilliant new book[2] by Caltech's Howard Lucas. The choice was surely pushed on Morgan by his younger colleagues, because he groaned about the more theoretical parts constantly. He started the course off by saying he did not understand the first chapter, was sure we would not either, and hence was skipping over it.

I remember almost nothing of Morgan's lectures, except once learning a structure for olive oil and something called Kondikoff's rule, which seemed to show up frequently, in one form or other, on the examinations. The rule was that, on the oxidation of an unsymmetrically substituted tertiary alcohol with chromic acid, the carbinyl carbon went with the smallest substituent group. In retrospect, this rule, which I never heard of again, is not unreasonable in the absence of Wagner–Meerwein rearrangements, because the most substituted alkene is expected to be favored by dehydration.

Fortunately, the organic laboratory taught by G. R. Robertson was a dream course, wonderfully organized, challenging, and with a high degree of competitiveness among the students. The emphasis of Robertson's course was on the physical principles of laboratory operations: how to understand extraction, crystallization, steam distillation, and fractional distillation. All of this information was given with remarkable clarity in the lectures and in the text,[3] as well as being emphasized in the actual manipulations.

There were relatively few graduate students in those days at UCLA to serve as teaching assistants in undergraduate courses. I was pleased when Crowell offered me a position as assistant in the analytical course. The pay was small, but sufficient so that I could abandon the bakery job and attend UCLA full-time.

Because there were relatively few research students and he had an avid interest, Crowell also gave me an opportunity to work on a different research problem. He proposed to set me up to work in his private laboratory, to which the only access was through his office. It was well equipped, a most pleasant place to work, and it became my headquarters for the next 2 years. Crowell wanted to determine the mechanism by which osmium compounds catalyzed the reaction of perchloric acid with hydrobromic acid. He became interested as an extension of some research he had done at Caltech in Don Yost's laboratory, which used ruthenium as a catalyst.

Crowell was a stickler for purifying reagents, and I prepared the hydrobromic acid by distillation from recrystallized sodium bromide. Here, for the first time, I could use equipment with ground-glass joints. Purification of the osmium was particularly important because of the possibility of contamination with ruthenium, which was already known to be a more effective catalyst than osmium. Crowell was pleased when I developed an alternate purification scheme in which the tetroxide was reduced with potassium hydroxide in ethanol to give the violet K_2OsO_4 salt. Osmium tetroxide was known to be toxic, and Crowell became very worried when there was evidence, by virtue of "seeing yellow spots before one's eyes", that I had ingested or inhaled some.

We used a pretty crude constant-temperature bath for the kinetics experiments; it was just a pan of boiling water and a siphon to keep it from going dry. Samples were sealed in ampules, and the bromine formed was determined by iodometric titrations. Measuring and analyzing kinetics before learning physical chemistry required quite a bit of independent study. After I had determined with some pride the kinetic order of the reagents, my morale was dealt a stiff blow by one of the graduate students who exposed my ignorance by asking about the kinetic salt effect and the activation parameters.

The results were written up and published jointly with Yost, so that my very first paper[4] (1940) provided a connection with Caltech. I did not meet Yost at the time, but Crowell reported that he was pleased with our findings.

I was having so much fun teaching and doing research that I decided not to try to graduate in four years, but to take an extra year and put off the inevitable crisis associated with undergraduate physical chemistry, which had a particularly grim reputation at UCLA. So I spent the academic year 1939–1940 as assistant in several chemistry courses, research, and qualitative organic analysis. Among the students whom I taught during my years as an undergraduate teaching assistant were George Pimentel, who became an extraordinarily fine spectroscopist, president of the American Chemical Society 1986–1987, and the leader of the National Academy of Sciences committee that produced the Pimentel Report[5]; Bruce Merrifield, winner of the 1985 Nobel prize in chemistry for peptide synthesis; and Jerome Hines, later to become a leading basso of the Metropolitan Opera.

I was becoming more and more enamored of organic chemistry, so, when I was looking for some research to do in the area, I naturally gravitated to G. R. Robertson. He told me about some of his ideas and then said, "But really, you shouldn't work with me, you should be working with Bill Young, who is at the forefront of his field."

William G. Young was 38 years old in the summer of 1940. A doctoral student of Howard Lucas at Caltech, he had already had as M.S. candidates Jerome Vinograd, Saul Winstein, and Stanley J. Cristol,

William G. Young and his wife Helen in 1950.

all of whom would later become members of the National Academy of Sciences. William Conger Morgan had died rather suddenly in the previous year, and Bill Young had been selected to succeed him as chairman, so he had moved his base of operations from his downstairs office and laboratory to the department office.

Young had been working for many years on allylic systems, substances that often undergo rearrangement readily and in many reactions give mixtures of products. Young got into this field as the result of an effort to prepare substantial quantities of pure *trans*-2-butene, **1**. He first established that 2-butenal, **2**, was wholly *trans* and envisioned a simple, straightforward synthetic sequence starting from **2** and yielding **1** (equation 1). When the final product turned out to be not pure *trans*-2-

$$\underset{\mathbf{2}}{(CH_3)(H)C{=}C(H)(CHO)} \longrightarrow (CH_3)(H)C{=}C(H)(CH_2OH) \longrightarrow (CH_3)(H)C{=}C(H)(CH_2Br)$$

$$\longrightarrow (CH_3)(H)C{=}C(H)(CH_2MgBr) \longrightarrow \underset{\mathbf{1}}{(CH_3)(H)C{=}C(H)(CH_3)} \qquad (1)$$

butene, but a horrendous mixture of 1-butene, *cis*-2-butene, and *trans*-2-butene, his natural impulse was to find out why.

The first complication was at the bromide step. Young and Winstein did excellent work unraveling the vagaries of the butenyl bromides, which were subject to facile equilibration and very difficult to prepare pure by the procedures known then for converting alcohols to bromides.[6] The observed rearrangements could be understood on the basis of intermediacy of the butenyl cation. The then-developing resonance theories of Linus Pauling attributed a special degree of stability to the butenyl cation **3** by virtue of its capability of being formulated as two electronic structures (**3a**, **3b**) with the same positions of the atomic nuclei.

As elementary as these ideas are today, they were novel in the late 1930s. Indeed, the whole qualitative concept of resonance presented a difficult problem for organic chemists. Vigorous debates continued for many years as to what Pauling meant by "resonating among several valence-bond structures."[7] For the butenyl cation (**3a** ↔ **3b**, equation 2), whether or not it was intuitive that it should have a special degree of stability as predicted by resonance theory, anyone could reasonably expect that it could combine with an anion, such as Br^-, at either C-1 or C-3 and give a mixture of products **4a** and **4b** (equation 2).

Young and Winstein found that, irrespective of bromide composition, the Grignard reagents from either **4a**, **4b**, or mixtures thereof,

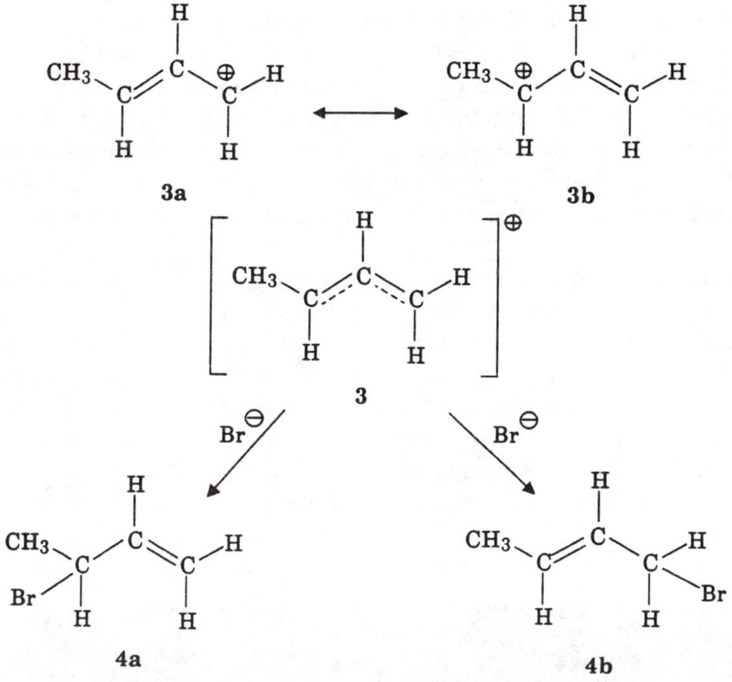

H H
CH_3 C ⊕ H CH_3 ⊕ C H
C C ⟷ C C
H H H H

3a 3b

H
CH_3 C H ⊕
C C
H H

3

$Br^{\ominus}$ $Br^{\ominus}$

H H
CH_3 C H CH_3 C H
C C C C
Br H H H H Br

4a 4b

(2)

gave the same mixture of butenes (about 56% of 1-butene, 27% of *cis*-2-butene, and 16% of *trans*-2-butene).[8] Today we have the luxury of rapid butene analysis by mass, IR, or NMR spectroscopy or by chromatography. In the absence of such methods, Young and co-workers used an extraordinarily tedious, but accurate, procedure wherein the mixture of butenes was converted to the corresponding mixture of dibromides. The dibromides, after purification, were assayed by density, refractive index, and reaction kinetics with iodide ion in acetone.

Little was known about the structure of Grignard reagents other than their extraordinary usefulness in synthesis, which had been recognized by the award of a Nobel prize to Victor Grignard less than 30 years earlier, in 1912. A question naturally arose about the degree to

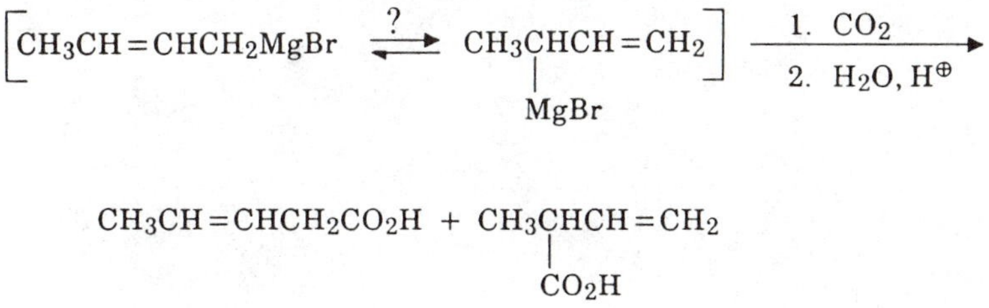

$$\left[CH_3CH{=}CHCH_2MgBr \overset{?}{\rightleftharpoons} CH_3\underset{\displaystyle MgBr}{\underset{|}{C}}HCH{=}CH_2\right] \xrightarrow[\text{2. } H_2O,\ H^{\oplus}]{\text{1. } CO_2}$$

$$CH_3CH{=}CHCH_2CO_2H + CH_3\underset{\displaystyle CO_2H}{\underset{|}{C}}HCH{=}CH_2 \quad (3)$$

which the composition of the butenes corresponded to the composition of allylic isomers in the butenyl Grignard solution. Methods to measure the isomer composition had to be indirect, gleaned from the nature of the products formed with various reagents. But, of course, with the myriad possibilities of rearrangement in reactions of allylic compounds, the product compositions resulting from reactions of the butenyl Grignard reagent might not necessarily represent the nature of the Grignard reagent itself.

During my interview with Bill Young, in the spring of 1940, about a possible undergraduate research problem, he told me about the butenyl Grignard reagents. Young said that Stan Cristol, as part of his M.S. research, had made a preliminary investigation of the carbonation of the Grignard reagents and believed that a mixture of carboxylic acids might have been formed (equation 3). Would I be interested in checking this out? Indeed I was, not because I understood much of the intricacies of allylic rearrangements, but because I had made a Grignard reagent in the advanced organic chemistry lab and was intrigued by them. In addition, I had the feeling that this was something I could do.

Saul Winstein in his UCLA office, 1950.

Early in the summer of 1940, as I was getting the materials together for the Grignard problem, Saul Winstein visited UCLA. He was fresh from a National Research Council postdoctoral year with P. D. Bartlett at Harvard University and ready to start an assistant professorship at Illinois Institute of Technology in Chicago. Saul possessed a good sense of humor, but was very intense, with a bulldoglike tenacity in thinking about and discussing chemical problems. After he obtained his Master's degree at UCLA, Saul carried on his Ph.D. work with Howard J. Lucas at Caltech—the same Howard Lucas with whom Bill Young had studied for his doctorate, and the man I would eventually succeed at Caltech as professor of organic chemistry.

Saul's thesis was a brilliant and perceptive attack on the stereochemistry of the reaction of the chiral 3-bromo-2-butenol isomers with hydrogen bromide.[9] A requirement for a Caltech doctorate in chemistry then (and now) was submission and defense of a number of propositions for doing original research. One of Saul's propositions evoked the possibility of S_N2' reactions of allylic derivatives, and that turned out to be the demise of my undergraduate research project on Grignard reagents. This was, in fact, fortunate because the Grignard problem needed much more time and experience than I had available in 1940.

At Harvard, Saul showed that the corresponding displacements of chloride from the butenyl chlorides, **5** and **6**, with ethoxide in ethanol

$$\underset{\mathbf{5}}{CH_3CH{=}CHCH_2Cl} + {}^{\ominus}OC_2H_5 \xrightarrow{S_N2} CH_3CH{=}CHCH_2OC_2H_5 + Cl^{\ominus} \quad (4)$$

$$\underset{\mathbf{6}}{CH_3CH(Cl)CH{=}CH_2} + {}^{\ominus}OC_2H_5 \xrightarrow{S_N2} CH_3CH(OC_2H_5)CH{=}CH_2 + Cl^{\ominus} \quad (5)$$

(crossed S_N2' pathways between **5** and **6** shown as not occurring)

gave the products expected from the S_N2 reactions and none of the S_N2' products, at least within the limits of his experimental error (equations 4 and 5).

He was anxious to find out whether, with acetate ion in acetic acid, the chlorides **5** and **6** would give S_N2' products. This became my undergraduate research project. As soon as I got started with the kinetics, it was clear that something was amiss, because it was just not possible to separate the reaction of a given chloride, **5** or **6**, cleanly into two processes, one independent and one dependent on the concentration of acetate ion. Something else seemed to be going on and, in my naiveté, I worked out in some detail a scheme whereby I could fit the data much better by assuming a concurrent rearrangement of **5** to **6** or vice versa. The process I suggested was later demonstrated for other systems by Winstein and co-workers[10] and christened "internal return".

The explanation was not acceptable to Saul. In a several-hour, detailed review, he zeroed in on the facts that I was using potassium acetate as my acetate source and that some potassium chloride was crystallizing from the acetic acid medium, thus changing the total salt concentration. He decided that what we needed was to use an acetate salt that gave a chloride salt soluble in acetic acid. Diphenylguanidinium acetate was chosen for the purpose and worked splendidly. Diphenylguanidinium chloride was soluble and the reaction was then cleanly separable into two parts, one dependent on, and the other independent of, the acetate-ion concentration. This was my first exposure to Winstein's tenacious way of digging into a problem until he was satisfied that he understood all of the extant information on it and had prepared a plan of action. It was an impressive performance, and I was to see many more of them in the future.

The results from the research on acetate formation from **5** and **6** in acetic acid and other solvents were decisive in that no evidence for the S_N2' mechanism was obtained. Within experimental error, we could account for all of the rearrangement products by intervention of the cation (**3a** ↔ **3b**). Young was so pleased with the results that I was delegated to present a paper at the Fall 1941 meeting of the American Chemical Society in Atlantic City, but that is part of other developments to be discussed shortly. The full paper was published in the *Journal of the American Chemical Society.*[11] It was fortunate that Saul diverted me from the butenyl Grignard reagent to S_N1 and S_N2 mechanisms, because it heightened my interest in such reactions and the rearrangements often attendant on the formation of carbocationic reaction intermediates, as well as exposed me to a great researcher at an early stage in his career.

In this era, the physical chemistry lectures at UCLA were given by James Blaine Ramsay, an extraordinarily dedicated man with a formidable reputation for making a tough course super tough. In his usual lecture, Ramsay would march in with a dark look, open the text, read a sentence or two, and then analyze the semantics and concepts in excruciating detail. Occasionally he would halt, affix with a stare one of those in the class who might be nodding off or otherwise failing to pay attention, and explode: "Roberts, what do you think of that?" It was hardly a fun-filled class, but we could never have had better lessons in how not to take the truth of printed words for granted. Ramsay inspired a very useful rigor of thought.

The physical chemistry laboratory course was wholly different. It was taught by Charles D. Coryell, a brilliant Caltech Ph.D. (under Pauling), who was almost the antithesis of Ramsay. He thought in broad general terms, was extremely friendly with the students, and wanted all

Charles Coryell, at the time of his appointment to MIT in 1946, taken at UCLA and autographed, "How do I look at Bill Young's desk?" Coryell was the leader of the research group at Oak Ridge that discovered in uranium-fission products the element of atomic number 61, for which his wife Mary Alice suggested the now-accepted name, promethium.

of us who were the least bit venturesome to break out of the standard routine of physical chemistry experiments. William G. McMillan, later to be a professor at UCLA, and I were laboratory partners.

One of the regular physical chemistry experiments was to involve calorimetry and uses of the Beckmann high-precision thermometer. For some reason, Coryell wanted to know the thermodynamic constants of sodium dithionite ($Na_2S_2O_7$). He proposed that we determine these from the heat of reaction of dithionite with silver–ammonia solution, a reaction used for analysis of dithionite. I had visions of elaborate apparatus and painstaking controls for determination of thermodynamic constants. Indeed, from Ramsay's lectures, it seemed that such work involved a dedicated priesthood of scientists obsessed with the need for ultimate accuracy. It was a shock to find that all we were going to do was to measure accurately the temperature rise when a weighed gram or so of sodium dithionite was added to 300 mL of ammoniacal silver nitrate in an ordinary thermos bottle (later called a "vacuum-jacketed calorimeter") set on a window sill of the laboratory so we could read the thermometer more accurately. We made three determinations, which agreed well, measured the heat capacity of the calorimeter itself, and dutifully calculated the heat of formation. Then Coryell took over. In a truly virtuoso performance, he parlayed our heat of formation and a host of literature data into a short paper[12] that dealt with dithionite free energy and entropy, as well as providing a brief discussion of the thermodynamics of the solvated electron.

In most of the experiments, McMillan and I had a symbiotic relationship—he was much better at theory and could figure out what we were supposed to be doing, while I, with a consummate knowledge of the stockrooms, would scrounge up the needed equipment or do the necessary glass blowing. Thomas L. Jacobs, an instructor in chemistry who came to UCLA via Cornell University and Harvard University, had made several acetylenic ethers. Somewhere along the line, he or Coryell suggested that their dipole moments might give interesting information about their electronic structures.

There was no instrument available for measuring dielectric constants, but the department had a lovely General Radio precision condenser. So we ventured off into the realm of electronics, of which I knew nothing, but where McMillan was quite knowledgeable. We found a wiring diagram for a Hartley oscillator and got that put together and working. The dielectric cell was a special problem. The literature usually specified concentric gold-plated brass cylinders for the condenser plates, and this was well beyond our means and time. Instead, we constructed a small Pyrex Dewar flask with tubes at the top and bottom so we could suck up liquids to fill the interstitial space, then silvered the inside and outside of the Dewar flask to provide the electrical plates. We measured the electrical capacity of the cell by adjusting the precision condenser, and thus the frequency of the oscillator, until we got a null audio interference with a table radio tuned to a particular AM station.

The off-null hetero beat frequencies gave wild squeals of changing pitch, and the experiments were best done in the evening. We cheated a little by using pure liquids, rather than measuring the dielectric constants of the usual range of concentrations in benzene, but correction by the Onsager equation gave quite acceptable dipole moments for several substances that had been measured previously by other methods. We were not very sophisticated in physical organic chemistry, hence our publication[13] of the results did not try to reach any very detailed conclusions about electron distributions or molecular geometries. But the relative ease with which we got significant results in this work kept me interested in such measurements for many years.

As usual for seniors, we had to start considering what was to happen after graduation, and postgraduate study looked very attractive. I went back to G. R. Robertson for advice. Robbie led me by the hand through the characteristics of the various first-rate schools in the country. My academic record was hardly prepossessing. Yes, I could blow glass, had a lot of teaching experience, and had done some research, but none of these achievements were on my scholastic record.

Robbie and I ruled out Berkeley, the University of Chicago, Caltech, Massachusetts Institute of Technology (MIT), and Columbia University as being too physical; the University of Illinois and Ohio State University as being too organic; and Harvard as being too choosy. We finally settled on the University of Wisconsin and Pennsylvania State College as being about right. I dutifully filled in and sent away the application forms, with no idea of what my reference letters would be like. I was turned down at Wisconsin, but Penn State offered admission and a teaching assistantship. Bill McMillan was a sure-fire applicant for physical chemistry and was accepted at Columbia. We decided to travel together to the East and the Atlantic City meeting of the American Chemical Society in the company of Francis Blacet.

Before leaving for graduate study, Bill McMillan and I were taken to Caltech for an afternoon by Charles Coryell. With Charles as our guide, we had a wonderful visit. I particularly remember meeting the biologist, James Bonner, then working on determining what made fruit ripen (avocados were his examples, and he was zeroing in on ethylene); A. J. Haagen-Smit's microanalytical laboratory; Verner Schomaker's electron-diffraction apparatus; and Edwin Buchman's attack on the synthesis of 1,3-cyclobutadiene. Edwin sketched out his approach and professed great patience for the difficult and untested steps ahead. Ten years after my earlier visits as a schoolboy, Caltech still looked like the Valhalla of chemistry, but I had no hopes then of getting any kind of a position there.

Graduate School at Penn State

In the summer of 1941, Atlantic City was in its heyday as a convention city. Because of the large number of attendees at American Chemical

Society meetings, one national meeting each year was usually held there. As Californians, we were amazed at the phalanx of gigantic, but aging, hotels, the wide and seemingly endless boardwalk, the hordes of people moving back and forth, and the lack of surf on the beach. My

Bill McMillan and the distinguished UCLA photochemist, Francis E. Blacet, at Independence Hall in Philadelphia in September 1941, before going to the Atlantic City meeting of the American Chemical Society.

presentation of our work on the butenyl chloride reactions seemed to go well in spite of shaky knees, although Bill Young told me that someone had complained to him that we had been too hard on one of the earlier workers in the field.

At Atlantic City, I had my first contact with Frank C. Whitmore at a Penn State get-together. Whether he remembered my application or not, Whitmore was certainly gracious when I identified myself as an entering graduate student.

Memory of the titans of any given era of modern chemistry tends to fade rapidly from generation to generation. Unless one is identified by name in connection with some particular discovery or technique, or with the Nobel prize (and perhaps not even that), one's most important contributions usually become an anonymous part of general chemical knowledge. Even such a titan as G. N. Lewis, who was extraordinarily influential for a long period, could probably only be identified by today's students through widely used "Lewis electronic structures." The problem of identifying achievements with people is a very difficult one for textbook writers, because you never know where to draw the line. Every fact or concept is associated with people, and mistakes in attributing priority are never forgiven.

The name of Frank Clifford Whitmore does not cut much ice nowadays, except with older Penn State alumni. Nonetheless, he was very influential in several important areas of organic chemistry and was certainly one of my personal heros. In 1941 Whitmore was 54, somewhat portly, balding, with a strong profile that would have served well as a figurehead on a clipper ship. He was ambitious, extremely vigorous and dynamic, and truly one of the all-time great lecturers on chemistry. He was a native of Massachusetts who worked his way through Harvard, mostly as an enormously successful and well-paid private tutor in chemistry. He graduated magna cum laude in 1911 and obtained his Ph.D. in 1914. After doing war research on poison gases, he held positions successively at Williams College, Rice Institute, University of Minnesota, and Northwestern University. He finally moved to Penn State in 1929, as dean of the School of Chemistry and Physics. I was much attracted to Whitmore by his dynamism and his efforts to achieve simplicity in the understanding of organic chemistry. One evidence of his dynamism was the increase in graduate enrollment at Penn State from 18 to more than 100 in a 10-year period.

At the time we first met, Whitmore had achieved considerable notice for his work on carbonium ion rearrangements, especially a theoretical paper published in the *Journal of the American Chemical Society*[14] calling attention to the commonality of the "open sextet" of electrons in rearrangement processes. He liked to tell how respective referees rated it as "unintelligible", "an excellent contribution", and "what everybody already knows". At this particular Atlantic City meeting, he talked about factors limiting yields in Grignard reactions, a subject about which I knew little, but which greatly interested me. I was wholly enthralled because the experiments seemed so cogent and the explanations so clear.

I think Whitmore suffered from the fact that much of his work was so clarifying that it became almost obvious. The Grignard work[15] was like that. Prior to Whitmore, had you asked a practicing organic chemist what to expect to get from isobutylmagnesium bromide and diisopropyl ketone, you would perhaps hear a suspicion that simple addition might not be favorable, and that would be it. Whitmore's work made it almost trivial to predict (correctly) much reduction, some enolization, and a little addition. Not a lot of hard thought was required. It could easily become a part of the working knowledge of anyone using Grignard reagents, with no special acknowledgment to Whitmore.

Finally the ACS meeting came to an end, Bill McMillan and I parted, and I set off for State College. My living arrangements were simple. I had a Spartan upstairs room in a private home, a block from the campus, and took my meals with the bachelor faculty and transients at the Faculty Club.

Despite the many portents of a world war ahead, my immediate goal was to get started in research, and the initial step was to have a talk with Whitmore. To my amazement, Whitmore's office looked like a chemical storehouse. There were 5-gallon cans, liter bottles, green bottles, brown bottles, bottles everywhere. Whitmore worked at an old, very plain oak desk. There were so many bottles and papers on top of the desk that he had to pull out the slide to have a place to write. In later life, Whitmore had many health problems usually ascribed to how hard he drove himself, but perhaps they were more likely caused by working in the midst of all of those chemicals. After Whitmore's death, whenever I encountered chemicals in a professor's office, I would deliver a lecture about the possible dangers, using Whitmore as an example. Nelson Leonard was one of those who took this seriously and was later grateful for having done so.

After I met Whitmore and heard his papers in Atlantic City, I was eager to start research with him. However, he was adamant that there was no space, even though I pleaded that I had come all the way from California to work with him. Finally he said I could have a place in the lab on a trial basis, but I would have to come up with a research problem that he found interesting before he would be willing to be my research supervisor.

Whitmore was well organized. When a student went to talk with him, he would pull out the slide on his desk, find a plain white pad, insert a carbon paper, and write down his understanding of each point discussed. At the end of the appointment, you got the original and he kept the carbon copy. He was strict about monthly research reports; they had a prescribed style, and he looked at them carefully. He almost never came into the laboratory unless he was escorting a visitor.

Whitmore's laboratory was an interesting place. It was very new, with black, oiled soapstone bench tops, each of which was equipped with a hood that had a built-in steam bath, level with the desk top. The large room was divided into two parts. One side was the domain of the very taciturn Russell W. Marker, who was just coming into fame for his studies of the steroids in Mexican plants, a development that eventually led to the founding of Syntex and to an enormous increase in the availability of steroids for medicinal purposes.[16]

When you first saw this laboratory, you had the feeling that something was missing; it looked different. If you worked there, you quickly realized why. Except in Marker's cubbyhole of an office, there were no laboratory stools, no chairs, and no writing desks. The only place to sit was on the floor. Because the laboratory was kept very clean, this was quite acceptable, even though it was strange to see a half dozen students sitting on the floor, backs to the wall, having a gabfest.

Penn State's Petroleum Research Laboratory was world famous for fractional distillation. Because almost all of Whitmore's work involved liquids, we were expected to set up efficient distilling columns. But not much was provided. My prior glass-blowing experience was of enormous benefit in making the small glass helices needed to pack a distilling column. I heated a soft-glass rod to just the right degree of softness and then wound it around a 3-mm mandrel driven by a slow-speed laboratory stirring motor. After the product cooled, the spiral was scratched with carborundum. The helices were then removed from the rod without breaking too many of them in half. The broken, too-thick, too-thin, and multiple-turn helices were then sorted out. The helices were carefully untangled from the usual interlocked mass and dropped, one by one, into the column. If you did all of this, you really appreciated your fractionating columns.

To heat our distillation flasks, we surrounded them with a coffee can insulated with asbestos paper with a split transite* cover and heated the whole thing with a small electric hot plate. The currents into the hot plate and the column heating jackets were controlled by rheostats made by winding flat nichrome wire around notched transite boards and making contact at the desired point, on the edges of the board, with spring battery clips. The result was inexpensive, effective, and hazardous.

On Sunday, December 7, 1941, at 1:30 p.m., my childhood friend and noted pianist, Eugene List, was to play on a broadcast with the New York Philharmonic. I was just settling in to listen when the broadcast was interrupted by the news of the attack on Pearl Harbor. With the outbreak of World War II, I had to reassess my position at State College. I was a long way from home, had just started my graduate research, and was wholly uncertain about what course Penn State would take. There was also the possibility of being drafted, which seemed much less attractive from Pennsylvania than from California.

Just after Christmas, I met with Bill Young at the National Organic Chemistry Symposium in Ann Arbor. He told me that there would be a war-research project starting up at UCLA, for which I could apply. I did, and, at the end of my first semester, I parted on good terms with Whitmore and returned to California. I had only been at Penn State for 4½ months, but the experience did much to shape my future interests.

---

* Asbestos board hardened with cement.

University of California at Los Angeles

War Work

I was not told much about the aims of our war project when we started. Ted Geissman and Bill Young were in charge, with Ted actually running the laboratory, which was in the subbasement of the chemistry

T. A. Geissman with his wife Lorraine at UCLA about 1949. In his later career, Geissman was best known for his research on the coloring matters of flowers.

building. There were just two hoods, a plywood inner door set at an angle so that hostile eyes could not peek in when the outside door was open, and four full-time occupants.

I was very fortunate in that the foreman, so to speak, of the laboratory was Maurice J. Schlatter, a Ph.D. graduate from Edwin Buchman's group at Caltech. He was a superb laboratory technician with experience in just those techniques I did not know about. In a few weeks, Schlatter had me proficient in carrying out multiple-step reactions on a small scale and using centrifuge cones as reaction, extraction, and crystallization vessels. Finally it was explained to us that the aim of our project was to produce oxygen in the field, or for aircraft in flight, by some simple process. The most attractive alternative at the time seemed to be the reversible oxygenation of "salcomine" (the cobalt salt of the diimine formed from salicylaldehyde and ethylenediamine, 7).

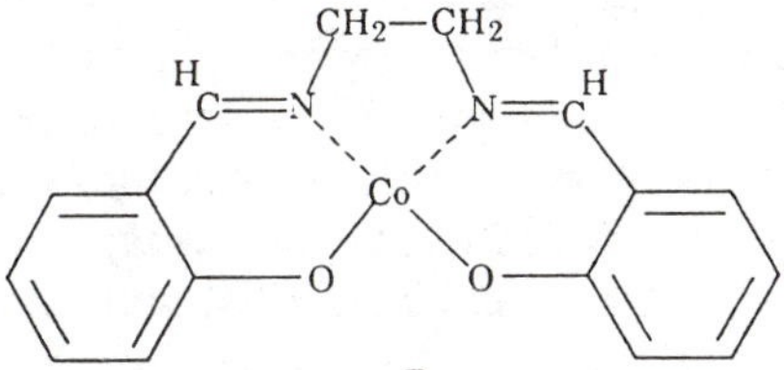

7

Salcomine, a light-brown solid, has the remarkable property of absorbing oxygen from the air at room temperature to form a black solid. When heated to less than 100 °C, this black solid gives up the oxygen and becomes light brown again. Each cycle is accompanied by a measure of irreversible oxidation. Our goals were to improve the lifetime of the material, to speed up both absorption and desorption of oxygen, and, if possible, to obtain a more favorable weight ratio of oxygen produced to carrier utilized.

For quite some time I made various salicylaldehyde derivatives, which were shipped off for testing to Berkeley, where Melvin Calvin was a principal investigator. There was not much feedback of results to me, which I took to mean that nothing we synthesized was better than what was already available.

The work assigned to us was ideal training for organic syntheses, and I learned a lot. I discovered the extensive research of Henry Gilman on metalation of anisole. Because we were especially interested in 3-substituted salicylaldehydes, I made a quite thorough survey of Gilman's work to see how we could utilize this kind of procedure in our syntheses. Not much came out of it for the purpose, but it did engender an appreciation of the breadth of Gilman's work and the acidity of aromatic hydrogens, which was to be part of a central theme in my later work.

I often had opportunities to discuss chemistry with Saul Winstein. Saul, with Tom Jacobs, was principal investigator in a program for synthesis of antimalarial drugs, and he worked hard at it. But his mind was often elsewhere, probing, questioning, and acquiring knowledge of organic reactions. Because Bill Young was overloaded with administrative work, I was one of the few around UCLA with whom Saul could discuss physical organic chemistry. Although I could contribute little more than questions, we spent many hours together in which he taught me much about what was understood at the time and about the nature of many critical future problems. It was an experience not unlike that provided by J. B. Ramsay in physical chemistry; Saul, like Ramsay, essentially refused to go on until he was sure of each step along the path.

In late 1943, Geissman was transferred to Philadelphia to work at a salcomine pilot plant. Not long afterward, there was an explosion in the plant. When Geissman ran around closing valves to safeguard the rest of the operation, he was terribly exposed to salcomine dust. He became very ill, with attacks of internal bleeding, and very nearly died. Later he regained much of his health, but for several years afterward he had catastrophic recurrences of internal bleeding, from which he made almost miraculous recoveries. He did a lot of fine research in later life, but never was able to live up to the promise of his early career.

Not long after Geissman's accident, the oxygen project was shut down because liquefication and fractionation of air to give oxygen was in fact practical in the field. We then returned to civilian pursuits. In my case, that meant beginning graduate work.

A Podbielniak fractionating column setup constructed by John D. Roberts for his Ph.D. research at UCLA, 1943.

Graduate School

I was now nicely positioned to begin the butenyl Grignard work in earnest, having acquired many needed skills at Penn State and on the war project. The major requirement was a suitable, highly efficient fractionating column with very low holdup. The Podbielniak Company made a vacuum-jacketed, 80-cm packed column with rather low holdup and a magnetic, actuated take-off. A complete unit, with controls, cost well over $1000, an almost unthinkable price in those days. However, the column itself could be purchased for some $250, which was thinkable. Bill Young bought the column on the strength of my promise that I would build a suitable control unit to use it efficiently.

It was my first foray into building a relatively complex apparatus. I was allowed to work in the physics shop under the tight supervision of their aged, but skillful, German instrument maker, who had grave and justified misgivings about my knowledge of metal working. Anyway, I planned and made a grand, highly automated column. Although it took all summer, the time was well spent because the apparatus worked from Day One, and the many distillations I had to carry out were quite automatic. The apparatus would run all day, with only occasional monitoring to determine which fractions of the distillate to combine, and the separations were excellent.

Our first objective was to repeat Stan Cristol's work on the carbonation of the butenyl Grignard reagent. As far as we could tell, there was only one acid formed (11), rather than the three possibilities suggested by the mixture of butenes formed from the Grignard reagent 8–10 with water (equation 6).[17] Direct interconversion of the *trans* form

$$\left[\underset{\mathbf{8}}{\overset{CH_3 \qquad H}{C{=}C}\atop{H \qquad CH_2\text{-}MgBr}} \overset{?}{\rightleftharpoons} \underset{\mathbf{9}}{\overset{CH_3 \qquad H}{CH{-}C}\atop{BrMg \qquad {=}CH_2}} \overset{?}{\rightleftharpoons} \underset{\mathbf{10}}{\overset{H \qquad H}{C{=}C}\atop{CH_3 \qquad CH_2\text{-}MgBr}} \right] \xrightarrow[2.\ H^{\oplus},\ H_2O]{1.\ CO_2} \underset{\mathbf{11}}{\overset{CH_3 \qquad H}{CH{-}C}\atop{CO_2H \qquad {=}CH_2}} \qquad (6)$$

8 with the *cis* form 10 seemed very unlikely on the basis of much other evidence that such processes are negligibly slow at room temperature.

Interconversion of either **8** or **10** through **9** does provide a mechanistically feasible route for the *trans* or *cis* interconversion.

So what was the formation of the single product **11** trying to tell us when water gave a mixture of butenes[8] and oxygen gave a mixture of alcohols[18] (albeit different compositions)?

The next experiment was directly inspired by Whitmore's work on Grignard additions to diisopropyl ketone,[15] which had so impressed me at the 1941 ACS meeting in Atlantic City. The crux of Whitmore's work was that he observed *no* addition of secondary Grignard reagents, such as isopropylmagnesium bromide, although the corresponding primary *n*-propyl derivative did add in reasonable yield.

Isopropyl- and *n*-propylmagnesium bromides are structurally analogous to the secondary and primary forms, **8** and **9**, respectively, of the butenyl Grignard. Therefore, a new working hypothesis was that **9**, like isopropylmagnesium bromide, would not add to diisopropyl ketone, **12**. However, if the equilibrium **8** ⇆ **9** ⇆ **10** existed and **8** (or **10**) would add as *n*-propylmagnesium bromide does, then all of the **9**

$$[8 \leftrightarrows 9 \leftrightarrows 10] + (CH_3)_2CH\text{—}C(=O)\text{—}CH(CH_3)_2 \;(\mathbf{12}) \xrightarrow{H_2O} CH_3CH{=}CHCH_2\text{—}C(OH)[CH(CH_3)_2]_2 \;(\mathbf{13})$$

$$\xrightarrow[?]{H_2O} CH_2{=}CH\text{—}CH(CH_3)\text{—}C(OH)[CH(CH_3)_2]_2 \;(\mathbf{14}) \qquad (7)$$

present would be converted to **8** (or **10**). Only the primary addition product **13** would be formed, and none of the secondary product **14** (equation 7).

Our hypothesis as to what should happen with the butenyl Grignard reagent and **12** did not survive unchanged. Both butenylmagnesium bromide and chloride added to **12** in excellent yields, of which 85% or more was **14**![19] Thus, we had unwittingly, and perhaps even unwillingly, made the first substances with three branched aliphatic groups connected to a single carbon atom. Interestingly, some **13** was formed but it was clearly a minor product. Quite a few years later, R.

Benkeser[20] at Purdue University discovered that the product ratio is actually time- and temperature-dependent. Apparently, addition corresponding to formation of **14** is fastest, but that corresponding to **13** is more energetically favorable. The longer you let the reaction mixture stand before adding water, the more **13** you get.

Why did these addition reactions proceed in such high yields when substances such as isopropylmagnesium bromide give no addition at all? Two possibilities came to mind: First, for some reason, the reactions competing with addition were slowed relative to addition, making addition the more favorable reaction. The second possibility was that the Grignard reagent has available to it some new and different way of adding to carbonyl groups.

The emergence of this second possibility as a leading contender for a viable working hypothesis now seems painfully slow. The more likely reason for our failure to grasp it at once was because so much of Young's earlier research had indicated that the butenyl Grignard reagent was a mixture of forms—a notion that did not die easily. Still, in hindsight, it is easy to decide that **8** (or **10**) is more stable than **9**. The primary isomer, **8**, of the butenyl Grignard reagent could react to give products that looked as though they were derived from **9** by the "ordinary" addition mechanism. This possibility had already been proposed by John R. Johnson in 1932 to account for the "abnormal" addition of formaldehyde to benzylmagnesium chloride.[21]

Johnson's legacy to organic chemistry is relatively small in terms of numbers of papers published. Nonetheless, he had a substantial influence through his advocacy of what he called "chelation" theory,[22] whereby many organic reactions could be correlated by considering that they occur "through an ephemeral cyclization within the primary reaction complex by means of a subsequent intramolecular coordination (chelation)."[23] The idea was not original with Johnson, but he developed it in detail, and it has become an important part of our understanding of organic reactions. The names "chelate" and "chelation" (derived from *chela*, a pincer-like claw) survive in several chemical contexts, but Johnson's chelate mechanisms are now most commonly known as cyclic mechanisms.

Johnson visualized benzylmagnesium chloride additions to a C=O bond, such as in formaldehyde, as involving an initial complexation, **15** (equation 8), but drawn to reflect, at least crudely, one of the

$$C_6H_5\text{—}CH_2MgCl + CH_2{=}O \rightleftharpoons C_6H_5(1,2)\text{—}CH_2\text{—}Mg\text{—}Cl \cdots O{=}CH_2 \quad (8)$$

15

possible geometrical configurations it might possess. Abnormal addition, when it occurs for formaldehyde, would proceed by a pathway involving **16** (equation 9). The abnormal reaction "arises as a result of

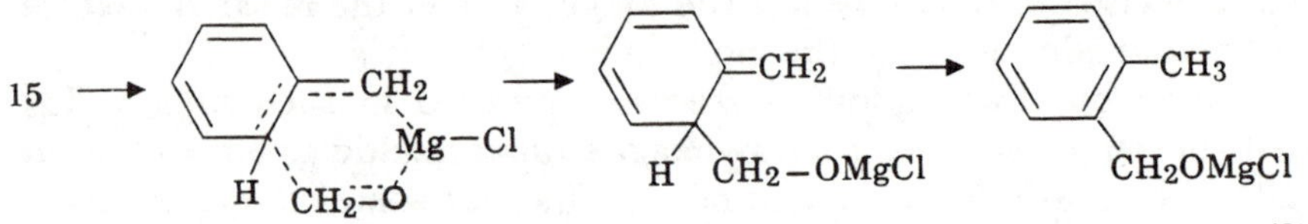

$$15 \longrightarrow \mathbf{16} \longrightarrow \mathbf{17a} \longrightarrow \mathbf{17b} \qquad (9)$$

(**16**: cyclic complex with ring, CH_2, $Mg-Cl$, H, $CH_2=O$; **17a**: $=CH_2$, H, $CH_2-OMgCl$; **17b**: $-CH_3$, CH_2OMgCl)

the ability of the Grignard reagent to forestall the normal reaction" by having the benzyl group become connected by its ring carbon atom (C-2) to the carbon of the carbonyl group to yield **17a** and then **17b**. Pathway **15** → **17a** is the "ephemeral cyclization" referred to by Johnson.[23]

Structure **17a** is rather unstable, involving destruction of the system of three cyclic conjugated double bonds of a benzene ring. Therefore, the corresponding reaction for the butenylmagnesium bromide, as **8**, can be expected to be much more facile by way of **18** (equation 10).

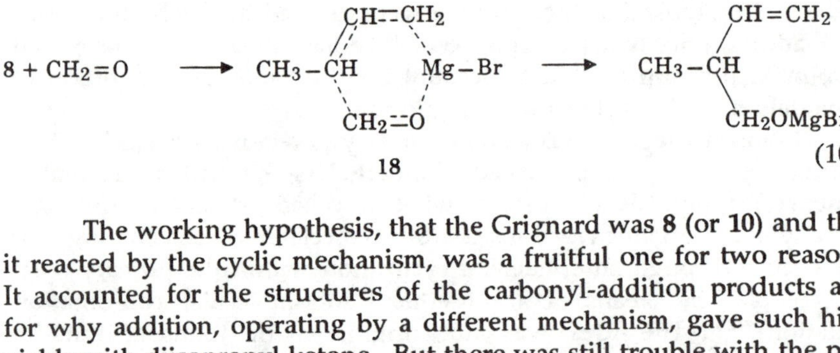

$$8 + CH_2{=}O \longrightarrow \underset{\mathbf{18}}{CH_3{-}CH(CH{=}CH_2)\cdots Mg{-}Br\cdots O{=}CH_2} \longrightarrow CH_3{-}CH(CH{=}CH_2)(CH_2OMgBr) \qquad (10)$$

The working hypothesis, that the Grignard was **8** (or **10**) and that it reacted by the cyclic mechanism, was a fruitful one for two reasons: It accounted for the structures of the carbonyl-addition products and for why addition, operating by a different mechanism, gave such high yields with diisopropyl ketone. But there was still trouble with the product mixtures formed with water and with oxygen.

That water might be classified as an unusual reagent was indicated by cleavage of the Grignard reagent with phenylacetylene as a proton donor. The product, analyzed for us by the then-newfangled infrared spectrometric technique at the Shell Development Company, was about 94% 1-butene and 6% *cis*-2-butene. This analysis is more consistent with the Grignard reagent being predominantly one form, rather than a mixture corresponding to the butene composition resulting from the reaction with water.[24] Unfortunately, the result does not tell us which is the dominant form of the Grignard reagent. So I ran through quite a few other reactions, which gave results generally consistent with our latest working hypothesis, but did not add much toward proving it. I even tried reducing the Grignard reagent with hydrogen, with the idea that removal of the double bond might tell which was the dominant form. Unfortunately, the reaction did not go.

The other interesting experiment involved the reaction of the Grignard reagent with acetomesitylene, **19**. The chemistry of this and

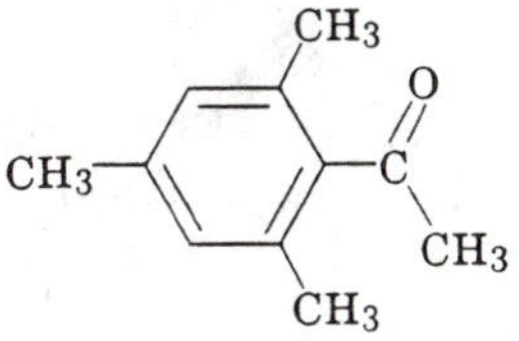

CH_3 O CH_3 C CH_3 CH_3

19

related compounds had been extensively investigated over the years by E. P. Kohler at Harvard and his student, R. C. Fuson, at the University of Illinois. The carbonyl carbon of **19** is rather well protected by the surrounding groups against ordinary addition involving Grignard or organolithium reagents. Indeed, no addition had ever been observed of such reagents. Instead, **19** acts as a proton donor to cleave the organometallic compound and yield a hydrocarbon and the enolate salt of **19**.[25]

It seemed possible that cleavage of the Grignard reagent with **19** might well occur *without* rearrangement, as shown by **20**–**21** (equation 11), where **8** is used as the Grignard structure and 2-butene would be the product. When I ran this reaction, I was quite annoyed to get a zero yield of butenes; a second run gave the same result. It was clear

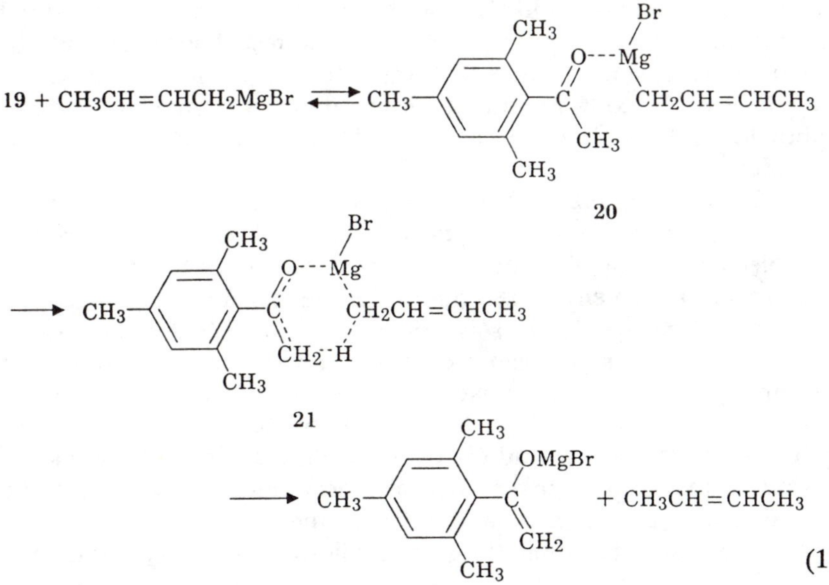

$$19 + CH_3CH{=}CHCH_2MgBr \rightleftharpoons CH_3\text{-}C_6H_2(CH_3)_2\text{-}C(CH_3){=}O\cdots Mg(Br)CH_2CH{=}CHCH_3$$

20

$$\longrightarrow CH_3\text{-}C_6H_2(CH_3)_2\text{-}C({=}O\cdots Mg(Br)CH_2CH{=}CHCH_3)CH_2\text{-}H$$

21

$$\longrightarrow CH_3\text{-}C_6H_2(CH_3)_2\text{-}C(OMgBr){=}CH_2 + CH_3CH{=}CHCH_3 \qquad (11)$$

that something unusual had happened, namely that a Grignard reagent had finally added to **19** with *no trace* of enolization. And so, what was the product? As usual, the product, **22**, was that expected of the cyclic mechanism operating from **8** (or **10**).[26]

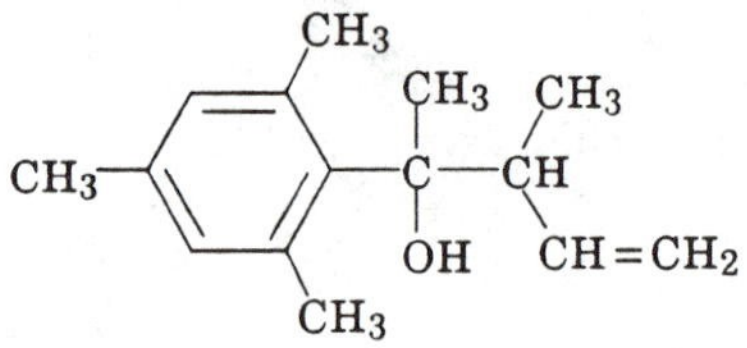

CH_3
CH_3 CH_3
CH_3—C—CH
OH $CH{=}CH_2$
CH_3

22

With acetomesitylene, the working hypothesis received a big boost. Addition occurred with this Grignard reagent and no other, not even with methylmagnesium bromide, and the product was the greatly crowded **22**. Obviously, this came about because of the special virtues accruing to the cyclic addition mechanism with **8** (or **10**).

This cyclic mechanism can also operate with benzyl Grignard and give "abnormal" products typical of addition to formaldehyde. Obviously, if normal addition is impossible, the benzyl Grignard, if it were to add, would have no choice but to add by the cyclic mechanism. So here was a chance to prove that the cyclic mechanism was responsible. I rushed to determine whether benzyl Grignard reagent would add at all and, if so, to demonstrate that only the abnormal product would be formed. The possible addition products are complex, but should be easily distinguished from one another by the fact that normal addition would insert a benzyl group that, on vigorous oxidation, would yield benzoic acid. In contrast, the abnormal addition product would yield *o*-phthalic acid. Addition occurred, and vigorous oxidation gave benzoic acid.[24]

So my great idea failed. Benzyl Grignard reagent could add to acetomesitylene in the normal way, without need for the cyclic mechanism. Needless to say, this finding left the structure of the butenyl Grignard reagent in something less than a definitive state. Some much more direct means was needed to solve the problem, and that means was not to be available for more than a decade. And so, some 11 months after starting my thesis research, I had built equipment, amassed a lot of data, generated a hypothesis about the central structural problem presented by the Grignard reagent, and discovered some startling new reactions. I was out of inspiration for new critical experiments, so it seemed to be a good time to wrap things up and write my thesis.

Not long after submitting an application for completion of the Ph.D., I got a summons to appear before Vern O. Knudsen, who was dean of the graduate school. Knudsen, a very distinguished acoustical physicist, was cheerful, well-tanned, and bald (today we would say "skin-headed"). His concern with me was my wholly nonexistent record of graduate courses. I agreed he was reading the record correctly and protested that it was a little late to be questioning my course record because I was already writing my thesis. "But surely you audited some courses that we could have you examined in, to generate some kind of academic record." I couldn't think of anything except graduate seminar, which wasn't very acceptable. He said he would talk to Bill Young. As far as I know, I got my Ph.D. without the appearance of taking courses.

UCLA had not been granting Ph.D. degrees for very long; indeed, their first Ph.D. degree was awarded to Gustav Albrecht in 1941. I am sure Knudsen was concerned that their fledgling Ph.D. program might be seriously set back if it was perceived to be shallow on requirements. I had hardly thought about taking courses when I was focusing so heavily on research. In retrospect, many things might have been easier later with some study of chemical physics, especially quantum mechanics, which was not a part of any of my studies at UCLA.

More on UCLA

Although I took no formal courses at UCLA, I learned a lot from the weekly graduate student seminars. Saul Winstein was running the seminar when my turn to talk came up. He asked (indeed, rather commanded) that I tell the group about the work done by Kirkwood and Westheimer[27] on substituent effects on the ionization constants of carboxylic acids. It was clear by the late 1930s that carboxylic acid ionizations offered an especially good way of systemizing the effects of substituents on reactivity. The work of Hammett[28] (to be discussed later) was especially relevant because, for the first time, it was possible to make genuinely quantitative correlations between reaction rates and equilibria of rather different kinds of processes. Saul wanted us all to understand the Hammett relationships better and, at the same time, was eager to have someone work through the Kirkwood–Westheimer papers, to see if it was likely to be important to his own research. It was a fortunate choice for me because I was pushed into something that had an enormous influence on my later work.

That some common inorganic acids are weak acids, while others are strong in water solution, has almost always been accepted as empirical fact. On the other hand, organic acids usually offer the possibility of small and incremental changes in structure. The Kirkwood–Westheimer approach is concerned with such changes. This

Frank H. Westheimer in his office at the University of Chicago, about 1950.

approach derives from some interesting speculations of the Danish chemist, N. Bjerrum, who was concerned with possible electrostatic effects of substituents on reactivity.[29] An example is trying to calculate the 80-fold-greater acid ionization of chloroacetic acid over acetic acid. This calculation depends on the fact that, with chloroacetic acid, the proton is in effect transferred from its bound position to infinity against (or assisted by) the electrostatic forces of a Cl–C dipole. In contrast, with acetic acid only a H–C dipole is involved.

The electrostatic energy change for taking a proton from the un-ionized acid to infinity according to Bjerrum is given by

$$\text{electrostatic energy change} = W_{\text{el}} = \frac{e\mu_{\text{CCl}} \cos \theta_{\text{CCl}}}{\epsilon r_{\text{CCl}}} \qquad (12)$$

where e is the numerical value of the charge on the proton, μ_{CCl} is the moment of the Cl–C dipole, cos θ_{CCl} is the angle that the line connecting the center of the dipole to the proton makes with the Cl–C bond, ϵ is the dielectric constant of the medium in which ionization occurs, and

r_{CCl} is the starting distance from the center of the dipole to the proton. Here, organic chemical reactivity is assessed in terms of simple physical laws. The formulation gains credence from the fact that if we substitute more chlorines on acetic acid (bigger dipole effect), the acid strength goes up, and if we make an analog of acetic acid with chlorines farther away from the ionizable proton, the acid strength goes down. Thus, $ClCH_2CH_2CO_2H$ and $ClCH_2CH_2CH_2CO_2H$ are weaker acids than chloroacetic acid, but still stronger than acetic acid, as expected for increasing r_{CCl}.

The Bjerrum formulation of the problem looks simpler than it actually is. One difficulty is assignment of values of r_{CCl} and θ_{CCl}. There are a number of possible geometrical arrangements, or conformations, of the chloroacetic acid, because of more or less freedom of rotation about the various bonds. The usual approach is to use reasonable average values of r_{CCl} and θ_{CCl} and get on with the calculation. Yes, but what are you going to use for ϵ, the dielectric constant of the medium?

Here Kirkwood and Westheimer made an important, even if approximate, step forward. They did what many entrepreneurs of science do, and that is to go in the right direction. This movement may not lead to a complete solution, but you hope to go at least as far as current knowledge and techniques will allow. If the step you make is really in the "right" direction, others will build on it. Such a step causes people to think and do experiments, even if the direction finally turns out to be a blind alley. Exploring blind alleys and marking them clearly is an important effort in science, even if it earns few "brownie points" for Nobel prizes.

Kirkwood and Westheimer attacked the dielectric-constant problem. Just how this collaboration between a theoretician and a physical organic chemist came about is not known to me. However, from experience with Frank Westheimer, I suspect he had the problem well defined and went to Jack Kirkwood to seek help on how it might be attacked. Their approach was to consider the acid undergoing ionization to be in a cavity of low dielectric constant, immersed in a medium of high dielectric constant, water. Obviously, in the absence of any kind of efficient calculator, it was necessary to use a model of feasible mathematical complexity. For the times, spherical or ellipsoidal cavities of uniform dielectric constant were appropriate.

As I read these papers for the first time (and was to reread them often, later), I was struck by the mathematical complexity of the problem. Things like Bessel functions were involved, about which I knew nothing. I hoped that Saul did not expect me to be able to explain how the equations were derived or even in any detail how the numerical calculations were made. The real point was that the treatment worked. It was possible to compute effective ϵ values on the order of 15 from the cavity model that could be used in the Bjerrum equation. My seminar on the Kirkwood–Westheimer papers appeared to go well. Saul seemed quite pleased, and I am sure he added the cavity theory to his mental tool box for dealing with new situations.

I had my final oral examination on October 20, 1944. It turned out to have the character of a seminar, and not many questions were asked. I would have liked to have been able to announce a more definitive solution to the structural problem presented by the butenyl Grignard reagent, but at least some exciting things had been found out and a reasonable working hypothesis was generated for others to follow or try to disprove (the more common incentive).

In October I was badly out of phase with the academic year. In order to get me back in phase, Bill Young arranged an instructorship for me—to teach analytical chemistry to Navy students in the wartime V12 training program in the spring of 1945. It was by no means clear what I would be doing after the instructorship was over. The main decision was, of course, between academic work and industry. Either choice meant moving elsewhere, because there was almost no very good industrial research in southern California, and my prospects for an academic position at UCLA or Caltech were not very promising at the time. Further, there was great uncertainty about what was going to happen in either sector when the transition from wartime to peacetime research occurred. Bill Young and Saul Winstein had consulting connections with the U.S. Rubber Company in Passaic, New Jersey. They had great faith in the basic polymer research programs being started there by Frank Mayo and Cheves Walling, former students of Morris Kharasch at the University of Chicago. These researchers were outstanding in free radical reactions, as well as physical organic chemistry in general.

As good as U.S. Rubber research was, E. I. du Pont de Nemours and Company was clearly the industrial leader in chemical research, with the development of nylon and neoprene as shining examples of how basic research could be turned into marketable products. I had an interview with the Du Pont recruiter, a man named H. W. Rinehart, in early 1945. It was my first experience in such affairs.

While I thought it went well, and I had found out a lot about Du Pont, Rinehart seemed to feel that somehow he was the one who was being interviewed. He told Bill Young later that I had made a poor impression by asking about how Du Pont operated, whether researchers kept their own hours, how projects were selected, and so on. I guess Rinehart felt I didn't appreciate that it was a privilege to be interviewed by Du Pont and to be invited for a visit later. And so I wasn't asked to visit Wilmington for a further interview.

The U.S. Rubber Company, with an expanding new program, and with Bill and Saul supporting me, was much more aggressive in offering a position. I did make an interview trip to Passaic and was enormously impressed with the programs initiated by Mayo, Walling, Matheson, and their colleagues. It appeared to be a wonderful opportunity to get in on the ground floor, but a combination of other events made me choose a different course, and it was fortunate that I did. Despite the superb progress the U.S. Rubber group made in delineating the mechanism of polymerization, of selectivity in copolymerization, and so on, the "bottom-line" people in management couldn't understand why better golf-ball covers were not coming out on a daily basis, and eventually the whole operation was dismantled. Mayo went to General Electric and Walling to Lever Brothers.

Although Bill Young was enthusiastic about the U.S. Rubber Company, he was also very encouraging about my trying to get into academic work. One way to start was to try for a National Research Council (NRC) fellowship, which was quite prestigious. Saul had an NRC fellowship to go to Harvard, and he also thought this was the way to go. The major part of the application was to offer a viable research proposal, and here Whitmore came to my rescue. The particular edition[30] of Whitmore's organic chemistry book that I purchased was a treasury of odd chemical facts. Whitmore seemed to include a lot of miscellany that just struck his fancy. For reasons not wholly clear to me now, I was becoming interested in small-ring compounds. Perhaps it was because not much physical organic chemistry had been done on them, and the ratio of the possibility of significant results to molecular size seemed favorable. Whitmore's book devoted considerable attention to small-ring compounds, and I remember particularly the statement that "cyclopropanol apparently does not exist."

The "existence" concept in organic chemistry has changed markedly over my lifetime. When organic chemists characterized their preparations by boiling point or melting point, by elemental analysis, and by degradative reactions, many only relatively reactive compounds could not pass muster. In 1945, the idea that substances could be stable and characterized only at –80 °C was not a part of the practicing chemist's thinking. Cyclopropene is an interesting example. It was made by the Russian chemists Demjanow and Dojarenko in 1923 and found to be stable only in the gas phase, to react with bromine in a flash of flame, and to form a silver nitrate complex in the same way that an acetylene derivative does.[31] This report of a compound easily expected on a classical basis to "nonexist" was difficult to accept at face value. However, the Dojarenko work has been amply confirmed; by me, among others.

An iffy synthesis of cyclopropanol hardly seemed to be enough for a proper proposal for the National Research Council. Although cyclopropyl chloride had been reported as a stable entity by the great Russian chemist Gustavson, precious little had been published about its properties. The possibility of making and studying the reactions of this compound seemed more attractive, particularly when coupled with a study of cyclopropylmethyl derivatives to further determine the degree to which cyclopropyl groups share properties with vinyl groups. Close similarity had been pointed out many times, with strong emphasis on bromine addition to cyclopropane. Such additions were probably regarded as the most characteristic reaction of vinyl compounds.

Simple extension of this line of reasoning would make cyclopropanol unstable like vinyl alcohol, cyclopropanone reactive like ketene, cyclopropyl chloride unreactive like vinyl chloride, and cyclopropylmethyl derivatives reactive like allyl derivatives. Such ideas made up the bulk of my fellowship proposal to the National Research Council.

Whether I could make it in academic circles as a researcher was only one question. There was also the problem of whether or not, with my hearing difficulties, I could be effective in the classroom. In most usual one-to-one activities, I could get by, aided by lipreading, but that was not going to work with large groups. Fortunately, just about the time that my need for them became acute, electronic hearing aids were coming on the market. Their development had been assisted by war research that required reliable small vacuum tubes. These aids made it possible for me to meet classes and answer questions, at least passably well.

Edith and I eloped to Las Vegas and were married on July 11, 1942. The date (7/11) seemed like a good omen for the future. We lived in a lovely apartment close by the UCLA campus, which, at $50 per month, was a real strain on our budget. Indeed, in the first few weeks of married life, our parents were probably surprised at how often we came to dinner—we were temporarily destitute! Edith worked as an underwriter for wartime cargo insurance in downtown Los Angeles and customarily had a hard run to catch the bus for the long, slow ride to the city. She seldom was able to get home before 6:30 or 7:00. If I was not working in the lab, I had the choice of either fixing dinner or trying to learn to play the piano that my sister-in-law generously left with us while her husband was overseas in the Marines. The piano playing was hard on our neighbors in the apartment beneath us. When their patience was exhausted, they would bang on the ceiling with a broom handle. Many nights I worked late and Edith would cook dinner in the laboratory over a Bunsen burner.

Edith and I developed a wonderful relationship with Bill and Helen Young. They often went on trips, and we had the opportunity to take care of their lovely home and cat. They were excellent at bridge, and we played with them often. Also, the UCLA faculty had an active dance group, with the Youngs, Winsteins, and Jacobs as active participants among the chemists. Saul was a superb dancer, and these parties seemed to have been a major source of relaxation for him. While I was not much shakes at dancing, we went to almost all of these affairs and got to know many of the other faculty members. Although I was ultralow on the UCLA faculty roster, Bill Young provided excellent opportunities to find out what was going on university-wide. Two or three days a week he would invite me to go with him to a Westwood restaurant to have lunch with a variety of his academic cronies.

While still at UCLA, I decided to jump the gun and go to work on that part of my National Research Council proposal that had to do with the chlorination of cyclopropane. It turned out that the dominating factor in cyclopropane chlorination is that chlorination of cyclopropyl chloride to form 1,1-dichlorocyclopropane proceeds much faster than formation of cyclopropyl chloride itself. Therefore, you need to have a substantial excess of cyclopropane over chlorine to keep the 1,1-dichloro compound from being the dominant product.

So what I did was to connect a cyclopropane cylinder to a home-built gas aspirator so that one volume of high-pressure gas from the cylinder would recycle three to four volumes of gas from the excess cyclopropane accumulating at the far end of the reaction system. The combined volumes went through a drying train, mixed with chlorine, were exposed to light from several photoflood lamps, ran through a water wash and a drying column, and then fractionally distilled to separate the cyclopropane. The cyclopropane was sent off to a flask, where it was kept boiling under a dry ice condenser while the vapor was drawn off by the gas aspirator and recycled. It was a rather impressive operation, and it allowed me to make several hundred grams of cyclopropyl chloride.[32]

Although I had been writing and submitting papers with Bill Young, I found things were a bit different when I sent in the cyclopropane chlorination work to the *Journal of the American Chemical Society*. The referees stomped on me hard. Arthur B. Lamb of Harvard, then the editor and a kindly gentleman if there ever was one, wrote a long letter to accompany the referee reports. He softened the blow and suggested a complete rearrangement of the presentation, in addition to showing how the paper could be shortened.

With cyclopropyl chloride in hand, I began to see if it could be converted to the Grignard reagent and thence to cyclopropanol. This was not easy. It took all of my prior experience, entrainment, fresh magnesium turnings, scratching, and iodine treatment, to get the reaction finally to go in small yield. I would guess that chlorobenzene is comparably difficult. But the Grignard reagent did form, and I was able to prove it was there by conversion to cyclopropanecarboxylic acid derivatives. I was also able to oxygenate it and make a crystalline cyclopropanol derivative.[33]

After I published the cyclopropane chlorination, Whitmore sent me a note that one of his people had also made cyclopropyl chloride, but was unable to get the Grignard reaction to go under any circumstances. I was pretty puffed up about the cyclopropanol synthesis, until Tom Jacobs said: "But that's not new. Cyclopropanol was reported by Cottle[34] in the *Journal* not so long ago." And so it was. He

$$C_2H_5MgBr + \underset{O}{\overset{CH_2}{|}}\!\!>CH{-}CH_2Cl \xrightarrow{FeCl_3} \xrightarrow{H_2O} \underset{CH_2}{\overset{CH_2}{|}}\!\!>CH{-}OH \qquad (13)$$

had formed cyclopropanol by a reaction one would hardly pick as a rational synthesis, but which gave about 45% yields (equation 13).

Charles Coryell had gone off to Oak Ridge and was not to return to UCLA. I needed a counterpart in physical chemistry to help me put together a sensible program to study the properties of the small-ring compounds I had made. Fortunately (again), there was just such a person at UCLA—Max T. Rogers, a temporary appointee who had earned his Ph.D. at Caltech doing work on electron diffraction and crystal structure.

Max lived a life of threescore and ten years, all of them underappreciated by chemists in general. Much of his difficulty came from the fact that he did not look the part of a dynamic, pioneering chemist. He was of medium height, sort of flat-faced, and with an upstanding shock of black hair that made him look a lot like Stan Laurel. But this was not the real Max. He was a very broad-gauge scientist with a wonderful sense of humor, which was all the more wonderful for its unexpected nature. He had not been told what was really expected of him, research-wise, at UCLA. He was in a sort of fill-in position for Coryell, and so he welcomed the opportunity to use his versatile talents to help work on my compounds. It was a zany collaboration. We interacted like a couple of Roman candles, worked outrageous hours to accomplish what we wanted to get done, and were full of laughter. Sometimes when things went wrong we would drop into our chairs and laugh and laugh—all the while trying to think of how we were to get around the obstacle, whatever it was.

We decided early that dipole moments were our best bet to determine the degree of interaction of the cyclopropane ring and attached substituents, relative to the same kinds of interactions for alkenes and larger-ring cycloalkanes.[35a] But the apparatus had to be more accurate than that used in the McMillan–Roberts undergraduate research. So we followed the designs of C. P. Smythe[35b] of Princeton University, the leading figure in dielectric and dipole measurements. Our detector was coupled to an oscilloscope so that we could revel in the beauty and variety of Lissajous figures*, as well as make blinking stop lights and rotating patterns galore. The calculations were laborious, but we had use of Coryell's magical Marchant electrical calculator, which would do long division automatically, albeit with much whirring and clanking. Square roots were more difficult, but Marchant provided a handy table with which quite good approximations could be made.

Max's wife, Ethel, was an attractive, vivacious woman. She, Max, Edith, and I celebrated the end of the war in Europe, VE Day, with a boisterous party at the Winstein's home. About 2 a.m. the Winsteins went to bed and the rest of us passed out on the living-room carpet, not leaving until well after breakfast.

The news of the first atomic bomb came to most of us as a great surprise. We knew, of course, that there was super-secret war research at Tennessee and somewhere in the wilds of the states of Washington and New Mexico. However, those of us at the low end of the totem pole and already out of war research knew nothing about the specifics of what was hoped to be achieved or the fantastic degree of progress toward those ends. As soon as possible, I got a copy of the Smythe report on the project and became as familiar as I could with the elements of what was involved.

In the early days of the Atomic Age (if Hiroshima can be said to mark time zero), we were immensely relieved to see the end of the war in the Pacific. The bloody battles at Guadalcanal, Coral Sea, Iwo Jima, and so on, which seemed to lead inexorably to a protracted, devastating invasion of the Japanese mainland, were very much on our minds, particularly because Edith's brother-in-law, a major in the Marines, had been captured in the Philippines and died on a Japanese prison ship.

The larger, longer-term implications of atomic bombs were much less in the forefront. We seemed to be assured by our political leaders, and it would appear that they were desperately trying to assure themselves, that the atomic bomb had a secret that would be impenetrable for a long time to come. Further, the horror of Hiroshima and Nagasaki was so great that surely there would be a way to control the atomic genii. For a few weeks we didn't get much done around the lab. We grouped and regrouped in various bull sessions, trying to understand the science and engineering involved, as well as the social and political issues.

---

* Beat frequency patterns formed when different frequencies are connected to the x and y axes.

For Edith and me, the die had finally been cast as to the next year. I heard from Paul Bartlett informally that the National Research Council fellowship board was going to act favorably on my application. He said that his colleagues in the selection process were not much impressed with my proposal, but that he had convinced them that there was possible merit therein. A chance to go to Harvard was not to be taken lightly, and we began to plan in that direction.

Still, I felt the need to lay the groundwork for a later return to the West Coast, and I called on both Stanford and Berkeley to see if anything could be set up. Berkeley, at this time, was a great chemical power. G. N. Lewis, W. Latimer, W. Giaque, T. D. Stewart, Melvin Calvin, Glenn Seaborg, W. Bray, and many others made up a fantastic array of chemical luminaries. I had been following the progress of S. Ruben, whose work on the path of carbon in photosynthesis using the short-lived radioisotope ^{11}C made an enormous impression on me. It was a real blow to me, even though I had never met Ruben, to hear of his tragic death from phosgene poisoning in the laboratory at Berkeley.

Bill Young arranged an audience for me with Wendell Latimer, who was then dean, or chairman. Latimer listened and sounded encouraging when I said I would like to carry on Ruben's tracer work on photosynthetic processes with the facilities at Berkeley. If he knew about Melvin Calvin's intentions to do the same thing, he didn't say anything. Of course, my credentials for starting such a program were nonexistent. I had never worked with radioisotopes and knew nothing about plant biochemistry. All I had was confidence that I could do the organic chemistry, and I did know this was an important problem. Latimer gave me the "We'll see; contact me again", which at least was not "No".

Another one of my fortunate happenings, which came around the end of my postdoctoral period at UCLA, was the availability of cyclopropyl methyl ketone in substantial quantities, at a very reasonable price. This substance, which was a bonanza as a raw material for my small-ring research, became available as the result of wartime synthesis of an antimalarial drug, atabrine. I purchased a kilo or more of the material to take to Harvard, because there was no assurance that it would remain commercially available.

When the time came to move to Harvard in 1945, we went by way of a National ACS meeting in New York. The meeting featured a lecture by Arthur C. Cope, lately of Columbia University and on his way to be head of chemistry at MIT. The subject was what is now known as the Cope rearrangement, and he had many beautiful examples involving shifts of allyl groups from carbon to carbon. I was enormously impressed by Cope's thoroughness and thoughtfulness. Although then only 36, he was clearly destined for leadership.

National Research Council Fellowship Year at Harvard

Edith and I arrived in Boston in September 1945, a most fortunate time to come to Harvard, but a most unfortunate time to find housing in Cambridge. At least the situation was dynamic. The war was over, there was an enormous influx of students under the GI bill, but there was also an efflux of workers whose wartime projects were closing down, Navy V12 students whose training was terminated, and so on. Harvard had a housing office with a System for listing rentals. Each day at 4 p.m., they put out the new listings for the day. There was a genuine mob scene as the lists were thrown out to the waiting wolves. Those with automobiles had their motors running and quickly sped off to what they hoped would be the choicer locations. The rest followed on bicycles or on foot. When the slower pokes finally got to a place on the list, it was either already rented or there was a line going in (and a line coming out) of something like a one-room apartment with bath upstairs, an electric roaster in a closet for a kitchen, and a high rent. After a couple of days of this kind of frustration, Edith took to ringing doorbells to find a place. It was not long before she found us an "apartment" in a large, rather dowdy, house on Wendell Street, within short walking distance of the chemical laboratories. We had the living room and kitchen of the original house as our part of the dwelling, with a bathroom widely shared with the upstairs residents. Although the apartment was furnished at least basically, we had a problem because an enormous theatrical trunk that contained the bulk of our earthly goods had been lost in shipment and could not be easily traced. Each arm of the hydralike New Haven Railroad passed the buck on to another arm, and we were both frustrated and worried because winter was coming and what we had in warm clothes and blankets was in the trunk.

Clearly, Wendell Street was not going to be satisfactory for the whole year, so Edith began a campaign to get us into housing owned by Harvard, which was the best thing going in Cambridge. The prize place was Holden Green, just a short way up Kirkland Street from Harvard Yard. Often called the "rabbit warren", Holden Green was a complex of apartments and small row houses that had been the gift of a wealthy lady who wished each of the married couples living therein to have appropriate training for real life. For this reason, each apartment had its own coal-burning furnace! But Harvard students, being both smart and unwilling to stoke furnaces, had found ways to get them converted to gas and regulated by thermostats. The apartments were mostly semifurnished with extraordinarily heavy and durable solid maple tables, chairs, and bureaus. The rents were reasonable and the intellectual atmosphere superb. The problem was to get in, and Edith was determined that we would.

Holden Green was run by the Harvard Housing Trust, with an office in Harvard Square, and the Trust had a System. The key was the List of Ten, who were assured of housing as it became available when one reached the head of the list. The problem was to get on the bottom. The System decreed that when a unit was made available to whomever was topmost on the list, the next person who phoned or came into the office got the 10th spot on the list. C. Gardner Swain, of whom I write more later, claimed to have spent $32 in nickels on phone calls before he got on the list. The Trust would not let you wait in the office, but you could come in any number of times a day to check on potential vacancies. Who devised the System, and why, was not known to us, but Edith got a book and undertook a marathon vigil in the hallway outside the Trust Office. After a few days, I believe, the woman in the office took pity on her and let her wait inside. Anyway, in a week, she got us on the list and then we only had to wait our turn to rise to the top.

The trunk was still missing, but, as you might have expected, at the beginning of December, just as we reached the top of the Trust's list, Edith was searching herself and finally found the trunk sitting on a loading dock of the New Haven Railroad in south Boston. We were fortunate enough to get a downstairs corner apartment with living room, bedroom, kitchen, bath, and coal furnace—the last of which we quickly converted to gas. Here we lived for the next seven and a half years. With but one bedroom, things became a bit cramped when our daughter Anne arrived on the scene in February 1951. The situation was almost ludicrous in the spring of 1953, when Edith's parents came to take care of Anne and Don (born in October 1952) while we went to

The downtown Boston skyline as viewed from MIT in 1948.

Europe. With four adults and two children in the apartment, my father-in-law observed that it would be ever so much simpler if we were all the same sex.

Once we were settled into Holden Green, Edith was able to use her insurance expertise to become assistant to Donald Warren, a partner in Field and Cowles, a large Boston firm that was an agent for the Insurance Company of North America, for which Edith had worked in Los Angeles. Although Holden Green was wonderfully located for me, it was less favorable for Edith to get to downtown Boston, and she had to take a bus to either Harvard or Kendall Square to get the subway to the city. In those days, Boston was pretty much a low-rise city. The skyline was dominated by the Custom House, the gold dome of the capitol on Beacon Hill, and, "out west", there were the Prudential and New England Mutual medium-rise towers.

At Harvard, in 1945, war research was winding down and there was a strong spirit of, "Let's get started again with the basics." The war research had brought new perspectives, techniques, and experiences. Paul Bartlett, because he was in a transition period, was very accessible. In fact, some of us were able to have lunch with him 2 or 3 days a week—episodes we looked forward to eagerly. At this time, Bartlett

Paul D. Bartlett, at his home in Weston, Massachusetts, about 1948.

was 34 years old and already recognized as a world leader in physical organic chemistry. I always think of him as a sort of typical Vermonter, but in fact, he was born in Butler, Indiana. Of good height, lean but not thin, with rugged features, he could pass as a woodsman, except for his heavy glasses that reflected earlier serious eye problems and severe near-sightedness. He seemed somewhat reserved and often spoke with abruptness, which masked a marvelous sense of humor that could erupt into warm and vigorous laughter. Paul liked to be precise; his words were carefully chosen and his sentences concise. He was not worldly in the fashionable sense, but worldly in the scholarly sense, with broad interests and high ideals. Paul married Mary Lula Court when he got his Ph.D. in 1931. Always known as Lou, she was highly intelligent, somewhat self-effacing, a wonderful person to be with, and a great complement to Paul.

Paul was not noted as a laboratory technician, but he had excellent ideas about the way to get experiments done. With the superb students he had, he was able to investigate some very difficult experimental problems. Bartlett's research was very important in bringing a strong emphasis to the "organic" in physical organic chemistry.

Harvard was a yeasty place then. Among the stars of the organic chemistry professorial faculty in 1945, besides Paul Bartlett, were Louis F. Fieser, R. P. Linstead, and R. B. Woodward. Linstead was just leaving to return to England to head organic chemistry at Imperial College, and his brilliant student, W. v. E. Doering, who had collaborated with Woodward on a much-publicized synthesis of quinine, had already departed for Columbia University.

The phalanx of graduate students and postdoctoral associates at Harvard during that period was absolutely outstanding in terms of future achievement. Among those I remember were George S. Hammond, Jerrold Meinwald, William G. Dauben (whose brother Hyp had left the year before for the University of Washington), Donald J. Cram, David Y. Curtin, Herbert Gutowsky, Harry Wasserman, C. Gardner Swain, Bernard Witkop, Harold Kwart, William Sager, J. E. Loeffler, Elliot Alexander, S. D. Ross, Abraham Schneider, Samuel Siegel, Blaine C. McKusick, Stanley J. Cristol, and Charles Heidelberger (the son of the famous immunochemist Michael Heidelberger). Of the group, 11 eventually became members of the National Academy of Sciences. Although not all of those listed are household words of chemistry today, each exerted a very timely and unique influence either at Harvard or, later, on my own work, as well as on chemistry as a whole.

It was a very highly competitive group, but a friendly one as well. We would often play touch football on the grass of the inner courtyard between the Converse and Mallinckrodt Laboratories. If Louie Fieser happened by during one of these sessions, he would doff his fur-collared greatcoat, put down his gold-crowned walking stick, and join in the game.

I was assigned a laboratory just across the hall from Bartlett's office, a well-worn room with a smallish double hood, but highly prized for being a corner laboratory with a view over a lovely large New England home toward Sanders Theater and beyond, to Harvard Yard. As I was outfitting the laboratory, a rather strange, wild-eyed apparition appeared with a wrench in hand, unscrewed the water aspirator from the water pipe on my lab desk top, and announced: "That is mine, the only good one at Harvard." Without another word, he bolted out the door. This was my introduction to Jack Loeffler, whom I did not see again for several years, and who became an imaginative and productive physical organic chemist at Florida State University.

The provocation occasioned by Jack Loeffler's decamping with the only good water aspirator was nothing compared to that engendered by my co-occupant of the laboratory, Elliot R. Alexander. Like

Elliot R. Alexander, then professor at the University of Illinois, not long before he died tragically with his wife in an airplane accident, 1951. Alexander was author of the first undergraduate-level book on reaction mechanisms, Principles of Ionic Chemical Reactions, *published by Wiley in 1950.*

me, Elliot was an independent postdoctoral associate who chose to come to work in Bartlett's laboratory. He was a Swarthmore graduate who did his Ph.D. work with Cope at Columbia, then worked for a year or two at the Central Research Department of Du Pont, where, I gathered, he had made a substantial impression. Elliot was an interesting and superdynamic person. Handsome, a natural athlete, and opinionated to the point of exasperation, he was not a laboratory partner who allowed one to pursue one's own course unchallenged. In fact, virtually every day in the laboratory was a debating day, and sometimes these debates got pretty heated. One sure thing about Elliot was that, although he had a lot of ideas, and good ones, about the things he wanted to do, his approach to science deeply reflected his background at Columbia and at Du Pont—conservative.

One of our biggest areas of contention was in structural theory—Elliot simply did not believe at that time that the concept of resonance was here to stay. After much argument, he usually came down to saying, ". . . you may be right, but I think in 10 years it will be replaced by some other theory." He was prescient in that molecular orbital theory provided a strong challenge a few years later, but it never quite displaced resonance for qualitative purposes.

Although Elliot seemed confident enough in his views to warrant the level of dogmatism he affected, there was a characteristic that possibly indicated some degree of insecurity. He was an intense nail biter, with his nails always chewed down to the quick. Without doubt, we led a stormy life as lab partners, but it lasted only for a few months. At the end of this time Elliot left for Berkeley for the other half of his postdoctoral year, and I had the laboratory to myself. When Elliot left he sold his Model A Ford roadster to Bartlett. To combat the winter cold on the long drive in from his home in Weston, Bartlett used to wear a war-surplus aviator's suit wired for internal electric heating. He was quite a sight in this rig.

The group of young turks who were then postdoctoral fellows at Harvard were attracted to R. B. Woodward like flies to honey. Woodward, perhaps the leading figure of organic synthesis in the 20th century, was already the young man of the hour because of his success with W. v. E. Doering on the synthesis of quinine. Woodward was a child prodigy whose performance in freshman chemistry was so outstanding that he was listed as student of the year on a plaque in the lecture hall at MIT. However, he lost interest or, more likely, patience, with the customary undergraduate routine and dropped out.

Some time thereafter, James F. Norris, the leading organic chemist at MIT, heard about a bright young man doing some outstanding research in the Food Technology Department on the constituents of coffee. Norris investigated, found Woodward, and made arrangements so that Bob could fulfill the undergraduate requirements by examination. Furthermore, he was allowed to do an essentially independent Ph.D. project. After this, he spent a summer as instructor at the University of Illinois and then moved to Harvard, first as an assistant. Later he was appointed to the Society of Fellows, a group of brilliant researchers in all academic fields who were encouraged to follow their own research agendas.

Robert Burns Woodward at his home in Belmont, Massachusetts, about 1950.

Throughout his life, Woodward was a heavy cigarette smoker. His fingers were heavily nicotine-stained, and he had a rather precise and mannered way of smoking. He dressed fastidiously, but monotonously—blue was his color. In the early days, he usually wore a gray sport coat with a white shirt and light-blue wool tie; in later years, he almost invariably wore a dark-blue suit, white shirt, and the light-blue tie. He affected blue automobiles and had the ceiling of his office painted dark blue. His students even painted his parking place light blue. He was a person of high intellect, and his general manner was imperturbable. He spoke relatively slowly in very well-chosen words and consistently used some uncommon pronunciations, such as "mole-ecule" rather than "moll-ecule".

Woodward's lecturing style was mannered, but extraordinarily clear. He greatly preferred to lecture at a chalkboard instead of using slides, and was absolutely meticulous in drawing organic structures. He simply would not abbreviate a single structure in detailing a sequence of synthetic steps. If the sequence involved modification of a steroid nucleus, he would draw, for each step, the whole steroid nucleus with the complete stereochemistry, slowly and carefully making each ring geometrically exact. Unlike many professors, he did not talk and draw at the same time.

Pedagogically, it was wonderful. By the time the lecture was over, the basic structures were firmly implanted in your mind and you understood what he was telling you. The problem was that, if the synthesis had many steps, just drawing molecular formulas was very time-consuming. As a result, the lecture could become very long indeed—3 or 4 hours was not unusual. Our postdoctoral group loved to congregate in Woodward's office after seminars, either for a rehash of the seminar or to listen to his views on current developments. Many of these sessions were socratic, with problems posed, discussed, and solved. Others were more delphic. Bartlett accused us of going to the horse's mouth, and observed (correctly), "He will practically neigh for you." We generated three axioms about Woodward: He never got drunk, he never got tired, and he never perspired. Each of these became less axiomatic on one occasion or the other, but they held up very well indeed for many years.

The other principal organic chemist at Harvard was Louis F. Fieser. Fieser's office was directly over Woodward's, but they had little in common. Fieser's passions were doing chemistry with his own hands, in his personal laboratory next to his office, and writing books with the help of his wife, Mary, who had been one of his Ph.D. students when he was on the faculty at Bryn Mawr. The Fiesers seemed to be a very smoothly operating team. They had no children, but they fancied Siamese cats and used sketches of them as decorations in their famous organic chemistry texts.

By modern standards, Louis was a rather old-fashioned chemist, although he had done some excellent physicochemical measurements on reduction potentials of quinones. Louis was especially proud of the synthetic sequence in his laboratory manual that led to the preparation of "Martius Yellow". It was on a relatively small scale, required preparation of seven compounds in all, and was regarded as an excellent test of preparative technique, at least of solid aromatic derivatives. Fieser made the sequence into a test of efficiency and speed in actual course work, so that the students were made to race against each other and him as well.

During the period while I was at Harvard, none among the talented graduate students and postdoctoral fellows then interested in physical organic chemistry seemed as impressively brilliant as C. Gardner Swain, a Bartlett graduate student who had moved up to postdoctoral fellow the year I came. He was later to spend time at Caltech before going to MIT as instructor.

Gardner was enormously enthusiastic, bright as a new silver dollar. He had an amazing talent for ferreting out simplicities in complex problems, which he turned to great advantage in the design of experiments. Gardner often did seem a bit unworldly. He was extremely boyish and almost too enthusiastic at times, especially when he described his physical feats, like skiing at Tuckerman's Ravine on Mt. Washington or climbing two or three peaks in the Presidential Range in one day. But it was stimulating to talk with him and experience the clarity with which he viewed science. He presented quite a contrast to Winstein, who tended to take the opposite track—feeling that nothing was really simple and trying to anticipate every subtlety.

In my final days at Harvard, I mostly tried to develop, from my store of cyclopropyl methyl ketone, syntheses of what I hoped would be interesting small-ring compounds for future study. For example, the haloform reaction to prepare cyclopropanecarboxylic acid, reduction to cyclopropylmethanol, and the Beckman rearrangement of the ketoxime to ultimately form cyclopropylamine.[33]

The Move to MIT

One day Arthur C. Cope dropped into my lab at Harvard and asked me if I would like an instructorship at MIT. This question really took me by surprise, particularly because, at the time, I had not really thought about staying in the East. My chips, as it were, were down on Berkeley. Furthermore, up to that point, I had fallen in with the general view around Harvard, fostered in part by Bob Woodward, that organic chemistry at MIT was "the pits". The description almost surely applied before Cope arrived on the scene. It was confirmed by my visits to MIT to attend the local American Chemical Society meetings, which were held in MIT's large lecture hall. Compared to UCLA's and Harvard's conventional university architecture, MIT, essentially one enormous interconnected set of buildings, seemed like a foreign world. It was somehow cold, with the mien more of a factory than of an educational institution. Further, there was a perception that MIT was dominated by engineering, for which science existed to provide educational services.

To my surprise, Paul Bartlett was strongly in favor of my going to MIT. He had a lot of faith in Cope and believed he could make MIT into a very attractive place. He understood that Roger Adams, "The Chief" at Illinois, had given Karl Compton, then president of MIT, a

(Left to right) John D. Roberts, C. Gardner Swain, Marguerite Swain, and Arthur C. Cope in MIT's large lecture hall (Room 10–250), awaiting the start of a meeting of the Northeastern Section of the American Chemical Society, about 1948.

strong pitch that the overall level of MIT chemistry was not first-rate and something should be done about it—preferably by taking on Cope as department head. Bill Young shortly told me that there was no hope for Berkeley, and that I should cast my lot with MIT, again because of Cope.

The unanimous enthusiasm shown for Cope and MIT by people for whom I had the greatest respect, along with Edith's liking for Cambridge and Boston, provided a strong impetus for accepting Cope's offer. When I visited MIT, it was clear that a very extensive renovation was in progress, with research labs planned that would have stainless steel bench tops and sinks—quite a departure from the norm of either black painted wood or oiled Alberene stone tops and heavy red or black tile sinks. Cope had been given rather a free hand to improve the situation at MIT. As head of the department, not as chairman, he had the power to appoint, apparently subject only to the approval of the higher-ups in the MIT administration.

The George Eastman Laboratory at MIT, 1949. The lefthand part of the building was for chemistry, the right for physics.

A few long-standing members of the MIT faculty, which was terribly inbred in the pre-Cope days, had been hired (perhaps more accurately, kept on) to teach freshman chemistry. As is not uncommon in a research university with excellent undergraduates, they had problems keeping up with the changes in chemistry and had been relegated to teaching quiz sections or just to keeping the books on grades or supplies. Rightly or wrongly, this plight of tenured professors gave me a strong prejudice against non-research-oriented professors teaching undergraduates in highly research-oriented universities.

Even though I was on the staff of MIT, Harvard allowed us to live in their housing near the chemistry department. This was great, because it meant I could conveniently go to the Harvard evening organic chemistry colloquia and the late Friday afternoon Bartlett physical organic chemistry seminars. MIT had no really comparable seminars. The Harvard meetings were fantastic because they were so frequently used by the speakers to announce new findings of historical importance. As just one example, Gilbert Stork gave the first exposition of his idea of the way that lanosterol and cholesterol were derived from squalene by cationic cyclizations. The Bartlett seminars were equally significant, and the discussions were superbly critical—perhaps almost too much so, if you judge from having a speaker interrupted before completion of a *first* long-winded sentence.

Moving to MIT just meant going off in a new direction each morning. Because we could not yet afford a car (my starting annual salary was $3200), I rode a bicycle or took a bus. My initial arrangements at MIT were not encouraging. As the low man on the academic totem pole, I was assigned to a quiz section in elementary organic chemistry and to a lab and lecture section, but I was also allowed to give an advanced organic chemistry course, of which more later.

My office–laboratory, because the renovation was not finished, was far away from the center of organic chemistry activity and in the second oldest MIT building on the Cambridge site, but it had my name painted on the door–an MIT custom of the day and perhaps today as well. Mail was delivered twice daily by a white-coated mailman. When the door was unlocked, he came in and put the mail on your desk, directly in front of you—in effect, challenging you to ignore it. Most of my mail was from John Wiley or was some kind of institute information that characteristically came in white envelopes, at the rate of 3–10 items in each mail delivery. When other things began to be important, I

Arthur C. Cope with muskelunge trophies in his den in Belmont, Massachusetts, 1948.

tended to ignore or heave out, unopened, those deadly white envelopes.

At the time Sheehan, Swain, and I arrived, the organic chemistry staff was more than a little inbred. MIT's premier organic chemists in earlier times were Samuel Mulliken and James F. Norris. Mulliken was very well known for his work on organic qualitative analysis and also as the father of Robert S. Mulliken, an outstanding theoretical chemist on the staff of the University of Chicago, who later received the Nobel prize for his contributions to molecular orbital theory. James F. Norris was primarily a physical organic chemist who measured rates of organic reactions, often interesting ones. Unlike James B. Conant, his neighbor at Harvard, he seemed incapable of knowing how to interpret the results.

Art Cope had the power to change the curriculum and who taught the courses, but he understandably chose to go a little slowly before throwing his new organic-chemistry-staff recruits into the fray. Although Cope showed restraint at first, it soon became clear to me that this man was a tough cookie. Of medium height and strong build, with a long face, curly blond hair, and strong jutting jaw, Art spoke carefully and quietly—indeed, he seemed mild and conciliatory, almost effete. But when you got to know him, you found out that behind that façade, the juices really flowed. When the going got rough, his long, slender, white hands seemed to flutter; he would speak more softly and sink deeper into his chair. You learned to watch for those danger signs. His graduate students labeled him the "iron fist in the velvet glove".

When I arrived on the scene at MIT, Cope was just getting started with a massive research effort on cyclooctatetraene, which was clearly one of the most marvelous problems of all time in organic chemistry. Cope decided to repeat the Willstätter synthesis of cyclooctatetraene, which was particularly pertinent at the time. Just a few years before the war, the idea emerged that Willstätter had not actually prepared cyclooctatetraene, but instead had obtained its well-known isomer, styrene, which gives a solid dibromide with a melting point near the one Willstätter reported for the dibromide of cyclooctatetraene.

The cyclooctatetraene skeptics included R. B. Woodward, who was said to have laid claim to being the first to have noticed the resemblance of the dibromide to that formed from styrene. Not at all explained by the skeptics was the yellow color Willstätter reported for cyclooctatetraene and its conversion by hydrogenation and oxidation to octanedioic acid. Also driving Cope was the news that the redoubtable German chemist Walter Reppe had in fact prepared cyclooctatetraene in one step by a catalytic tetramerization of acetylene under pressure. As more detail came in on Reppe's work, cyclooctatetraene began to look like a chameleon, reacting with various reagents to give products with quite different ring systems. We could hardly wait to see how much of it could be confirmed.

The other important new faculty member in organic chemistry, when I started at MIT, was John C. Sheehan. This extraordinarily capable organic chemist had obtained his doctorate with W. E. Bachman at the University of Michigan while working on a war project aimed at improving the preparation of the brisant military explosive, RDX. He had then gone to Merck, where he worked on penicillin. When he came to MIT, his prime research goal was to synthesize penicillin, then a

John C. Sheehan, professor of chemistry at MIT, renowned for his successful synthesis of the antibiotic penicillin and for precursors that have allowed the preparation of a variety of penicillin analogs with the important property of at least transitory resistance to deactivation by bacterial β-lactamases.

most difficult undertaking. This synthesis had eluded Bob Woodward, who had made himself one of the ranking experts on penicillin. Because of its relatively low molecular weight and extraordinary antibiotic activity, it had attracted wide attention, particularly from the legendary Robert Robinson of Oxford. The penicillin story was complex, and there was intense competition to determine the unusual structure and develop a synthesis more efficient than fermentation.

John's personality is such that, despite his success with penicillin and other difficult synthetic problems, it is sometimes hard to take him seriously as a great chemist. Woodward looked and acted the part. John, in contrast, always seems more like an Irish political figure. He plays his chemical cards close to his chest, and his imaginative, shrewd chemical judgment is often not appreciated as it should be. One characteristic that separated him from the Woodward afficionados was that he was more frankly guided by instinct than by physical organic theory. He used theory ably, but his papers seldom reflected the fact, and he resisted presenting ex post facto rationalizations as a priori reasoning. John and I generally had excellent relations with Art Cope, who treated us much more as equals than would normally be expected, considering our differences in age and position. But then Art was, at the outset, relatively isolated from the senior incumbent members of the department.

Often one thinks about the things that might have been. In 1948, MIT had such an opportunity brought about by a visit from Carl Djerassi, now professor at Stanford, but at the time, working for Ciba in New Jersey. Carl was an emigrant from Austria, who obtained his undergraduate degree at Kenyon College in 1942 and his Ph.D. at the University of Wisconsin in 1945. An extraordinarily talented and entrepreneurial person, at the time of his visit he was applying the Wohl–Ziegler (*N*-bromosuccinimide) reaction to steroids and wrote a comprehensive *Chemical Reviews* article[36] about it. It seemed clear that Carl was looking for an academic position and Cope was very impressed with his talent, vigor, and enthusiasm. However, Art hesitated, I think, because he was concerned about adding still another person who might rock the boat in the organic chemistry group. He was having plenty of trouble already with collisions between the older professorial members and the rest of us. Nonetheless, it would have been really interesting to have had Carl at MIT, and it might well have changed the course of much of steroid chemistry. At MIT, Carl may not have wanted to become affiliated with Syntex, and without Carl, it is unlikely that Syntex would have developed into such a world power in steroid chemistry.

Research at MIT: Carbocations

Getting started in research at MIT was not easy. I had no real lab space for students, if any came along; no research funds of my own; and, as I have said, I was located well out of the mainstream of the organic chemistry activity. My initial plans were to continue with cyclopropane derivatives and other small-ring compounds, apply the work I had done at Harvard to the synthesis of cyclopropene, and start looking into the electrical effects of substituent groups on reactivity and dipole-moment measurements. A little later, I decided to investigate the use of ^{14}C as a tracer for organic reaction mechanisms. The electrical-effect work was to be directed to the interpretation of substituent-group influences, as measured by their σ constants. In effect, I intended to wade into the interpretative thicket that Hammett chose to skirt. The groups that looked promising for study, at the outset, were trimethylsilyl, trifluoromethyl, trimethylammonium, and cyano.

My plans for the use of ^{14}C as a tracer were the direct result of Art Cope's asking me if I had any problems that might warrant support by MIT's new Laboratory for Nuclear Science and Engineering. I wrote up something for Art about studying reaction mechanisms with ^{14}C. Lo and behold, before long he came back with funding for my research. It was very generous funding indeed for a new instructor, $44,000 for the fiscal year 1947–1948. In fact, it would be a generous sum for a grant made today, when corrected for inflation and not burdened with overhead and staff benefits. I had no idea how to spend such an enormous sum, but Art had some ideas for expenditures that would generally benefit the organic chemistry group, and I was pleased to accommodate. Because it was clear that I did not need so much support in subsequent years, the level of support was halved, but that was still very generous.

A procedure for ^{14}C analysis was worked out by my first two students at MIT, E. W. Holroyd, Jr., and Winifred Bennett.[37] Neither was at the top of the class. Holroyd was an M.S. candidate studying under the GI bill, a man of great cheerfulness and doggedness, but not well prepared for graduate work. Winifred did her senior thesis with me and was determined to do a synthesis project, preferably good for a Nobel prize and definitely not one directed to mundane small-ring compounds. Desperate to get someone working with me, I kept her interest up by devising a route to synthesize the unknown pentalene.

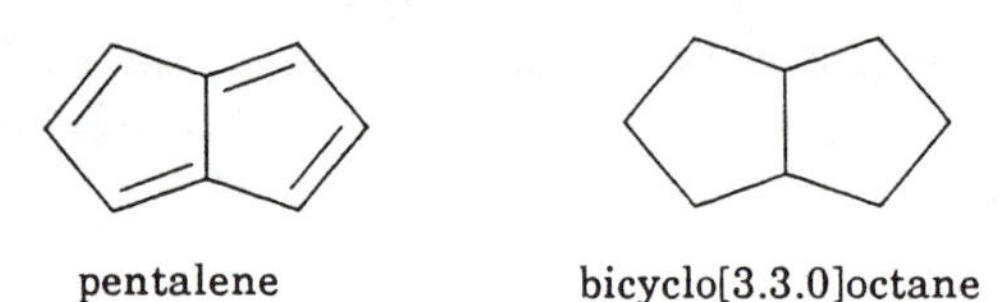

Unfortunately, the basic bicyclo[3.3.0]octane ring system was not very easily accessible. The pentalene problem was a rather hot one in 1946. There was considerable division of theoretical opinion as to whether it would be like cyclooctatetraene or quite different. My plan was to get into the bicyclo[3.3.0]octane system by converting readily available dicyclopentadiene to a dicarboxybicyclo[3.3.0]octane (equation 14).

From the diacid, various routes could be envisioned to insert the double bonds of the pentalene structure. We could not make the oxidation step work in a practical way. If nothing else, the effort wasted on this project convinced me that there was not much to be gained by putting senior thesis students to work on complex multistep syntheses, and I vowed to eschew glamor for problems that I was reasonably sure would give tangible results.

H H $\xrightarrow[H_2SO_4]{H_2O}$ HO H H

$\xrightarrow[Pt]{H_2}$ HO H H HO_2C $\xrightarrow{[O]}$ HO_2C H H HO_2C (14)

Our work on small-ring compounds and carbocation (carbonium-ion) rearrangements turned out to be much more fundamental and important. These programs started out quite separately and, later, achieved such substantial overlap that it almost became hard to tell one from the other. Getting students was easier by this time because I had moved into my spanking-new third-floor office–laboratory, complete with stainless steel benchtops and hood. This change of venue brought me into the center of organic chemistry activity. Art Cope's office was a few yards away, I was next to the microanalytical laboratory, and, across the hall, my large laboratory was contiguous with those occupied by co-workers of Cope and Sheehan. With ample funds, I could even think about postdoctoral fellows.

My first postdoctoral associate was Donald J. Cram, who just finished his Ph.D. with Louis Fieser at Harvard and wanted a summer job before taking off to UCLA as a postdoctoral fellow. We had gotten to know the Great Cram and his charming wife, Jean, rather near the end of our Harvard stay, and had hit if off well right from the beginning. Flamboyant, immensely energetic, and vocal, Cram exuded apparent self confidence that, in fact, served to mask a measure of insecurity. He and Jean were fun to be with, having had a host of amusing experiences, and they were full of outrageous chatter. One of Don's more revealing statements to me was: "My trouble is that my subconscious is too closely connected to my mouth." At my urging, Cram had already signed up to do postdoctoral work at UCLA, but first came his period of travail as my postdoctoral fellow, hired for the twin objectives of mak-

Donald J. Cram and his first wife, Jean, in Cambridge, about 1947.

ing methylenecyclobutane and deuterioacetic acid, the latter of which I wanted for some purpose now unclear.

During most of the summer of 1948, Edith and I were away in California, visiting our families and friends at UCLA and Caltech, while Don was at work in my new laboratory–office. On our return, I was no end put out by the fact that the last step in the preparation of methylenecyclobutane had gotten away from him. My beautiful new stainless steel hood was badly stained, and permanently so. To make matters even less satisfactory, I had no methylenecyclobutane to show for the money and damage.

There was more success with the deuterioacetic acid project. The idea was to make it by adding heavy water to carbon suboxide and then decarboxylating the resulting deuteriomalonic acid (equation 15). Don

$$O=C=C=C=O + 2D_2O \longrightarrow DO_2C-CD_2-CO_2D \xrightarrow{\text{heat}} CD_3CO_2D + CO_2 \tag{15}$$

made this work. It also sparked his interest in the chemistry of carbon suboxide, which led to some later nice publications for him at UCLA.

I carried out the synthesis of cyclopropene and 1,2-dibromocyclopropane myself. It went very smoothly, but broke no new ground except for the new route to cyclopropylamine from cyclopropyl methyl ketone. I was quite impressed by the avidity of cyclopropene for bromine. If the reaction was allowed to go too fast, combination occurred with a flash of flame. Iodine also added readily, to give a colorless diiodide.

I wanted very much to get someone to start with methylenecyclobutane and synthesize cyclobutanone so that we could convert it to cyclobutane (by the then-fashionable Wolff–Kishner reduction) and thence to cyclobutyl chloride for study of solvolysis rates and products. However, students have their own ideas of what they want their theses to cover. So Charles William Sauer, my first Ph.D. candidate to sign on the dotted line, although willing enough to synthesize cyclobutanone, was more interested in seeing whether it could be converted to cyclobutadiene.

Charlie Sauer was round-faced, a bit plumpish, rather formal in bearing and dress, with some tendency to flutter when hard pressed, and possessed of a rather shy sense of humor. He was a careful and meticulous laboratory worker, a quality that was excellent for the cyclobutanone project, and he carried out with distinction new syntheses of not only cyclobutanone, but also cyclobutylamine, cyclobutene, and cyclobutane.[38] Charlie's ambitions to get to cyclobutadiene were foiled at several junction points. Then, because he was running out of material and dreaded starting all over again, way back, with the preparation of methylenecyclobutane, he started to say, ". . . time to write my thesis." So another cyclobutadiene synthesis straggled away in the sandy wastes.

One of the joys of being a professor is when an exceptional student comes along and wants to work with you. Such a person was Robert H. Mazur, who did his undergraduate work at MIT and in later life became famous as the discoverer of NutraSweet at the G. D. Searle Company in Chicago. He was an intensely serious person with a deep interest in mathematics. It was surprising that Bob Mazur decided to get a Ph.D. in organic chemistry. Already balding in 1948, he seemed to be hidden behind his thick-lensed glasses. You could hardly see his eyes, but he had a ready smile and a fine sense of humor. He was dedicated, skillful, and meticulous—qualities that were to be important in later years when the veracity of his results was brought into question by H. C. Brown and others. The research that Bob did, when pub-

Robert H. Mazur as a graduate student at MIT in 1949. He later did outstanding research at G. D. Searle Company on peptides, which led to the discovery of NutraSweet.

lished, became one of the most-cited articles in chemical history.[39] It was especially important as the opening of the Pandora's box of an extraordinarily difficult and subtle problem—a problem concerned in an important way with what we mean when we write chemical structures on paper. This problem has resisted experimental resolution even to this day, some 40 years later.

Bob Mazur's primary objective was to investigate the reactions of cyclopropylmethyl compounds that, in reactions where carbocations could be expected to function as intermediates, had a pattern of giving mixtures of cyclopropylmethyl and cyclobutyl derivatives. Thus, the Russian chemist Demjanow[40] reported that both cyclopropylmethylamines and cyclobutylamines with aqueous nitrous acid give mixtures of cyclopropylmethanol, **23**, and cyclobutanol, **24**. Bob found that their

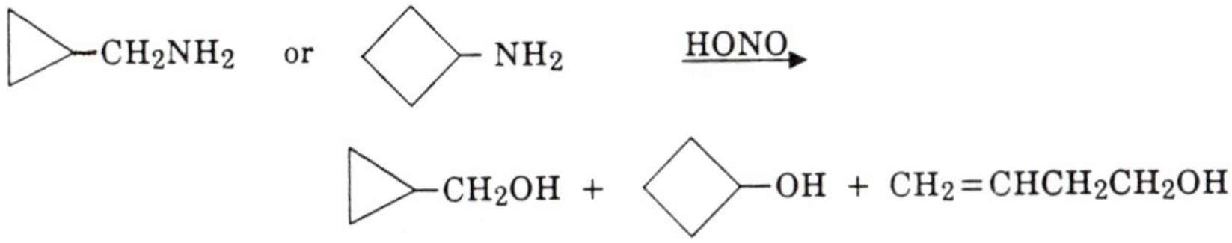

$$\triangleright\!-CH_2NH_2 \quad \text{or} \quad \diamond\!-NH_2 \xrightarrow{HONO}$$

$$\triangleright\!-CH_2OH + \diamond\!-OH + CH_2{=}CHCH_2CH_2OH \qquad (16)$$

23	24	25
~48%	~47%	~5%

report was incomplete. In fact, 3-buten-1-ol, **25**, was also formed (equation 16).

The relative percent yield figures given for the alcohols formed are probably not terribly accurate. Because gas–liquid-phase chromatography was not yet available, Bob Mazur analyzed the alcohol mixtures, using somewhat catch-as-catch-can procedures. At least by this time we had entered the modern instrumental age, with a Baird double-beam infrared spectrometer available for general organic chemistry purposes, and we used that instrument very hard and productively. Complications arose in many of Mazur's studies of reactions of cyclopropylmethyl compounds. From reaction to reaction, he found that sometimes kinetic control was dominant, at other times thermodynamic control, and at still other times combinations of these factors prevailed. In analyzing the results, it is important to remember that cyclopropylmethyl compounds are *much less stable* than the corresponding cyclobutyl compounds, and both are *very much less stable* than the 3-butenyl derivatives (*see* structures **26–28**).

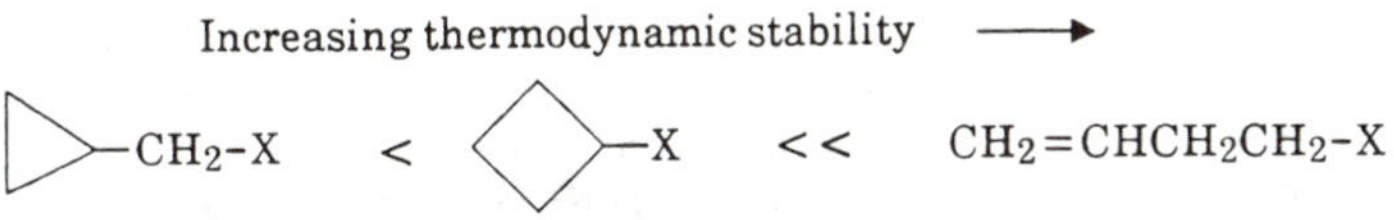

Increasing thermodynamic stability ⟶

$$\underset{26}{\triangleright\!-CH_2\text{-}X} \;<\; \underset{27}{\diamond\!-X} \;\ll\; \underset{28}{CH_2{=}CHCH_2CH_2\text{-}X}$$

As we got more and more into this research, we began to get more and more tangled up with Saul Winstein's interests. None of us foresaw the potential overlap at the outset, because so little was known about the interrelationships of the rearrangements that occurred in compounds embodying the structural features of **26–28**. Saul regarded *i*-sterol rearrangements as being within his purview of carbocation reactions. The *i*-sterol rearrangement is well illustrated by ionization of **29** by removal of X as X^-, which would be expected to form the cation **30**. However, if the ionization is carried out in methanol, the product is **31**, called 3,5-cyclo-6-cholesteryl methyl ether (equation 17).

$$\mathbf{29} \xrightarrow{-X^{\ominus}} \mathbf{30} \xrightarrow{CH_3OH} \mathbf{31} \qquad (17)$$

I remember Saul telling me, when I was still at UCLA, how very interesting these observations (some of them made by Byron Riegel at Northwestern) were to theories of carbocation reactions. One of the striking features of the *i*-sterol rearrangement is that nowhere does a cyclobutyl derivative appear to be formed. This fact was probably important in keeping us from recognizing that there was a relationship between reactions involving **26–28**. The *i*-sterol rearrangement, of course, correlates with the aforementioned formation of some **25**, in the reactions of cyclopropylmethyl- and cyclobutylamines with nitrous acid. Occurrence of **23–25** in the alcohol mixtures might seem easily explained by rapid equilibration of the carbocations, **32–34** (equation 18). The ratios of the alcohols formed would be expected to depend on the concentrations of each cation and the rates at which each reacts with water.

$$\underset{\mathbf{32}}{\text{cyclopropyl-}\overset{\oplus}{C}H_2} \rightleftharpoons \underset{\mathbf{33}}{\text{cyclobutyl}^{\oplus}} \overset{\oplus}{\rightleftharpoons} \underset{\mathbf{34}}{CH_2{=}CHCH_2{-}\overset{\oplus}{C}H_2} \qquad (18)$$

The then-contemporary discussions of the ratios of products to be expected from carbocationic reactions usually emphasized only relative carbocation stabilities as the determining factor and did not take into account the relative rates of the carbocations with the solvent. In processes where kinetic control is operative, proper account must also be taken of the *relative stabilities of the reactants*. With most acyclic substances relative stability is often neglected, because the thermodynamic stabilities of the reacting isomers are not very different compared to the energy differences that correspond to solvolysis rate differences. With strained molecules undergoing solvolysis, relative stability can be a very important factor.

The concept of a "driving force" in solvolysis reactions for formation of a more stable entity is not a new idea. Ionizations of isomeric allylic halides proceed much more rapidly than would be expected if the nondelocalized carbocations were the initial products. This concept was first suggested in 1939 by Nevell, de Salas, and Wilson,[41] who studied the solvolysis of isobornyl chloride, **35**, and camphene hydrochloride, **36**.

Rearrangement of **37** to **38** is the result of a Wagner–Meerwein 1,2-bond shift converting a secondary cation into a tertiary cation. The rearrangement is expected to be energetically favorable and, under kinetic control, to lead to formation of the tertiary alcohol, **39** (equation 19). The solvolysis of **36** appeared to Wilson to be extremely facile. He formulated cation **38** with an unusual bridged structure, **40**, wherein the

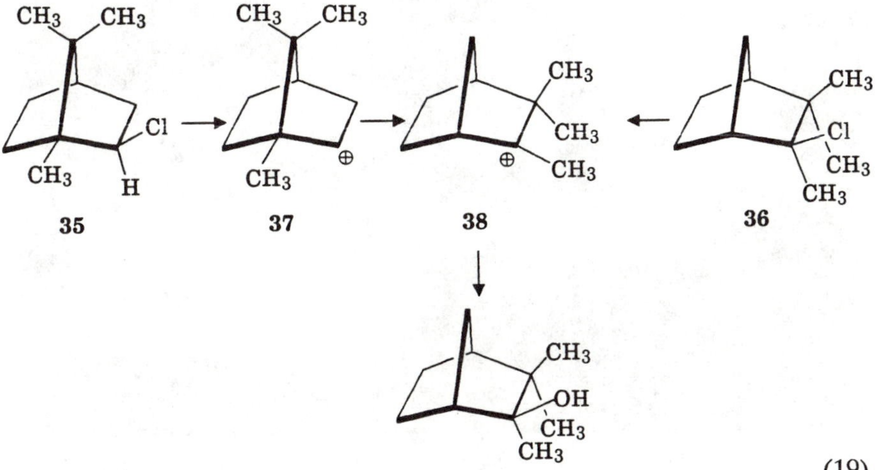

CH_3 CH_3 CH_3 CH_3

Cl → ⊕ → ⊕ ←

CH_3 H CH_3 CH_3 CH_3 Cl CH_3 CH_3

35 **37** **38** **36**

↓

CH_3 OH CH_3 CH_3

39 (19)

CH_2 at position 6 is partially bonded to both C-1 and C-2. This additional bonding, and the electron delocalization implied by it, could confer extra stabilization, analogously to the stabilization of the allylic cation. The suggestion of structure **40**, according to Christopher Wilson's own recollection to me later, when he was a professor at Ohio State, was made in an off-hand and tongue-in-cheek way. Despite having unleashed the first salvo in the famous (or infamous) war over "nonclassical" carbocations, which lasted for some 35 years, Wilson seemed mostly a bemused bystander during its height.

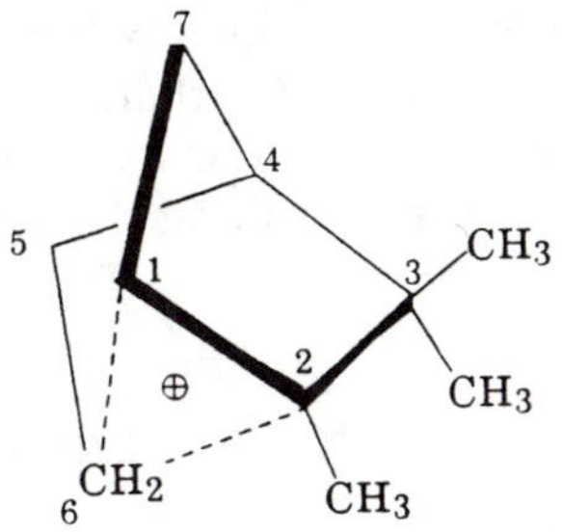

7 4 5 1 3 CH_3 2 ⊕ CH_3 $_6CH_2$ CH_3

40

Christopher Wilson in his office at Ohio State University, about 1950. When at University College, London, Wilson was the first to propose and offer rate-enhancement data as well as stereochemical evidence for a bridged nonclassical cation.

At the outset, nonclassical cations seemed to be invoked whenever one encountered unexpected behavior in a carbocation reaction. Later, there was more or less general consent on a narrower definition based on the idea of having unusual bonding. In particular, such bonding related to the type predicted for H_3^+. This ion, known to exist in the gas phase, is predicted to be much more stable in the form of an equilateral triangle, **41**, than with a linear arrangement of its three hydrogens, **42**. An important early analogue of this kind of bonding was provided by "bromonium" cations, which turned out to be much more stable in the bridged form, **43**, than in the linear form, or what Winstein liked to call the "open" form, **44**. Winstein's doctoral research (p 17) was very important in establishing the significance of **43** as a reac-

H—H (⊕) H (triangle) **41**

$[H{-}H{-}H]^{\oplus}$ **42**

$CH_3CH{-}CHCH_3$ bridged by $Br^{\oplus}$ **43**

$CH_3CH(Br){-}\overset{\oplus}{C}HCH_3$ **44**

tion intermediate. That achievement, of course, made it natural for him to be interested in other possible bridged carbocations. The camphanonium cation as structure **40** is bridged in analogy to **41** and **44**; the carbons at the cationic centers of **37** and **38** are open cations analogous to **44**.

Mazur's first problem was in the preparation of the pure cyclopropylmethyl, cyclobutyl, and 3-buten-1-yl halides. The last two were simple. Pure cyclobutyl chloride was obtained by photochemical chlorination of cyclobutane by a vastly simpler procedure than the one I had used earlier for cyclopropane. In fact, the procedure was modeled after a chlorination of toluene that I saw being carried out at a Caltech Open House in the early 1930s. The 3-buten-1-yl chloride came easily from the corresponding alcohol and thionyl chloride.

Cyclopropylmethyl chloride (and cyclopropylmethyl bromide) presented problems. Reactions that were noted for producing chlorides without rearrangement from alcohols by way of silyl ethers gave products containing substantial amounts of cyclobutyl chloride. Apparently, even a *tendency* for carbocation formation in a reaction became competitively favorable and led to some rearrangement product.

The only source of cyclopropylmethyl chloride that did not contain cyclobutyl chloride was vapor-phase, free-radical, photochemically induced chlorination of methylcyclopropane. However, the product mixture so obtained, besides containing what appeared to be ring-chlorinated methylcyclopropanes, afforded nearly equal amounts of cyclopropylmethyl chloride and 3-buten-1-yl chloride. This free-radical substitution process, unlike carbocationic reactions, gave apparent conversion of the cyclopropylmethyl radical to the 3-buten-1-yl radical without intervention of the cyclobutyl radical. Further, the cyclobutyl radical does not give rearrangement products.

Some further mechanistic relationships are revealed by the facts that cyclobutyl halides can be converted to Grignard reagents and that these reagents react without rearrangement to give cyclobutyl derivatives. In contrast, cyclopropylmethyl or 3-buten-1-yl halides, when converted to Grignard reagents, react to give 3-buten-1-yl derivatives. This pattern of rearrangement offers the prospect of determining whether or not a carbocation is involved in particular reactions by determining if a cyclobutyl reactant gives any cyclopropylmethyl product, or vice versa, in that reaction. No interconversion implies no carbocationic intermediate.

We knew that cyclopropylmethyl- and cyclobutylamines gave very similar product mixtures with nitrous acid, and it became a critical issue as to whether 3-buten-1-ylamine would behave in the same way. The reaction products, actually a quite complex mixture, were difficult to analyze (equation 20), but Mazur was equal to the task. He found, for the alcohol mixture, that about 25% of the alcohol product is clearly

$$CH_2{=}CHCH_2CH_2NH_2 \xrightarrow[H_2O]{HONO}$$
45

$$CH_2{=}CHCH_2CH_2OH + CH_2{=}CHCH(OH)CH_3 + HOCH_2CH{=}CHCH_3$$
46 **47** **48**

$$+ \text{cyclopropyl}{-}CH_2OH + \text{cyclobutanol (H, OH)} \qquad (20)$$
49, 14% **50**, 13%

the result of ring closure. It is very important that the approximately 1:1 ratio of **49** to **50** corresponds quite well to the ratio observed for cyclopropylmethyl- and cyclobutylamines with nitrous acid. But there is a large fraction of 3-buten-1-ol, **46**, and two new products, 3-buten-2-ol, **47**, and 2-buten-1-ol, **48**, which result from 1,2-hydrogen migrations. The approximately 2:1 ratio of **47** to **48** formed turned out to be comparable to the ratios found by Mazur in the deamination of 3-buten-2-yl- and 2-buten-1-ylamines with nitrous acid.

The pattern of products that arise from 3-buten-1-ylamine, **45**, and nitrous acid makes it obvious that there was something wrong with the earlier working hypothesis that the 3-butenyl cation was in equilibrium with cyclopropylmethyl and cyclobutyl carbocations. One would expect that much more 3-buten-1-ol, **46**, would be formed along with **47** and **48**, the products of hydrogen migration, when one started with cyclopropylmethylamine or cyclobutylamine.

The rates of solvolysis of the various chlorides in 50% water–50% ethanol (by volume) indicated that cyclopropylmethyl chloride was exceptionally reactive for a primary halide; the conventional order of reactivity is primary << secondary <<< tertiary. It was a surprise and a thrill to find that a cyclopropane ring could activate a primary halide for solvolysis more than a double bond. Cyclobutyl chloride also turned out to be quite reactive: 40 times more reactive than cyclohexyl chloride and about three times more reactive than cyclopentyl chloride.

$$\text{cyclopropyl}{-}CH_2Cl \; : \; \text{cyclobutyl}{-}Cl \; : \; CH_2{=}CHCH_2CH_2Cl$$
$$1 \qquad 0.037 \qquad \sim 0.0004$$

This finding was not only a surprise, but a shock. My studies at Harvard of the solvolysis of cyclopropyl chloride had shown it to be very unreactive indeed, and surely several thousand times less reactive than cyclohexyl chloride. By the arguments advanced about the effect of angle strain on carbocation stabilities, the cyclobutyl carbocation should have much greater angle strain than cyclobutyl chloride. This difference in angle strain should markedly reduce cyclobutyl chloride's solvolytic reactivity. But here it was, highly reactive, and thus it presented another puzzle.

There were more surprises than the unusual reactivities. Mazur found that the rate of solvolysis of cyclopropylmethyl chloride dropped off rather rapidly in the manner associated with internal return.[10] Here, internal return was occurring in 50% ethanol, a most unlikely candidate for an internal-return-favorable solvent. To make things more interesting, the overall internal-return rate was fully 25% of the total reaction rate. Most of this gave cyclobutyl chloride, but substantial amounts of 3-buten-1-yl chloride were also formed. Further, internal-return rearrangement of cyclobutyl chloride to 3-buten-1-yl chloride also seemed to occur in 50% ethanol. In the more favorable internal-return solvent, acetic acid, internal return was about twice as fast as solvolysis with cyclopropylmethyl chloride; cyclobutyl chloride was formed about twice as fast as 3-buten-1-yl chloride. With cyclobutyl chloride, internal-return isomerization to 3-buten-1-yl chloride occurred at almost the same rate as solvolysis.

There was no evidence of isomerization to the allylic chlorides corresponding to 1,2-hydrogen migration expected if $CH_2{=}CHCH_2CH_2^+$ were formed. Here again, it appears that $CH_2{=}CHCH_2CH_2^+$ is not the intermediate that is the precursor to the 3-buten-1-yl chloride. It is relevant that aqueous hydrolysis of cyclopropylmethyl and cyclobutyl chlorides gave alcohol mixtures with infrared spectra identical to those from deamination of the corresponding amines.

The super solvolytic activity of cyclopropylmethyl chloride suggests, as one possibility, formation of an unusually stable carbocation. An alternative would be driving force supplied by a significant relief of the angle strain of the cyclopropane ring through formation of the 3-buten-1-yl cation, rather than the cyclopropylmethyl cation. This cation would have some 10 kcal less angle strain, as judged by the 10-kcal difference in free energies of formation of cyclopropane and propene. If all of this relief of strain were realized in the transition state, ionization would be accelerated by as much as a factor of 10^7. This is almost the degree of acceleration required to make a primary chloride as reactive as a somewhat sluggish tertiary chloride.

The bad news for this suggestion is that the 3-buten-1-yl cation gives the wrong products. It seems inconceivable that it would have sufficient positive charge at C-3 and C-4 to yield essentially equal quantities of cyclobutanol and cyclopropylmethanol, with only a small amount of 3-buten-1-ol. Direct formation of the cyclobutyl cation with relief of strain inherent in the cyclopropane ring is a more realistic possibility. However, the relief of strain would be very substantially smaller because conversion of methylcyclopropane to cyclobutane is associated with only about 1 kcal relief of strain. And how can the cyclobutyl cation be so organized as to possess sufficient positive charge at C-2 and C-4 to permit attack of the solvent, put back the angle strain lost in the ionization, and yield as much cyclopropylmethanol as cyclobutanol? How can there be enough positive charge at C-3 to allow formation of as much as 5% 3-buten-1-ol?

We seem to be left with the alternative that an unusually stable cyclopropylmethyl-like cation is formed in the ionization of cyclopropylmethyl chloride. Saying this does not preclude the possibility that more than one carbocation is involved in the formation of the various products. In Mazur's work, we postulated that two carbocations were involved, one formed from cyclopropylmethyl chloride and the other from cyclobutyl chloride, which itself is unusually reactive. Why is it so stable? Relief of strain? Electron delocalization? Binding in the carbocation of a type not available for stabilization of more conventional carbocations?[39]

To really understand the relationships between reactivity, carbocation stabilities, and stabilities of the starting materials and products, it is very important to map out an energy diagram for each system. I had made a serious mistake by not doing that when thinking about the relative solvolytic reactivities of nortricyclyl and dehydronorbornyl derivatives.[42] Later, Winstein and Kosower[43] made a beautiful analysis of the problems involved with derivatives of this kind and published, to my knowledge, the first complete energy diagram with reactants, intermediates, and products of the type discussed earlier.

This approach had been used schematically before with reference to $C_4H_7^+$. As an example, Figure 1 is taken from a slide that I used at a symposium on reaction mechanisms, held on September 7, 1951, at the 75th anniversary celebration of the American Chemical Society in New York. The Winstein–Kosower approach was very different in being quantitative. Although the Winstein–Kosower paper was published in 1959, Herbert C. Brown attributes the concept of such diagrams to Goering and Schwene[44], whose paper came out more than six years after that of Winstein and Kosower. The charitable explanation is that Brown never read the Winstein–Kosower paper, but I'm little inclined to be charitable in this particular situation.

Because this same redoubtable Herbert C. Brown will be a major figure in some of what is to follow, it is appropriate to detail my first contretemps with him. This occurred shortly after he moved from Wayne State University to Purdue, and the issue was one that involved

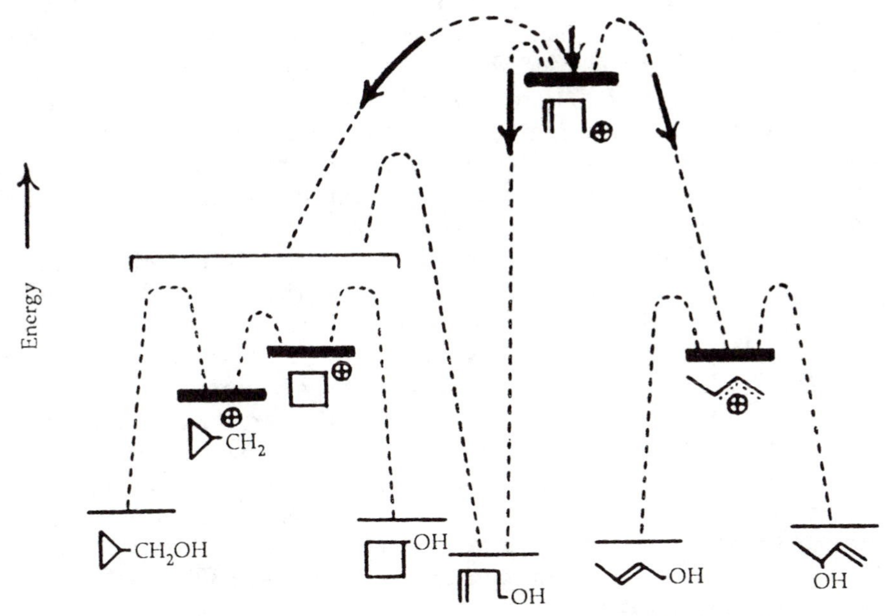

Figure 1. Slide drawing made by the author for a talk on nonclassical cations at the 75th Anniversary Meeting of the American Chemical Society in September 1951. The drawing illustrates the relative thermodynamic stabilities and rate barriers that would be expected if classical cations were involved in cationic interconversion reactions of cyclopropylmethyl, cyclobutyl, and 3-buten-1-yl derivatives.

Mazur's work on cyclopropylcarbinyl chloride. In my NRC postdoctoral work at Harvard, I had shown that cyclopropyl chloride was an extremely unreactive halide (p 70). The possibility that cyclopropyl derivatives might be unreactive also occurred to Brown, as the result of his fascinating studies of structural influences on the dissociation constants of complexes of boron compounds with amines. Herb had the bright idea of using these dissociation constants as indicators of what might be expected for rates of carbocation formation in S_N1 solvolysis reactions, where the carbocations to be formed were structurally analogous to the boron moieties of the complexes. Although correlations between equilibrium constants and rate constants are known sometimes to be less than satisfactory, there seemed no good reason to suspect that substantial difficulties would be involved in making the comparisons that Brown wanted to make. Indeed, because his dissociation measurements were made in the gas phase, they had the bonus of possibly shedding some light on the role of solvent in causing differences in S_N1 reactivity.

Brown had an amusing penchant for simple designations of the structural influences he used to explain his results with amine–boron complexes. He used *F* ("front") strain for interferences between large groups on boron and nitrogen that would assist breaking of the B–N bond of such complexes; *B* ("back") strain for mutual interferences between large groups on boron that were expected to increase dissociation, because then the large groups could get farther away from one another as the boron becomes planar (the reverse of the Bartlett–Knox effect); and he later proposed *I* ("internal") strain for the bond-angle effect, which was the same effect I was looking for with cyclopropyl chloride. As you might guess, the chemists of the time had a field day with the "*FBI*" strains, and Brown loved every minute of it. However, in the long run, his savoring of the limelight was certainly diminished by the fact that at least one of the correlations he was hoping to find turned out to be very poor. Surprisingly, despite the simplicity of the concepts involved, the *FBI* designations never really caught on and are seldom heard today, although *steric hindrance*, which Brown certainly helped to make an important concept in physical organic chemistry, is still with us in spades.

Herb Brown got involved with cyclopropylmethyl chloride as the result of trying to establish the importance of the *I*-strain effect. Embarking for the first time into relative solvolytic reactivities, he reported at the Chicago meeting of the American Chemical Society in 1950, solvolysis rates for a series of tertiary cycloalkyl chlorides and suggested that his results supported the *I*-strain concept. The reactivity he reported for 1-chloro-1-methylcyclopropane seemed very high to me on the basis of any reasonable extrapolation from what I had observed for cyclopropyl chloride. So after his talk, I asked Herb how he had prepared his sample. He said it was by liquid-phase chlorination of methylcyclopropane, which certainly could be expected on the basis of the knowledge of the time to yield 1-chloro-1-methylcyclopropane. However, Mazur had found that chlorination of methylcyclopropane gave predominantly cyclopropylmethyl chloride (p 69). So I told Herb that I thought he was actually solvolyzing cyclopropylmethyl chloride. He got rather angry and protested that the identity of the product had been proven by making a solid derivative through the Grignard reagent, and the derivative had the right melting point. I suggested that he test the solid derivative to see whether it contained a double bond, because Mazur had found that cyclopropylmethyl halides gave 3-buten-1-yl derivatives in Grignard reactions (p 69). Brown apparently took this suggestion seriously and never published a formal paper on the solvolysis of 1-chloro-1-methylcyclopropane. In later life, he liked to point out that he had never had to retract any of his published experimental work, but perhaps he did not count meeting abstracts among his published works.

Along with Mazur's work on C_4H_7 derivatives,[39] we were pursuing the use of ^{14}C as a tracer for organic reaction mechanisms, particularly on what I came to call "isotope-position" rearrangements. These are reactions in which the primary structural change is a shift in the position of an isotopic label. Part of my interest in such rearrangements stemmed from some brilliant and interesting studies by Otto Beeck and co-workers at the Shell Development Company in Emeryville, California. During the period in which these studies were made, this Shell laboratory was one of the preeminent industrial research organizations in the world. It had an absolutely fantastic record of basic research accomplishments, especially as related to the use of advanced techniques of instrumental analysis. The laboratory had a very fine group in mass spectrometry, headed by David P. Stevenson, and this group was involved with Beeck in many very imaginative and important experiments.

The Shell group's experiment with isotope-position rearrangements used ^{13}C as a label for butanes and involved water-promoted aluminum bromide as a catalyst (equation 21).[45] The reactions studied

$$^{13}CH_3CH_2CH_2CH_3 \xrightarrow{H_2O \cdot AlBr_3} \underset{\mathbf{51}}{^{13}CH_3-CH(CH_3)-CH_3} + \underset{\mathbf{52}}{CH_3-{}^{13}CH(CH_3)-CH_3} + CH_3{}^{13}CH_2CH_2CH_3 \qquad (21)$$

involved both isomerization and isotope-position rearrangements. The result of the Beeck experiment was, at first glance, a shocker. Specifically, gaseous $^{13}CH_3CH_2CH_2CH_3$ was put over solid water-promoted aluminum bromide and the isotope locations in the mixture of butane and isobutane were analyzed as a function of time. Even the initially formed isobutane was a statistical mixture of isobutane-1-^{13}C, **51**, and isobutane-2-^{13}C, **52**. Statistical, here, means in a ratio of 3:1. Some $CH_3{}^{13}CH_2CH_2CH_3$ was also found.

For this chemistry, particularly relevant and indeed wonderful experimental results were published in 1944 by Bartlett, Condon, and Schneider.[46] This was one of the all-time great findings in organic chemistry. Bartlett et al. showed that an extremely fast reaction can occur between a hydrocarbon and a carbocation, provided, of course, that the reaction is not thermodynamically unfavorable. A specific example is given in equation 22.

The initial systems we chose for study of isotope-position rearrangements were simple enough, the possible conversion of **53** to **54a–54d**, equation 23. Perhaps they were ridiculously simple, because

$$CH_3{-}\underset{\substack{|\\ CH_2\\ |\\ CH_3}}{\overset{\substack{CH_3\\ |}}{C}}{-}H \;+\; \oplus\underset{\substack{|\\ CH_3}}{\overset{\substack{CH_3\\ |}}{C}}{-}CH_3 \;\rightleftarrows\; CH_3{-}\underset{\substack{|\\ CH_2\\ |\\ CH_3}}{\overset{\substack{CH_3\\ |}}{C}}\oplus \;+\; H{-}\underset{\substack{|\\ CH_3}}{\overset{\substack{CH_3\\ |}}{C}}{-}CH_3 \qquad (22)$$

$$^{14}CH_3{-}\underset{\substack{|\\ OH}}{\overset{\substack{CH_3\\ |}}{C}}{-}CH_2{-}CH_3 \xrightarrow[-\,H_2O]{HCl} \underset{\textbf{53}}{^{14}CH_3{-}\underset{\oplus}{\overset{\substack{CH_3\\ |}}{C}}{-}CH_2{-}CH_3} + \overset{\ominus}{Cl} \longrightarrow$$

$$\underset{\textbf{54a}}{^{14}CH_3{-}\underset{\substack{|\\ Cl}}{\overset{\substack{CH_3\\ |}}{C}}{-}CH_2{-}CH_3} \;+\; \underset{\textbf{54b}}{CH_3{-}\underset{\substack{|\\ Cl}}{\overset{\substack{CH_3\\ |}}{C}}{-}CH_2{-}^{14}CH_3} \;+$$

$$\underset{\textbf{54c}}{CH_3{-}^{14}\underset{\substack{|\\ Cl}}{\overset{\substack{CH_3\\ |}}{C}}{-}CH_2{-}CH_3} \;+\; \underset{\textbf{54d}}{CH_3{-}\underset{\substack{|\\ Cl}}{\overset{\substack{CH_3\\ |}}{C}}{-}^{14}CH_2{-}CH_3} \qquad (23)$$

none of the traditional organic chemists would have thought such rearrangements very plausible under any conceivable conditions. Of the possible isotope-position isomers **54a–54d** that might be formed from **53**, only **54a** is expected on the basis of applicability of the *principle of least structural change* to organic reactions.

This principle evolved in the early days of structural organic chemistry as a sort of guiding idea, one that suggests (or hopes) that life is simple and that organic reactions will occur with minimal structural disturbance. The usefulness of this notion is that, in trying to assign structures to the products of organic reactions, you start by believing that nothing very striking happened and consider other possibilities only when forced to do so. Today, with our knowledge of so many processes that violate the principle, it seems rather absurd to think that such a notion would ever have received much credence. Still, even as late as 1945, it must have been comforting to be told to take the simple approach, that the product of a reaction is likely to resemble the reactants in a reasonable way, and that it is not necessary to consider that all of the bonds in the molecules are disrupted and reassembled at random.

To carry on this research, I had two postdoctoral fellows. The first, Robert E. McMahon, got his Ph.D. with McElvain at the University of Wisconsin. He was short, chunky, and dogged, but cheerful and hardworking, and later worked in industry. Bob's co-worker was Jack S. Hine, with a Ph.D. from the University of Illinois. Hine was wholly different; tall, slender, extremely intelligent, and well read. Jack already had some imaginative ideas about how to establish the intermediacy of dichlorocarbene, CCl_2, as an intermediate in the hydrolysis of chloroform, $CHCl_3$. These ideas were to help establish an excellent reputation for him in the field of physical organic chemistry.

McMahon and Hine found that there was no isotope-position rearrangement in the reaction shown in equation 23, which, in principle, could lead to the isotope-position isomers **54a–54d**. Only **54a** was formed, which means that the carbocationic intermediates do not have a long enough lifetime to undergo the rearrangements formulated in equations 24–26. This finding was surely no surprise to traditional organic chemists.

Matters were much more interesting when the chloride, **54a**, was put in contact with solid aluminum chloride, which is expected to generate carbocations readily and reversibly, with much longer lifetimes

$$^{14}CH_3-\underset{\oplus}{\overset{CH_3}{C}}-CH_2-CH_3 \xrightleftharpoons{\overset{\ominus}{H}:\text{ shift}} {}^{14}CH_3-\underset{H}{\overset{CH_3}{C}}-\underset{\oplus}{CH}-CH_3 \xrightleftharpoons{\overset{\ominus}{CH_3}:\text{ shift}}$$

$$^{14}CH_3-\underset{\oplus}{CH}-\overset{CH_3}{CH}-CH_3 \xrightleftharpoons{\overset{\ominus}{H}:\text{ shift}} {}^{14}CH_3-CH_2-\underset{\oplus}{\overset{CH_3}{C}}-CH_3 \qquad (24)$$

$$^{14}CH_3-\underset{\oplus}{\overset{CH_3}{C}}-CH_2-CH_3 \xrightleftharpoons{\overset{\ominus}{H}:\text{ shift}} \oplus{}^{14}CH_2-\underset{H}{\overset{CH_3}{C}}-CH_2-CH_3 \xrightleftharpoons{\overset{\ominus}{CH_3}:\text{ shift}}$$

$$CH_3-{}^{14}CH_2-\underset{\oplus}{CH}-CH_2-CH_3 \xrightleftharpoons{\overset{\ominus}{CH_3}:\text{ shift}} CH_3-{}^{14}CH_2-\underset{H}{\overset{CH_3}{C}}-\overset{\oplus}{CH_2}$$

$$\xrightleftharpoons{\overset{\ominus}{H}:\text{ shift}} CH_3-{}^{14}CH_2-\underset{\oplus}{\overset{CH_3}{C}}-CH_3 \qquad (25)$$

$$ {}^{14}CH_3-\overset{\displaystyle CH_3}{\underset{\oplus}{C}}-CH_2-CH_3 \;\underset{}{\overset{\overset{\ominus}{C}H_3\text{: shift}}{\rightleftarrows}}\; {}^{14}CH_3-\overset{\displaystyle CH_3}{\underset{\displaystyle CH_3}{C}}-\overset{\oplus}{C}H_2 $$

$$ \overset{{}^{14}\overset{\ominus}{C}H_3\text{: shift}}{\rightleftarrows}\; CH_3-\overset{\displaystyle CH_3}{\underset{\oplus}{C}}-CH_2-{}^{14}CH_3 \qquad (26) $$

than would be expected in solvolytic reactions. The reaction was not very clean because hydrogen chloride is liberated, and the combination of hydrogen and aluminum chlorides causes substantial polymerization. The trick is to quench the reaction before all of the chloride is destroyed, and then look to see whether any of the other isomers, **54b**–**54d**, are formed. This stratagem showed that extensive rearrangement occurred in accord with equation 24 (or equation 26) and very little, if any, by way of equation 25.

With evidence that rearrangement by equation 24 appears to be much faster than that by equation 25, it was interesting to try to compare the rate of reactions according to equation 24 with that according to equation 26. This comparison is possible because, when proper account is taken of statistics, each isotope-position rearrangement from **54a** to **54b** by equation 25 can be equated to a conversion of **54c** to **54d**. Rearrangement of **54a** to **54b** via the route of equation 26 does not convert **54c** to **54d**. Knowing that rearrangement of **54a**–**54b** to **54c**–**54d** is slow, we can measure **54a** → **54b** and **54c** → **54d** in a single experiment with a mixture of **54a** and **54c**. Although we expected that equation 26 would not be an important process, we obtained evidence that **54a** went to **54b** only slightly faster than **54c** went to **54d**.[47]

With the information that isotope-position rearrangements could be observed with *t*-pentyl derivatives, George R. Coraor, a postdoctoral fellow from the University of Illinois, ran a Bartlett, Condon, and Schneider exchange with ^{14}C-labeled *t*-pentyl bromide, isopentane, and aluminum bromide, so that the labeled carbocation would pick up a hydride ion from the isopentane and form labeled isopentane. The location of the ^{14}C in the isotope was established by photochemical bromination to *t*-pentyl bromide (a nonrearranging process) and degradation by the procedures used earlier for *t*-pentyl chloride. With a residence time of 0.2 s, essentially *complete* isotopic equilibration was observed. Statistical distribution of ^{14}C in the various possible positions of isopentane corresponded to the isotope-position isomers of *t*-pentyl chloride, as shown in **54a**–**54d**. In about 0.005 s, the isomerizations of equation 24 and equation 26 were complete. There was some interconversion, as expected by operation of equation 25.[48]

Two conclusions can be drawn from the work, with which perhaps not all might agree. First, rearrangement is more rapid than hydride transfer; this fact accounts for the Shell results with ^{13}C-labeled butane. Second, these experiments foreshadowed the rapid shuffling of groups observed much later by nuclear magnetic resonance spectroscopy of carbocations in strongly acidic solutions ("magic" acid and its analogs[49]). Certainly, 0.005 s is pretty fast on the NMR time scale, but there has been little, if any, acknowledgment of Coraor's careful and pathbreaking experiments in this area.

The experiments on isotope-position rearrangements described so far involve changes between carbocations of quite different energies, as in the examples of equations 24, 25, and 26. Many rearrangements involving formally identical carbocations can be detected by isotope-position rearrangements. Such rearrangements have a special interest because of the possible existence of intermediate bridged structures, which might have special stabilization analogous to that postulated for **41** and **43** (*see* p 68).

An earnest and dedicated graduate student, Joel Yancey, looked for rearrangements that might be associated with methyl bridging, starting with ^{14}C-labeled pentamethylethanol, **55**. Conversion of **55** to the corresponding chloride and degradation (equation 27) showed that, at

$$(CH_3)_3C{-}C(CH_3)(OH){-}{}^{14}CH_3 \ (\mathbf{55}) \xrightarrow[-H_2O]{H^{\oplus}} \left[(CH_3)_2C \overset{CH_3}{\cdots} C(CH_3)({}^{14}CH_3) \right]^{\oplus} \xrightarrow{Cl^{\ominus}} (CH_3)_3C{-}C(CH_3)(Cl){-}{}^{14}CH_3 \ (\mathbf{56a}) + (CH_3)_2C(Cl){-}C(CH_3)_2{-}{}^{14}CH_3 \ (\mathbf{56b}) \quad (27)$$

0 °C, where there was negligible exchange of the product (**56a**–**56b**) with radioactive chloride, no more than 15% rearrangement occurred in 1 min. This result again shows that normally one can expect little rearrangement in essentially irreversible reactions where the lifetime of the cations is not long.[50] Hydrolysis of the chloride in water resulted in no additional rearrangement in going back to the alcohol, **55**, from **56a** and **56b**.

One might wonder why *formation* of the chloride gives 15% rearrangement but *hydrolysis* of it gives little, if any. The answer must be that formation of the chloride from the alcohol, 55, is actually inefficient. The cations are formed reversibly from the alcohols and are able to equilibrate to some degree before going off to the chlorides. On the other hand, in hydrolysis of the chlorides, the cations react efficiently with water to give the alcohols without much isotope-position equilibration. We were also able to show that longer aggregate lifetimes, such as those obtained in highly reversible carbocation reactions (using zinc chloride as catalyst), cause complete isotope-position rearrangement of **56**.[50] This rearrangement was later beautifully confirmed by NMR rate measurements.[51]

With this background of work on isotope-position rearrangements, it was perhaps not surprising that, in early spring of 1950, examination of the $C_4H_7^+$ carbocation came to mind. By then it seemed clear that whatever intermediate gave the cyclopropylmethyl products was in equilibrium with whatever intermediate gave the cyclobutyl products, or else these intermediates were one and the same thing. A critical experiment was to determine whether a ^{14}C label in the α-CH_2 position of cyclopropylmethyl derivatives would stay there, get mixed between the three CH_2 carbons, or, perhaps, even penetrate to the CH carbon. Thus, equation 27 and formation of a bridged *bicyclobutonium* cation, **57**, (with a structure discussed in detail in my talk at the 1951 ACS meeting) would yield an entity that, considering the charge distribution over its various resonance structures, **58a–58c**, could reasonably be expected to

$$\text{cyclopropyl-}^{14}CH_2X \xrightarrow{-X^\ominus} \left[\begin{array}{c} ^{14}CH_2 \\ CH_2 \cdots CH \\ CH_2 \end{array} \right]^\oplus \equiv \begin{array}{c} ^{14}\overset{\oplus}{C}H_2 \\ CH_2 \quad CH \\ CH_2 \end{array} \leftrightarrow \begin{array}{c} ^{14}CH_2 \\ CH_2 \quad \overset{\oplus}{C}H \\ CH_2 \end{array} \leftrightarrow \begin{array}{c} ^{14}CH_2 \\ \overset{\oplus}{C}H_2 \quad CH \\ CH_2 \end{array}$$

57 **58a** **58b** **58c**

$$\mathbf{57} \xrightarrow{H_2O} \text{cyclopropyl-}^{14}CH_2OH \; + \; \begin{array}{c} ^{14}CH_2 \\ CH_2 \quad CHOH \\ CH_2 \end{array} \; + \; ^{14}CH_2{=}CHCH_2CH_2OH \qquad (28)$$

react with water to give the observed product mixture of cyclopropylmethanol, cyclobutanol, and 3-buten-1-ol with minimal rearrangement of the label (equation 28).

The situation would be very different for equilibrating unbridged structures, **59a–60b**, which we formulate here with conventional structures, but which might also be in equilibrium with bridged structures such as **57**. The only restriction with respect to bridged structures would be that none of them be so stable relative to the other cations that equilibration would not occur because the system would get stuck in a particularly stable species, perhaps **57**. The consequences of equilibration are shown in equation 29, wherein it will be noted that

$^{14}CH_2X$ $\xrightarrow{-X^{\ominus}}$ $^{14}CH_2^{\oplus}$ ⇄ ^{14} ⊕ ⇄ ^{14} $CH_2^{\oplus}$ ⇄ ^{14} ⊕

59a **60a** **59b** **60b**

$^{14}CH_2OH$ ^{14} —OH ^{14} —CH_2OH ^{14} —OH

61a **62a** **61b** **62b** (29)

59a does not go directly by a Wagner–Meerwein process to **60b**. Depending on the rates of the equilibration reactions and the lifetimes of the cations, anything from complete equilibration to formation of essentially only **61a** and **62a** could result.

Bob Mazur carried on experiments of this kind by using the reaction of cyclopropylmethyl-1-^{14}C-amine and nitrous acid to avoid the possibility of internal-return rearrangement. The work was not easy, but Bob was quite experienced and skillful by then. He found that the cyclic product was a bit short of complete equilibration, for which the ratios of **61a** to **61b** and **62b** to **62a** should be 1:2. Mazur's experimental ratios were **61a**:**61b** = 46:54, and **62b**:**62a** = 28:72.[52] These are high degrees of rearrangement for irreversible carbocation reactions where there is no enthalpy change, as for **59a** isomerizing to **59b** or **60a** isomerizing to **60b**. The progress of equilibration for **59a** → **59b** is 54/67 × 100 = 81% complete, and that for **60a** → **60c** is 28/33 × 100 = 85% complete. These are not in the expected direction for operation of equation 29, but there could have been some contribution for a nonrearranging nucleophilic displacement involving the diazonium cation precursor of **59a**.

At first I was pretty well satisfied that the evidence showed the reactions were proceeding in accord with equation 29. However, a flat-out equilibration of cyclopropylmethyl and cyclopropyl cations was hardly in accord with the unexpected solvolytic reactivities of the corresponding halides. Bridging seemed to be the answer, but it seemed strange that a bridged cation, such as **57**, occupying a reasonably stable energy well, would want to equilibrate so rapidly as to give a near-equilibrium mix of products **61a**–**61b** and **62a**–**62b**.

I was in a quandary and, in one of my frequent visits to Harvard for seminars, I had a brainstorming session about it with Bob Woodward in his office. Here, as in earlier sessions, I was mightily impressed by Bob's fertile imagination and his desire to look far beyond the bounds of trivial or humdrum solutions to problems. In the process, virtually every reasonable way of producing the observed results was critically examined. The concept of the *tricyclobutonium* cation, **63**, emerged as first choice. This cation had the virtue of producing, in one fell swoop, an equilibrium distribution of ^{14}C among the CH_2 groups. If it reacted with water at the CH position, it would give cyclobutanol; at the CH_2 positions it would give cyclopropylmethanol or, conceivably, even 3-buten-ol.

The **63** structure was so unprecedented and bizarre that I christened it, then and there, a "nonclassical" carbocation. This designation caught on like gangbusters, for no obvious reason. Indeed, it

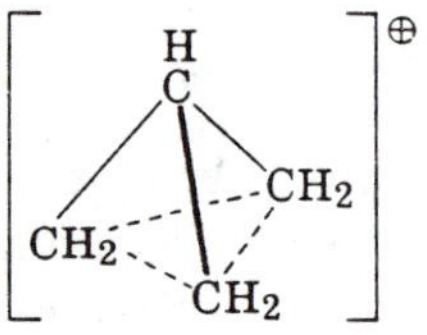

H C CH2 CH2 CH2 ⊕

63

became a generic description for unusual bridged organic intermediates, whether cations, radicals, or anions. It was not a very precise term, and perhaps that was part of its appeal. When Paul Bartlett wrote his review[53] of the nonclassical cation situation in 1965, he had this to say about my creation: "After protesting for years against the inappropriate name 'nonclassical ions', I have been overruled by general usage and am employing the term because of its extreme familiarity. For the present purpose, an ion is nonclassical if its general ground state has delocalized bonding σ electrons." Then, in a footnote, he says: "Familiarity and contempt are still not entirely separable. I quote from *Chemical Abstracts*, **59**, 912b (1964): 'Only a dark, undistillable resin remained upon removal of the solvent. This suggests that a nonclassical carbonium ion intermediate is involved in the mechanism!' "

The idea of the tricyclobutonium ion was exciting in its simplicity and disturbing at the same time. Two difficulties were apparent: First, with its very constricted bond angles, why should it be so stable? Second, would there be enough positive charge at the CH position to allow formation of nearly equal amounts of cyclobutanol and cyclopropylmethanol?

While I was in this indecisive frame of mind, two things happened. First, I received the abstracts of the 1951 American Chemical Society meeting in Boston, in which my old friend and fellow graduate student, Samuel Siegel, reported that cyclopropylmethyl benzenesulfonate was highly reactive in solvolysis in ethanol. He suggested formation of the tricyclobutonium carbocation, **63**, as the responsible intermediate. One problem with the Siegel report[54] was that no rearrangement was detected in the reaction, the sole product being cyclopropylmethyl ethyl ether. This finding was quite at odds with Mazur's observation of extensive rearrangement in such reactions. We were able to show much later[55] that the lack of rearrangement was surely the result of an S_N2 reaction between the benzenesulfonate and ethoxide ion. The discrepancy arose because the reaction product was isolated under quite different conditions (i.e., a much higher ethoxide ion concentration than used in the measurement of the reaction rates).

Siegel's report stirred competitive juices, because Mazur's work offered the possibility of providing experimental confirmation of his speculations. However, neither Sam nor I had any good reason to offer why the tricyclobutonium cation could be regarded as a specially stable intermediate. In fact, Sam told me that "it just looked so good to form the two bonds which lead to the pyramidal intermediate."

The other thing that happened was an encounter with the era's Peck's Bad Boy of chemistry, Michael J. S. Dewar. This man is in many ways a genius, broad of interests, quick of intelligence, very outspoken, and possessing a gargantuan humor. Michael, who was an Oxford D.D.

M. J. S. Dewar with Paul Bartlett after his talk on π-complexes to the Bartlett seminar group at Harvard in 1951. This was a notable encounter between two of the world's most creative chemists.

Phil., achieved instant fame by suggesting the seven-membered ring tropolone structure for colchicine in 1945. He was quick to discover simple molecular orbital theory and wrote a most provocative book called *The Electronic Theory of Organic Chemistry* (1949),[56] of which I will say more later. Dewar loves to take pot shots at anybody's sacred cows and then bursts into a sort of shy, nervous, embarrassed laughter. No statement seems too outrageous for him to make. Relatively short, built like one of those round-bottomed, regain-upright dolls, with a profile like New Hampshire's "Great Stone Face", in no time at all Dewar was a force to be reckoned with.

On his first trip to Cambridge, Dewar spoke at a Bartlett seminar at Harvard, with Swain and me there from MIT, and Cheves Walling from Lever Brothers. It was standing room only, as he talked about π-complexes and their consequences. Needless to say, the discussion was vigorous and long—a great occasion and a wonderful introduction to a most stimulating chemical colleague. At the time, Michael had just gone to Queen Mary College in one of the most bombed-out sectors of London and was trying to get a broad-gauge research program going under very difficult circumstances.

I told Michael about my problems with the tricyclobutonium ion and, quick as a flash, he said he could explain by molecular orbital theory why it was so stable. Indeed, he saw what I didn't see—that it could be formulated as another example of the three mutually overlapping orbital, two-electron system, such as the center orbitals postulated for the Walsh model of cyclopropane.[57] Indeed, the Walsh orbitals do just what one would like, in predicting a low-energy, stable orbital system made up of the three σ-overlapping orbitals, one from each CH_2 group.

Figure 2 from our original publication shows Dewar's formulation quite clearly. With this way of accounting for the stability and the rearrangement results themselves, my doubts about publishing support for the "tricyclobutonium ion" faded, and it did not take long to get a Communication to the *Journal of the American Chemical Society*.[52] The structure, as formulated in Figure 2, has all of the CH_2 hydrogens in *one* plane, a matter that will be of some importance later.

The tricyclobutonium communication created quite a stir, but in the cold gray light of dawn, I still had my doubts about how easily it could give cyclobutyl derivatives with water and other nucleophilic agents. Of course, the tricyclobutonium ion could be postulated as an additional intermediate, or else as a transition state for equilibrating isotopic labels between bicyclobutonium ions of the kind illustrated by 57, but this made the overall picture complex again. I was determined to try to resolve these questions by further experimentation, but more of this later.

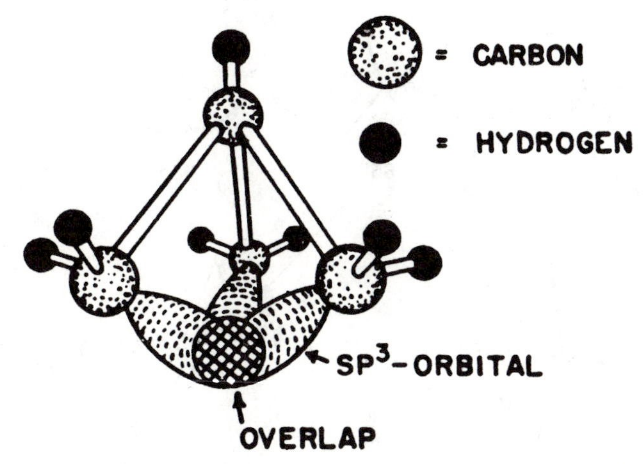

Figure 2. A drawing of the favorably overlapping atomic orbitals that might make for a stable tetrahedral structure for the $C_4H_7^+$ carbocation.[52]

Other carbocation research that we carried on at MIT was on the norbornyl and dehydronorbornyl systems. As we progressed in our studies of isotope-position rearrangements, it became clear that we could investigate bridging by using ^{14}C-labeled norbornyl derivatives.

My point man on this project was Chuck Choi Lee, who was sent to MIT from Edmonton, Alberta, by Reuben B. Sandin, an old-line organic chemist who nurtured quite a few successful graduate students at Harvard, MIT, and Caltech. Chuck was from mainland China via Hong Kong, spoke excellent English, and was short, rotund, jolly, and a bear for work, efficient work. As Winstein had shown, solvolysis of the *exo*-tosylate derivatives was complicated by internal return, and this process was expected to lead to isotope-position rearrangement on its own. To assess the importance of internal return, we investigated both the tosylate solvolysis[58] and the reactions of ^{14}C-labeled *exo*- and *endo*-norbornylamines with nitrous acid.[59] This work was done with the aid of William H. Saunders, a postdoctoral fellow from Northwestern who later had a distinguished career in physical organic chemistry at the University of Rochester.

Our results showed a pattern of ^{14}C distribution that required substantial intervention of hydrogen migration concurrent with the Wagner–Meerwein carbon shifts expected from equation 30. For reasons not obvious to me now, but perhaps following Woodward's philosophy of going for the far-out, I proposed that the hydrogen-shift part of the reaction occurred by equation 31, with involvement of a sort of super nonclassical cation, 67, which I called the "nortricyclonium cation". In equation 31, the starting materials and reaction products are

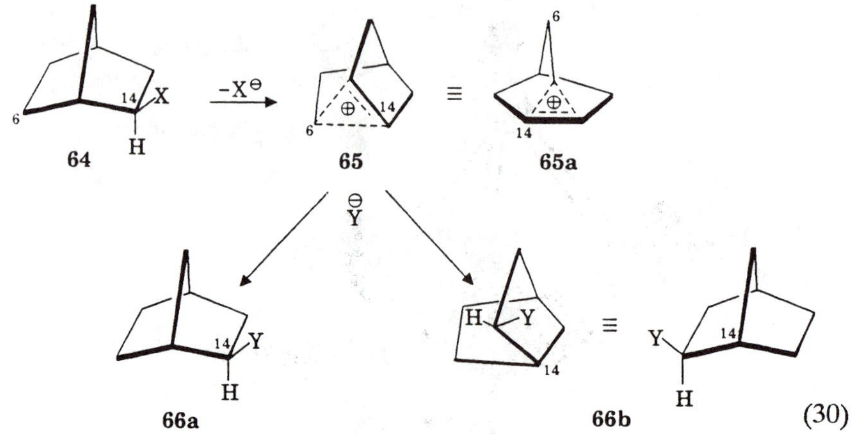

64 $\xrightarrow{-X^{\ominus}}$ 65 ≡ 65a

$\xrightarrow{Y^{\ominus}}$ 66a + 66b

(30)

$\xrightarrow{-X^{\ominus}}$ [67]$^{\oplus}$ $\xrightarrow{Y^{\ominus}}$

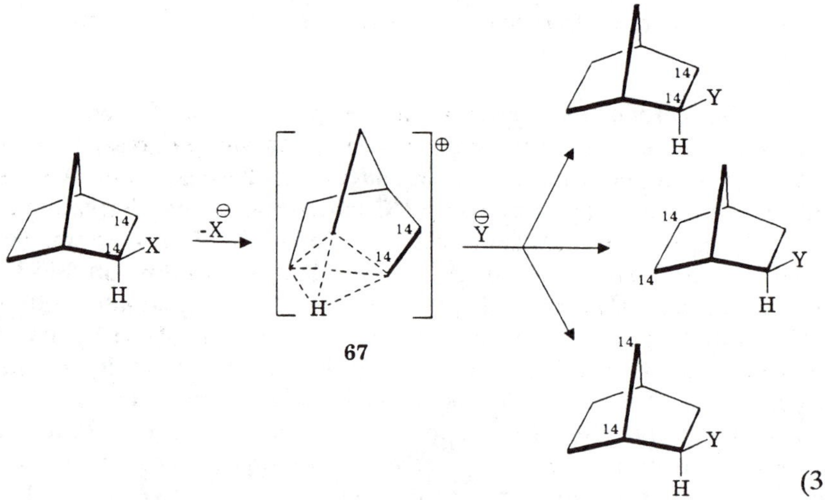

(31)

shown with the pattern of ^{14}C labeling that was actually used in Chuck Lee's experiments. My idea was that **67** represented a "center-protonated cyclopropane" derivative, with the proton bound primarily to the center-directed orbitals of the Walsh cyclopropane formulation.[57] Others have formulated the hydride shift as occurring via **68**, an "edge-protonated cyclopropane" derivative, and theoretical support has been adduced for **68** over **67**.[60]

The story of the discovery of the extraordinary reactivity of the *anti*-7-norbornenyl derivatives, **69**, is of some interest here. F. O. Johnson, a graduate student, started work on compounds related to **69** in early 1949; the project was continued by R. A. (Rudy) Carboni in mid-1952 and published in 1954.[61] Not much was known of such substances,

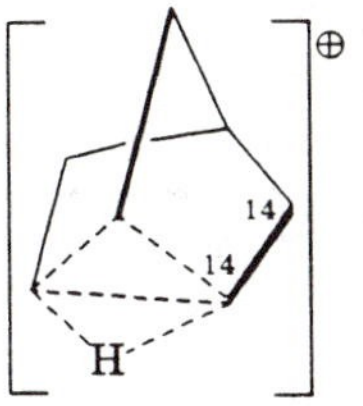

68

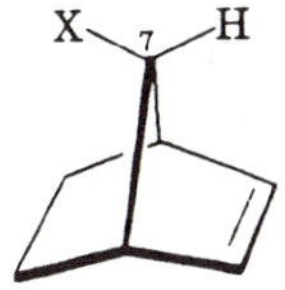

69

how to synthesize them, or even, in the absence of such procedures as NMR, how to establish their structures. I will not go into detail here about the reaction mechanisms or how we proved the structures of the products. We were able to prepare two isomeric 7-chloro-*exo*-norborneols, **70** and **71**, by equations 32 and 33. Dehydration of **70** and

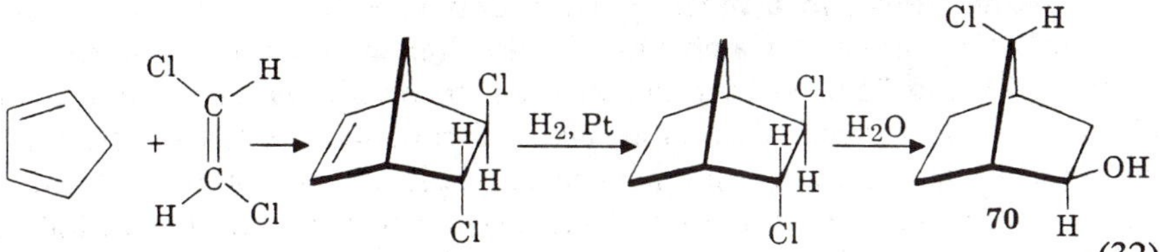

(32)

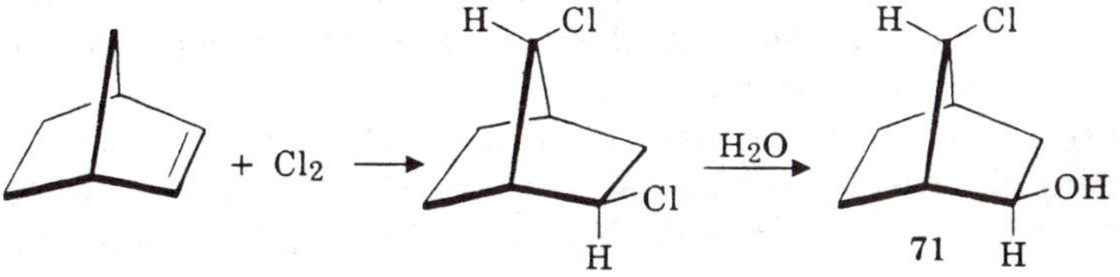

(33)

71 is not easy, but by pyrolyzing the 2-naphthoate esters, small amounts of more or less pure *anti*- and *syn*-7-chloronorbornenes, **72** and **73**, respectively, were obtained (equation 34). These substances were reduced by hydrogen to the same 7-chloronorbornane, **74**.

Interestingly, although Carboni mentions in his thesis abstract that **72**–**74**, including the *anti* compound **72**, are singularly unreactive in solvolysis, there is no evidence in either his thesis or his laboratory notebook that he actually did more with **72** than hydrogenate it to **74**. I was on sabbatical leave at Caltech and later in Europe when most of the work was done and had to omit saying anything about the reactivity of **72** when we published our account of all of the other work involved. There can be no question that we actually had **72**. Because of the method of synthesis, its infrared spectrum was different from that of **73**, and the infrared spectra obtained for **74**, prepared from either **72** or **73**, were essentially identical. And so we dropped the ball on a great opportunity.

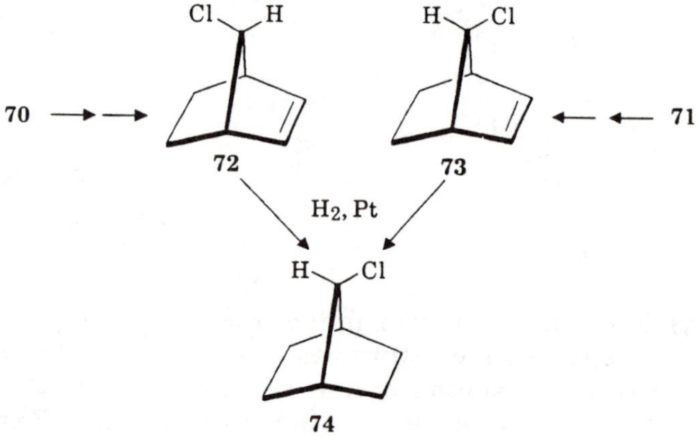

Cl H H Cl

70 →→ 72 73 ←← 71

H_2, Pt

H Cl

74 (34)

Nonetheless, in a meeting with Bob Woodward and Saul Winstein in Bob's office in Cambridge, I told them about the gross unreactivity of **73** and **74**, and then asked them to predict what to expect for **72**. There was a lot of discussion, pro and con, about whether the orbitals of the cation, **75**, would be correctly placed or not for extensive delocalization of the double-bond electrons into the vacant *p*-orbital at C-7 of the cation. Believe it or not, even though I knew then how to calculate the degree of overlap to test the proposition, the consensus was that the orbitals were just *not* close enough, *not* properly oriented, to result in effective stabilization.

The next year after our first publication, while W. G. Woods was reinvestigating the preparation and solvolysis of **72** at Caltech, Winstein, Woodward, and co-workers[62] published data showing that *anti*-7-*exo*-2-dibromonorbornane, **76**, and *anti*-7-norbornenyl toluenesulfonate (**69**,

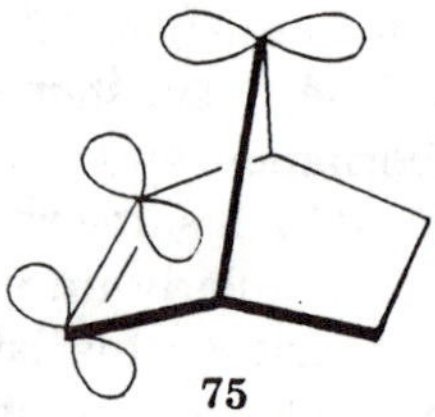

75 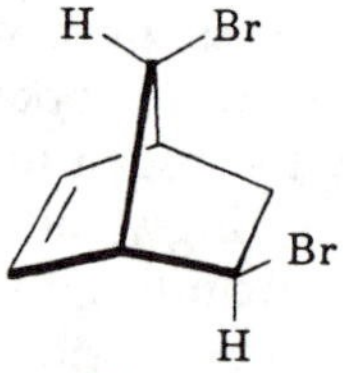 H Br Br H 76

with X = 4–$CH_3C_6H_4SO_3$–) showed extraordinary solvolytic reactivity. Woods[63] also found this awesome reactivity for **72**. He was able to show by Hückel molecular orbital calculations that the orbitals of **75** were, in fact (and, of course, *after* the fact), so situated as to be very effective for electron delocalization. Clearly, this structure is best designated as a nonclassical *bis*-homocyclopropenyl cation, **77**.

Even the many orders of extraordinary reactivity of the *anti*-7-norbornenyl derivatives were not enough to be wholly convincing in

the battle over nonclassical cations. One might argue that **77** is actually classical, or possibly that **77** is better formulated as **78**, a cyclopropylmethyl-type carbocation. But cations like **78**, by Herb Brown's definition, are classical—even if the reasons for their stability are not understood. Structure **77**, if it were the only, or an isolated, example of the cause of extraordinary solvolytic reactivity, might have been the prime focus of the nonclassical ion controversy. However, it was really too strong a bastion to attack effectively. As any military strategist knows, you should not mount a full-scale attack on your opponent's strong points; better to try to outflank them.

The nonclassical-cation controversy ranks among the classics of chemical disagreements.[53] It was fought out, nationally and internationally, almost to the present, through seminars, meetings, referee reports, papers, books, and private communications. The early stages are wonderfully documented by Paul Bartlett's annotated reprint collection.[53] Paul's comments are perceptive, pithy, and full of good humor. Unfortunately, not all of the protagonists were amused and, at times, even Paul lost patience. Participation in the battle was remarkably one-sided. Herbert C. Brown held a myriad of the top physical organic chemists at bay for many years, as a veritable Horatio at the bridge, determined to keep nonclassical cations from being widely enough accepted to make it to the land of "classicality". The rallying cry was often Occam's razor—the appeal to reason that holds that nothing should be made more complex than it needs to be. Whether Bishop William of Occam intended or did not intend it to be so, his razor does have two edges. Contriving, at all costs, to keep something seemingly simple can make it uncomfortably complex. In the final analysis, that is exactly what Herb did, and that is what did him in. More about that later.

Electrical Effects of Substituents at MIT

Although the structure of $C_4H_7^+$, the norbornyl cation, and the potential existence of other nonclassical cations were fascinating and engrossing, I got involved in a number of other interesting research problems at MIT. One was a rather systematic attempt to learn more about the electrical effects of substituent groups along the lines of the earlier work of Hammett[28] and of Kirkwood and Westheimer.[27] As I began to study Hammett's important book, *Physical Organic Chemistry*, I was greatly impressed by the sharpness of Hammett's focus on the reactivity problem and by the way in which he swept aside the details relating to what was going on with individual substituent groups to concentrate on the overall correlations that could be made. Because of the importance of the Hammett treatment to what follows, I will digress briefly to discuss the principles that it involves.

In brief, the Hammett treatment shows that the rates and equilibria of many reactions of 3- and 4-substituted benzene derivatives can be related to one another very simply. Hammett separated the reactivity problem into two parts: one that is characteristic of the nature of the substituent itself and another that is characteristic of the reaction. For the example of the 4-nitro group, the substituent component of the reactivity, $\sigma_{4\text{–}NO_2}$, is defined as the logarithm of the ratio of the ionization constant, $K_{4\text{–}NO_2}$, of 4-nitrobenzoic acid to the ionization constant, K_H, of benzoic acid in water at 25 °C, so that $\sigma_{4\text{–}NO_2} = \log(K_{4\text{–}NO_2}/K_H) = 0.778$. A very large number of Hammett σ constants have been determined in this way. They can also be determined indirectly from substituent effects on other reactions, but the values so obtained are always somewhat suspect, because σ is defined as being the influence on acid ionization constant of a substituted benzoic acid relative to benzoic acid itself. Because the free energy of ionization is $-RT \ln K$, the value of $\sigma_{4\text{–}NO_2}$ is related to the difference in free energy of ionization between 4-nitrobenzoic acid and benzoic acid.

The reaction-specific component is defined as ρ, where ρ is the statistically determined slope of a straight line correlating the logarithms of the rate or equilibrium constants of a given reaction (same concentrations, solvent, and temperature) with different 3- and 4-substituents. It should be clear that ρ reflects the sensitivity of the reaction to changes in substituents. At least two different substituents are needed to obtain a ρ value. For example, for the hydroxide-induced hydrolysis of substituted ethyl benzoates, a ρ can be obtained from just the rates $k_{4\text{–}NO_2}$ of ethyl 4-nitrobenzoate and k_H of ethyl benzoate. Thus, $\rho\sigma_{4\text{–}NO_2} = \log(k_{4\text{–}NO_2}/k_H)$. The numerical value for ρ so obtained is 2.017 for rates of ester hydrolysis when determined at 30 °C in 50% ethanol–water with 1 M sodium hydroxide. The ρ value that fits some 12 different substituted esters with a standard deviation of 0.067 is 2.498. Changes in the reaction conditions will almost invariably change ρ, and it will usually be smaller, the farther removed the substituent group is from the reaction center.

Now, if ρ is defined as unity for the ionization of benzoic acids in water at 25 °C, we can write the simple Hammett equation, $\log k_x = \rho\sigma_x - \log k_H$, which has great beauty in that k_x (or K_x) can be calculated with the aid of relatively short tables of σ_x values, of k_H (or K_H) values for rate (or equilibrium) constants for unsubstituted compounds, and of ρ values. The goodness of fit of calculated and experimental values is usually excellent. In fact, serious deviations, for 3- and 4-substituted benzene derivatives can usually be taken as an indication that one should look for special complexities in the reaction mechanism. Thus, the compounds in one range of σ_x values may be reacting by a quite different mechanism than those in another range.

Hammett found that his treatment failed when he applied it to 2-substituted benzenoid derivatives. He attributed this to the fact that, with these substances, the reaction center was so close to the substituent that the size and shape of the substituent could be as important as its electrical effect in determining the relative reactivity. The simple $\rho\sigma$ relation can be modified to take into account such factors, but then matters become much more complex. Hammett preferred to stick with substituents in the 3- and 4-positions. Hammett relationships have been extremely important in physical organic chemistry. Hammett pretty much ducked the problems of what about the substituent determines σ and what about the reaction determines ρ, but there have been plenty of smaller fry, including me, to try to tackle those problems as well as to try to expand, modify, or even supplant his treatment.

At the outset, my work on the Hammett relationships was carried almost exclusively by undergraduates. The process was greatly facilitated by the fact that, in those days, all MIT undergraduates had to do a senior thesis. This requirement provided me with a prolific supply of excellent co-workers, because those MIT seniors were very good indeed. I got them to study the electrical effects of substituent groups on reactivity and dipole moments. This type of work has the advantage of involving relatively simple syntheses and experience with physical methods. Further, a senior thesis could be complete with investigation of one, two, or five different substituents.

Determination of substituent (σ) constants by the Hammett equation seemed to be the best way to connect our new data with existing literature data. We devoted a lot of attention to determining such constants, and complemented the interpretation of their magnitudes, when appropriate or possible, with dipole-moment data. The substituents we investigated included: $(CH_3)_3Si$–[64,65] (from the interest engendered by Whitmore in his work with Leo Sommer); cyano,[66] hydroxyl, and carboethoxyl[67] (the literature data were incomplete); CF_3–;[68] and $(CH_3)_3N^+$–.[69]

Another project was to show that the Hammett equation, at least as applied to benzoic acids, was independent of steric hindrance as long as that hindrance could be regarded as constant throughout the series being studied. We did this by showing that a good Hammett relation

John D. Roberts in his laboratory at MIT determining the dipole moment of 1,5-dichloro-1,4-cyclooctadiene (see pp 135–136), 1947.

could be obtained for a series of 4- and 5-substituted 2-methylbenzoic acids. Whatever the effect of the 2-methyl group on the reactivity of the substituted benzoic acids, it was constant and was also far enough away from the 4- and 5- substituents not to influence their conjugation (if any) with the aromatic ring.[70,71]

The work on the CF_3– group had quite an impact, because it gave strong indications by both σ-constant and dipole moment data that the CF_3– group has the kind of qualities characteristic of a conjugating substituent group, such as nitro or cyano. My interpretation of the results involved the concept of C–F no-bond resonance, or hyperconjugation, in which electrons are donated to the C–F bond and can involve

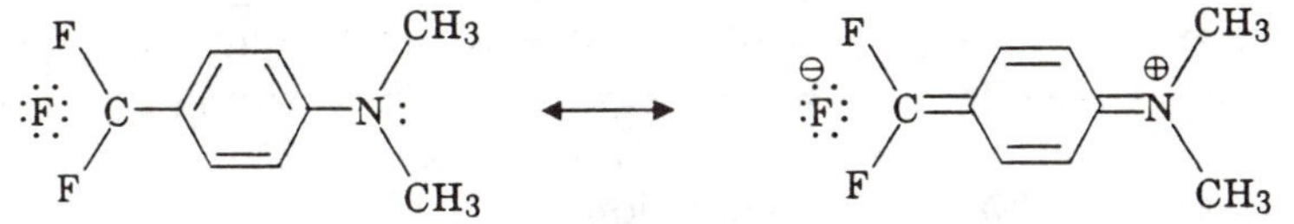

79 (35)

conjugation with a strongly electron-donating 4-substituent like the dimethylamino group, as in **79** (equation 35).

There were subsequent trenchant attacks on C–F hyperconjugation, most notably by Streitwieser and co-workers[72] on the basis of experiments designed to determine whether removal by base of the C-1 hydrogen of **80** and its conversion to **81** occurs readily or with difficulty (equation 36). Steric inhibition of resonance suggests that resonance structures such as **81b** should be unfavorable. Thus, if there were rapid

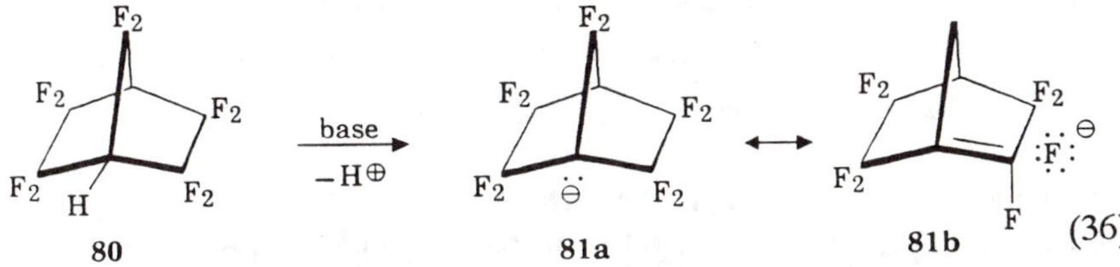

80 base, $-H^{\oplus}$ 81a ⟷ 81b (36)

proton loss from **80** relative to an analogous compound where steric inhibition of resonance would not be important, this would argue against the importance of C–F hyperconjugation. The reported high relative reactivity of **80** seemed decisive, and C–F hyperconjugation became discredited, if not disreputable.

Later work cast some doubt on the validity of Streitwieser's study of **80**, and a rather recent examination of the crystal structure of the CF_3O^- anion[73] has demonstrated bond lengths in accord with the concept. However, the postulation of hyperconjugation as represented by **79** is disputed on the basis of theoretical calculations made by Paul R. v. Schleyer.[74] If I understand Paul correctly, he does not believe his calculations support the idea of any kind of important hyperconjugation between substituents and an aromatic ring. This position was espoused long ago by Dewar,[75] albeit not wholly consistently, and in a somewhat different form by Charles C. Price[76] with respect to virtually all conjugation of substituent groups.

Perhaps so, but I still feel that if something acts as if it has double-bond character, smells as if it has double-bond character, and tastes as if it has double-bond character, you might as well go on pretending that it has double-bond character. That's Occam's razor for you. Anyway, controversy, although by no means as intense as with the structure of $C_4H_7^+$, has dogged C–F hyperconjugation for years, since it was first proposed by Brockway.[77]

The work on the σ constants of the $(CH_3)_3N^+$– group reinforced the idea that σ-constant and dipole-moment data indicated important hyperconjugative effects with the CF_3– group, by showing that electrostatic polarization of a position in an aromatic ring by the $(CH_3)_3N^+$– group does not behave like conjugation.[69]

The Hammett equation is not applicable to reactions involving *ortho* positions or *ortho* substituents, but it might apply to the *para:meta* ratios in the electrophilic substitution of substituted benzenes. This is of interest because *para* substitution in halobenzenes is supposed to be facilitated by a strong resonance contribution of forms such as **82** (equation 37), of which **82d** will be seen by inspection not to be possible for

$$C_6H_5Cl + HNO_3 \xrightarrow{-HO^{\ominus}} \mathbf{82a} \longleftrightarrow \mathbf{82b} \longleftrightarrow \mathbf{82c} \longleftrightarrow \mathbf{82d} \qquad (37)$$

meta substitution. If **82d** is important, then the Hammett equation should not apply to *para:meta* ratios in reactions like nitration.

This question was investigated by Janet Sanford, my first female doctoral student, with the idea of measuring the very small expected amount of *meta* nitration reasonably accurately by the isotope-dilution technique. For chloro-, bromo- and iodobenzenes, the measured ratios of *para:meta* reactivities were 77, 52, and 33, respectively. Overall, we found a poorer Hammett relation than usual, with a very large ρ value of about 7, and with *para* reactivities from 4–25 times greater than expected. These observations were in accord with stabilization of the nitration intermediate by those substituents studied, such as **82d**, that would be expected to have significant contributions of resonance structures.[78]

A research problem regarding the Hammett equation, which turned to gold in the long run, was actually difficult to get graduate students to accept. My idea was to replace 4-substituted benzoic acids, **83**, by 4-substituted bicyclo[2.2.2]octanecarboxylic acids, **84**, and to see whether a Hammett-type relation held with these compounds (certainly expected). Even more, I wanted to determine the values of σ and ρ for comparison with those obtained with **83**. I had this idea for quite a while but was unable to sell it because the syntheses were sure to be difficult. Also, there seemed a very good chance, with the saturated rings of **84**, that the effects of substituents might be too small to be interesting.

I remember my brilliant physical chemistry colleague, Walter Stockmayer, encouraging me to push on with the problem because he was appalled that I was not able to get someone to work on it. His enthusiasm was so great that I started work myself with the initial steps of the preparation and was able to make a small amount of the diester, **85**. At about this point, another excellent graduate student came along

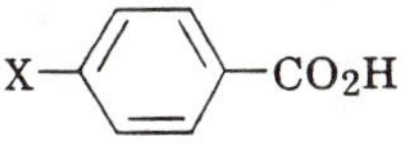

X—C_6H_4—CO_2H

83

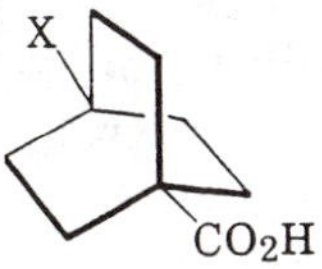

X, CO_2H

84

$C_2H_5O_2C$— —$CO_2C_2H_5$

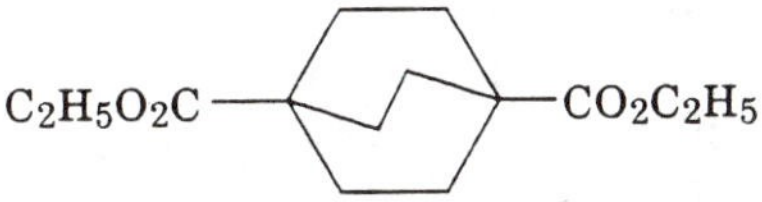

85

who wanted to work on the problem. Of course, I was pleased to oblige. Walter Moreland, later to have a fine career at Pfizer Company was a handsome, very modest, and quietly efficient student who carried out the syntheses and physical measurements with great aplomb.

Our first indication that the electrical effect of a substituent through the bicyclooctane ring could be substantial was when Walt found that alkaline hydrolysis of the diester, **85**, went quite cleanly to the monoester, **86**, rather than giving a sloppy mixture of **85**, **86**, and the dicarboxylic acid (equation 38). It seemed quite clear that the hydrolysis of the ester group of **86** was inhibited by the negative charge of the carboxylate group at the other end of the molecule. This finding was most encouraging, and Walt was able to synthesize bicyclo[2.2.2]octane-1-carboxylic acids with six different groups at the 4-position, **84**.[79]

$$85 \xrightarrow[\text{fast}]{\overset{\ominus}{OH}} \overset{\ominus}{O_2C}\text{—[bicyclooctane]—}CO_2C_2H_5 \xrightarrow[\text{slow}]{\overset{\ominus}{OH}} \overset{\ominus}{O_2C}\text{—[bicyclooctane]—}\overset{\ominus}{CO_2} \quad (38)$$

86

As expected, the reactivities of the members of this series gave a superb Hammett correlation. Quite precise σ constants could be assigned to each substituent, which we called σ' constants. The really big surprise was the fact that the corresponding ρ' constants, which measure the response of each reaction to the electrical effects of the substituents, had magnitudes not much different from those reported for benzoic acids under the same conditions. Nonetheless, there were substantial differences in the σ and σ' values as expected, because, in the saturated system, the resonance effects of the substituents are not expected to be operative. This is especially true of the OH group, which is strongly electron-donating in the 4-position of benzoic acids, but is rather strongly electron-attracting as the X group of **84**. There was a substantially better correlation of σ'constants with the Hammett σ constants obtained for the 3-position than for the 4-position, which indicates that 3- substituents on the benzene ring exert their influence more by inductive than resonance effects.

Because of my graduate seminar (*see* pp 35–38), I was particularly interested in just how well our substituent effects could be correlated by the Kirkwood–Westheimer theory, which treats the ionization of organic acids as occurring in a cavity of low dielectric constant. Acids like **84** should have quite a bit better defined cavities than benzoic acids. It turned out that the results were not as good as I had hoped.

In general, the Kirkwood–Westheimer treatment predicted that the effects of the substituents would be much smaller than was actually observed. Thus, for some reason, the effective dielectric constants calculated by their equations were too large. Further, there were some considerable discrepancies between the relative effects expected for various substituent groups. With respect to these matters, Frank Westheimer offered a commentary, which is given here in part. (The rest is included in our original publication.[80])

> . . . the data for the bicyclooctane system clearly show that equations based on our cavity model somewhat underestimate the effect of substituents on ionization constants. It should be noted, however, that whereas our equations underestimate the observed effects by a factor of 2, the older Bjerrum–Eucken equation underestimates them by a factor of 16; the newer approximation is therefore better than the older one by almost an order of magnitude.

Moreland's research on the 4-substituted bicyclo[2.2.2]octanecarboxylic acids was quite a hit and has been greatly extended by others. Particularly important has been the work of Robert A. Taft, Jr.,[81] who took our data and that of a number of other workers to separate resonance and inductive effects and generate a very much wider quantitative relationship between reactivity and structure than the benzoic derivatives correlated by Hammett.

X ← CH_2 ← CH_2 ← CH_2 ← CO_2H

87

One of the questions that came up repeatedly in connection with studies of reactivity in organic molecules was the relative importance of the so-called "field" and "inductive" effects of substituent groups. The Kirkwood–Westheimer treatment, in its simplest form, is a field-effect treatment (i.e., the electrical effect of a substituent can be calculated on the basis of the electric field it produces at a particular distance, across a medium with a particular effective dielectric constant). The inductive effect operates in a somewhat different way, by having the substituent exert its influence along a chain of bonds. In effect, the electrons are polarized in successive bonds, as might be depicted for an electron-attracting substituent by **87**. The attenuation of the electrical effect with distance from the substituent is suggested in **87** by diminishing line weights for the arrows. The zigzag arrangement of the bonds is intentional, to give credence to the idea that the inductive effect, in its simplest form, travels along the bonds and not through space.

It would have been nice to be able to synthesize **88a** and **88b** to test the relative importance of these influences. Presumably, with the

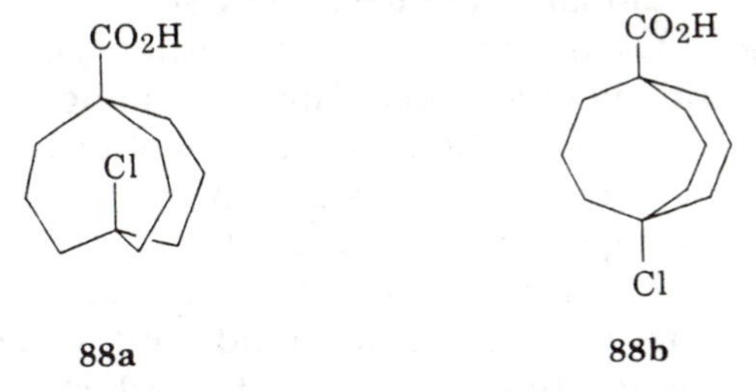

88a 88b

"inside" isomer, **88a**, the negative end of the C–Cl dipole would be closer to the carboxyl group than the positive end. This should, if the field effect alone is operative, make **88b** a weaker acid than the parent compound with no chlorine substituent. On the other hand, the "outside" isomer should be a stronger acid than either the unsubstituted or the "inside" isomer. If only the inductive effect operates, both isomers should have the same acid strength because they have the same number of bonds between the carbon carrying the chlorine and the carboxyl group. Paul Bartlett said, about these postulated electrical effects, "The field effect travels by plane and the inductive effect by train."

In an era when the synthesis of compounds like **88a** and **88b** was not very thinkable, we settled for a relatively inelegant substitute, the 2-, 3-, and 4-substituted phenylpropiolic acids, the critical 2- and 4-chloro isomers of which are shown as **89** and **90**. My graduate student,

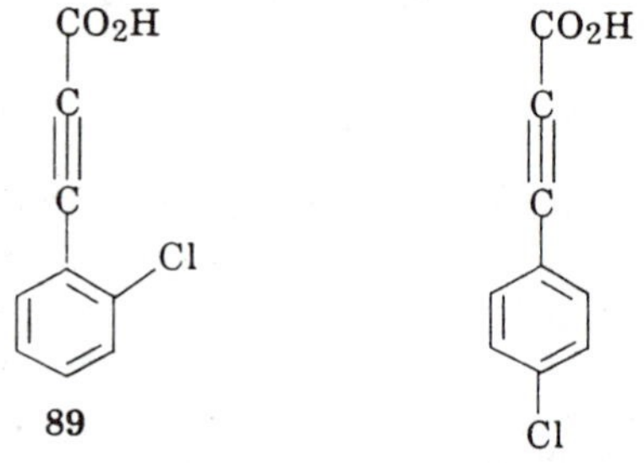

CO_2H C C Cl

89

CO_2H C C Cl

90

Rudy Carboni, found that with derivatives like **89** and **90** carrying electron-attracting substituent groups, a 2-substituent did not behave as though it were essentially electroneutral or opposite in effect to the 4-substituent, as might be predicted from the angle that the dipole of the 2-substituent makes to the carboxyl group. Instead, such substituents have an effect comparable to, or less than, the 4-substituent group, even if it is much closer to the carboxyl. We concluded from such comparisons that the inductive and field effects must have fairly comparable influences.[82]

We studied these effects in solvents of differing dielectric constant to see if we could make the field effect more prominent. In measurements of the rates of reaction of the acids with diphenyldiazomethane, the substituent effects were larger (meaning that ρ is larger) in dioxane of dielectric constant 2 than in ethanol of dielectric constant 24. However, the order of reactivity with the different substituents did not change with this change in solvent, as might be expected for a sizable field effect. With the dielectric constant 2, there should no longer be a pronounced cavity factor, and the magnitude of the electrostatic effect should change rather drastically relative to the inductive effect.

Sometimes you find a research problem that is sheer fun. One such problem was our study of the reactions of diphenyldiazomethane (DDM) with various acids in different solvents. We got into this because I was looking for a convenient reaction to use for the measurement of σ constants. In Hammett's book[28] I happened to run across mention of work by James F. Norris, one of my predecessors at MIT, who had studied the reaction rates of some organic acids with bis(4-methylphenyl)diazomethane in nonpolar solvents.[83] When I looked this up, I noted with pleasure that such diazo compounds have striking permanganate colors, in contrast to the colorless reaction products.

This color change provided a wonderful opportunity for measuring reaction rates spectroscopically. I bought a Beckman DU ultraviolet–visible spectrophotometer with a thermostatically controlled cell holder and began measuring the rates of DDM reactions in ethanol as solvent. The bulk of the work was done by Warren Watanabe, one of Frank Westheimer's doctoral students. Warren was one of the most meticulous laboratory workers I have ever had work with me. He recorded hundreds and hundreds of readings on the Beckman spectrophotometer. It was a pleasure to see him take a reading, pause, and then go back and verify it before writing down the value. For years afterward, I used Warren's notebooks as models of how laboratory records should be kept.

The chemistry was fascinating because the reactions were quite complex. DDM reactions are subject to general-acid catalysis, and furthermore, DDM seemed to be especially effective in reacting with undissociated acids HX to give $(C_6H_5)_2CH{-}X$. Then, depending on the nature of X, the product often undergoes a solvolytic reaction with ethanol to generate diphenylmethyl ethyl ether and regenerate HX. This sequence generated interesting rate curves, which were beautifully susceptible to exact analysis. So we wound up studying solvolyses, along with measuring reactivities of carboxylic acids.[84] In the course of this work, I got a very useful education on how to analyze complex kinetics, acid catalysis, and deuterium isotope effects[85–87] (which, although long known, were just coming into general prominence for use in determining reaction mechanisms, thanks to Frank Westheimer).

Some Other Small-Ring Chemistry

Another truly joyous research project, which bore fruit in many directions and opened up a new area of chemistry, owed much to a very unusual accident—in the right place and at the right time.[88] The hero of this episode was G. Bruce Kline, energetic and determined. Bruce started his graduate career in 1948, with his eye on making cyclobutadiene or any substituted cyclobutadiene, short of the stable dibenzo derivative known as biphenylene.

Short and sweet was Kline's ideal. The simple straightforward way, at first, seemed to be to dimerize alkynes with appropriate functional groups. Such reactions might occur photochemically (transition-biphenylene

metal complexes of conjugated cyclic polyalkenes were not yet known). Bruce tried to dimerize diphenylacetylene photochemically, but got a naphthalene rather than a cyclobutadiene (equation 39). Harmon's reports[89] suggested an analogous try with tetrachloroethylene (equation 40). The first step of this rather reasonable-looking sequence failed when the tetrachloroethylene was heated to 300 °C for 12 days (equation 41). Another prospect was to dimerize 1,2-dichloro-1,2-difluoroethylene, **91**, and eliminate chlorine from the cycloadduct (equation 42). The initial problem was to obtain **91**, which was not commercially available.

$$2\, C_6H_5C{\equiv}CC_6H_5 \xrightarrow{\text{sun-light}} \text{1,2,3-}(C_6H_5)_3\text{-naphthalene} \tag{39}$$

$$2\, CCl_2{=}CCl_2 \xrightarrow{?} C_4Cl_8 \xrightarrow[?]{Zn} C_4Cl_4 \tag{40}$$

$$6CCl_2{=}CCl_2 \longrightarrow 3CCl_3\text{-}CCl_3 + C_6Cl_6 \tag{41}$$

$$\underset{\mathbf{91}}{2CFCl{=}CFCl} \xrightarrow{?} \underset{\mathbf{92}}{(FCl)_4C_4} \xrightarrow[?]{Zn} C_4F_4 \tag{42}$$

About this time, 1950, I had become associated with Du Pont as a consultant and had the opportunity to visit the Jackson Laboratory of the Du Pont Organic Chemicals Department. Chlorofluoro organic compounds were made for sale by Kinetic Chemicals, a subsidiary, at the nearby Chambers Works. I was assured that I could get

$CCl_2F{-}CCl_2F$ from Kinetic Chemicals, and that this, on treatment with zinc, would give the desired **91**. We ordered several hundred grams and made the chlorofluoroalkene. It did indeed dimerize at 180 °C in useful, but hardly exciting, yields. On the basis of the work of Coffman at Du Pont,[90] we figured that an alternative short cyclobutadiene synthesis might be achieved by equation 43. Again, addition went well in a

$$C_6H_5C{\equiv}CH + CFCl{=}CFCl \;(\mathbf{91}) \longrightarrow C_6H_5{-}\overset{}{C}\underset{CClF}{\overset{CH}{\diagup\!\diagdown}}CClF \;(\mathbf{93}) \xrightarrow{Zn} C_6H_5{-}\text{(cyclobutadiene with H, F, F)} \quad (43)$$

sealed-tube reaction at 150 °C, in reasonable yield, almost the first shot out of the box!

Problems came when we tried to treat the cycloadducts with zinc. Nothing seemed to work the way that Bruce expected. The material made by dimerization, equation 42, only reacted with one equivalent of zinc, and so on. Then we noticed that what we thought was **92** had the same melting point as the dimer from $CF_2{=}CCl_2$. Furthermore, the infrared spectrum of the starting alkene was correct for $CF_2{=}CCl_2$ and not $CFCl{=}CFCl$!

When I complained to Kinetic Chemicals about sending CCl_3CF_2Cl in place of CCl_2FCCl_2F, the person who replied had the gall to say that they knew it was not what we ordered, but they decided it didn't matter because we were going to use it as a solvent! Well, that scotched the particular approach by equation 43, because what we really had synthesized was **94** and/or **95**. Bruce was not to be turned off by this setback. Because CCl_2FCCl_2F was not commercially available, he synthesized it, treated it with zinc, and obtained **91**. But now it turned out that **91** would not dimerize, and it added poorly to phenylacetylene. Further, the adduct did not behave any better with zinc than did **94** (or **95**), giving only starting material or polymer.

Discouraged with this result, Bruce spent some months trying to dimerize unsaturated nitriles and introducing double bonds into 1,2,3,4-tetraphenylcyclobutane, neither of which ever got very far. He was quite a bit put out by his lack of success, and I suggested that, at least, he had a new compound in the $CF_2{=}CCl_2$–phenylacetylene adduct, which he could make in any desired quantity. Why didn't he find out something about its chemistry? Perhaps he could hydrolyze the halogens and get a ketone. Anyway, he needed to show whether it was **94** or **95** (equation 44).

Almost a year to the day after Bruce had first made the adduct, he began to investigate its chemistry. When he started, he was sure the

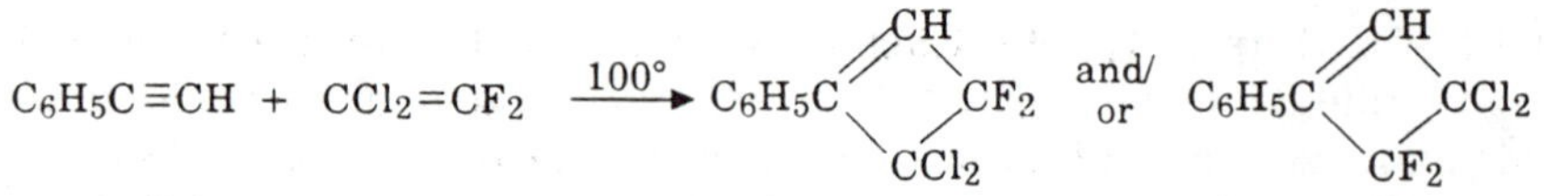

$$C_6H_5C{\equiv}CH + CCl_2{=}CF_2 \xrightarrow{100°} \text{94 or 95} \quad (44)$$

94: C_6H_5C / CH / CF_2 / CCl_2 (cyclobutene; and/or)

95: C_6H_5C / CH / CCl_2 / CF_2 (cyclobutene)

structure was **95**. Attempted hydrolysis by boiling in 50% water–50% acetone liberated no chloride ion. However, I pointed out that the hydrolysis of phenyltrichloromethane was known to be catalyzed by strong acid. Why did he not see what sulfuric acid would do? He reported that it did react, turned color, bubbles formed, and "a strong odor of HCl was apparent." When the compound was poured into water, a light-yellow solid was formed, melting point (mp) 79.0–80.2 °C after purification. The solid was unsaturated and had a carbonyl band in the infrared. A few days later, he came in with a long face because the analysis for $C_{10}H_6OF_2$ was 10% low in carbon and 0.5% low for hydrogen. I suggested he recalculate for $C_{10}H_6OCl_2$, and that came out wonderfully well.

Oxidation gave benzoic acid, so the compound was not **96**. Then zinc-amalgam reduction gave 3-phenylcyclobutanone, so the product was clearly the cyclobutenone, **97**. Furthermore, dilute sodium hydrox-

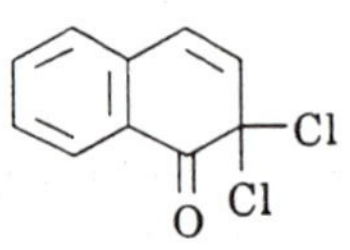

96 (naphthalenone with Cl, Cl, O) **97** (C_6H_5, H, =O, Cl Cl cyclobutenone)

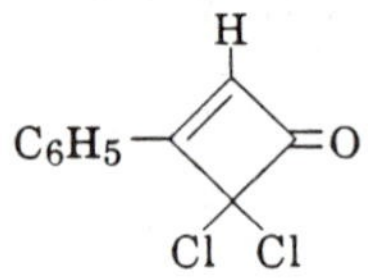

ide at the steam-bath temperature caused ring opening to a carboxylic acid, **98**, by a reaction analogous to the haloform reaction (equation 45).

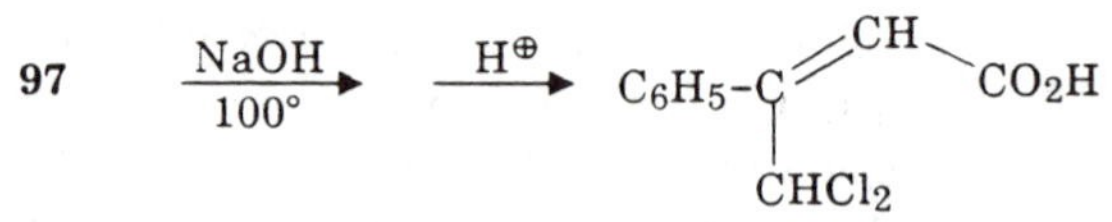

$$\textbf{97} \xrightarrow[100°]{NaOH} \xrightarrow{H^{\oplus}} C_6H_5\text{-}C(CHCl_2){=}CH\text{-}CO_2H \quad (45)$$

98

The structure of **98** was established by ozonization and by reduction to 3-phenylbutanoic acid. Bruce was making the adduct, **94**, in quantity now, but because the heavy glass tubes often blew up, it was economical to make several small runs at a time. He added a pinch of hydroquinone in hopes of inhibiting polymerization.

One day, he decided to add a milliliter of triethylamine to hold down the hydrogen–halide concentration, in case acid gases were causing decomposition or were responsible for the pressure in the tubes that caused them to burst. Hydrolysis of the cycloadduct, **99**, with sulfuric acid gave a ketone of mp 80 °C, **100**, the same as **97**, but which yielded a different carboxylic acid, **101**, on cleavage with sodium hydroxide (equation 46). Apparently, the triethylamine caused an anionotropic rearrangement of **94** to **99** under the conditions of the addition reaction.[91]

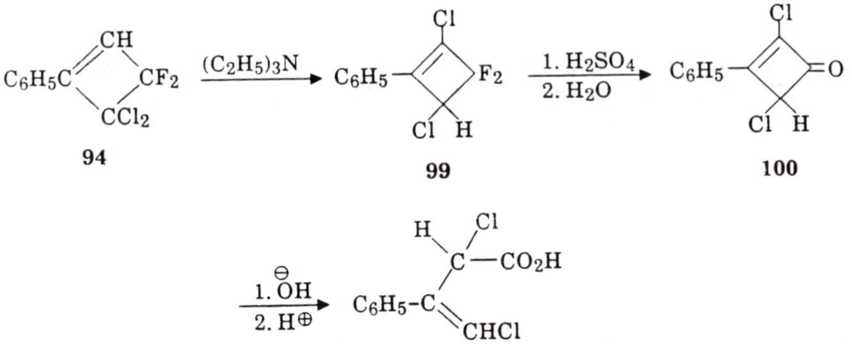

$$\underset{\mathbf{94}}{C_6H_5C(=CH)CF_2CCl_2} \xrightarrow{(C_2H_5)_3N} \underset{\mathbf{99}}{C_6H_5\text{-}C_4(Cl)(F_2)(Cl)(H)} \xrightarrow[2.\ H_2O]{1.\ H_2SO_4} \underset{\mathbf{100}}{C_6H_5\text{-}C_4(Cl)(=O)(Cl)(H)} \xrightarrow[2.\ H^{\oplus}]{1.\ \overset{\ominus}{O}H} \underset{\mathbf{101}}{C_6H_5\text{-}C(=CHCl)\text{-}C(H)(Cl)\text{-}CO_2H} \quad (46)$$

The cyclobutenone, **100**, is a fascinating substance, because it is just a proton shift away from a hydroxycyclobutadiene, **102** (equation 47). Well, it doesn't happen; **100** seems perfectly happy as it is. This

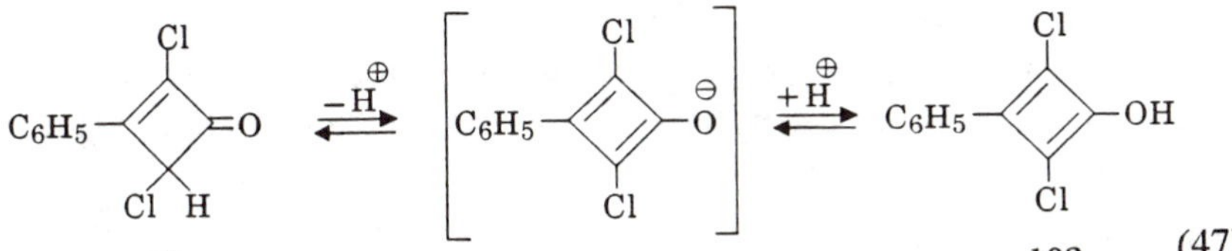

$$\underset{\mathbf{100}}{C_6H_5\text{-}C_4(Cl)(=O)(Cl)(H)} \underset{}{\overset{-H^{\oplus}}{\rightleftharpoons}} \left[C_6H_5\text{-}C_4(Cl)(Cl)\text{-}O^{\ominus}\right] \overset{+H^{\oplus}}{\rightleftharpoons} \underset{\mathbf{102}}{C_6H_5\text{-}C_4(Cl)(Cl)\text{-}OH} \quad (47)$$

tells us something about the stabilization of the cyclobutadiene system. Several years later, an excellent Swiss postdoctoral fellow, Erwin Jenny, did a wonderful job of looking into the propensity of **100** to be converted to **102**. Our procedure was to make **100** optically active by selective destruction of one enantiomer. Any conversion of **100** to **102** should cause loss of optical activity, because **102** is expected to be planar and achiral.

Interestingly, **100** loses its optical activity on heating, but it does so *without breaking* the bond to the hydrogen attached to the –CHCl– carbon. This was demonstrated by finding that there is no H–D exchange when the loss of optical activity occurs in CH_3CO_2D. Furthermore, there was no evidence that chloride ion caused the loss of optical activity by reactions such as that shown in equation 48. If you know that racemization is occurring without breaking the C–H bond or the C–Cl bond, then it must be breaking a C–C bond. An excellent way to

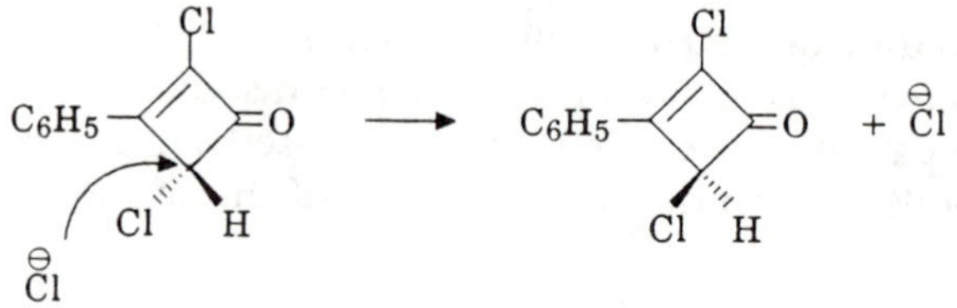

(48)

have this happen is by reversible formation of a vinylketene, 103 (equation 49). It was rather surprising for this reaction to be reversible. We

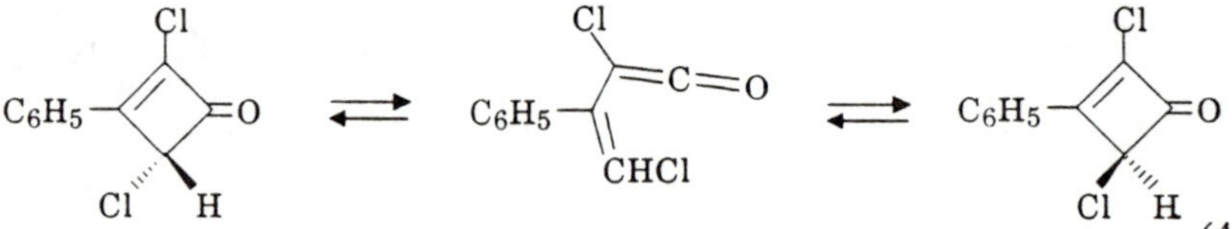

103 (49)

showed that reversible formation was taking place by trapping the ketene, 103, with ethanol to yield the corresponding ethyl ester. The rate of formation of ester equaled the rate of loss of optical activity in the absence of ethanol.[92] It was a really fun research problem!

Just as interesting was the discovery by postdoctoral fellow E. J. Smutny[93] that the adduct of $CFCl{=}CF_2$ with phenylacetylene, 104, could be hydrolyzed in good yield to the stable cyclobutadienoquinone, 105, by 92% sulfuric acid.

The structure of 105 is quite well established by oxidation with hydrogen peroxide to phenylmaleic anhydride, 106 (equation 50). Here,

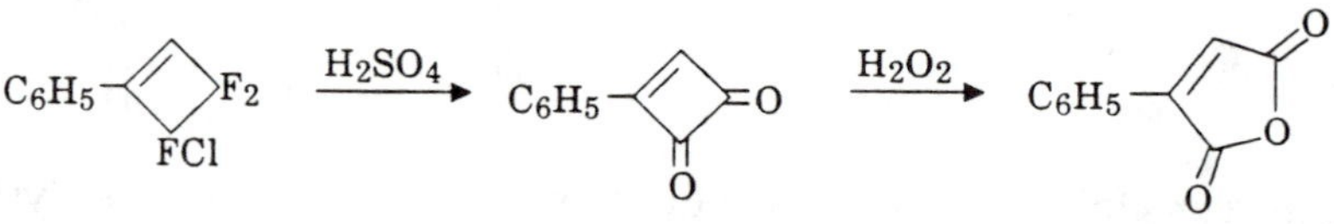

104 **105** **106** (50)

in 105, we have the quinone of the cyclobutadiene system. It is significant that 105 is quite stable and the corresponding *o*-benzoquinone is much less so. Compounds like *o*-benzoquinone react in some manner or other to try to regain benzene-ring stabilization. None of the cyclobutadienoquinone derivatives we looked at did anything but try to avoid being converted to cyclobutadiene derivatives. Clearly, this is not a matter of angle strain, because 105 must have about the same angle strain as cyclobutadiene.

We found many wonderful reactions of 105[94] and the corresponding cyclohexenyl analog.[95] Among these were substitution with halogens, 107, and conversion of the halides to the hydroxy and amino derivatives 108a and 108b, respectively (equation 51). The halo-

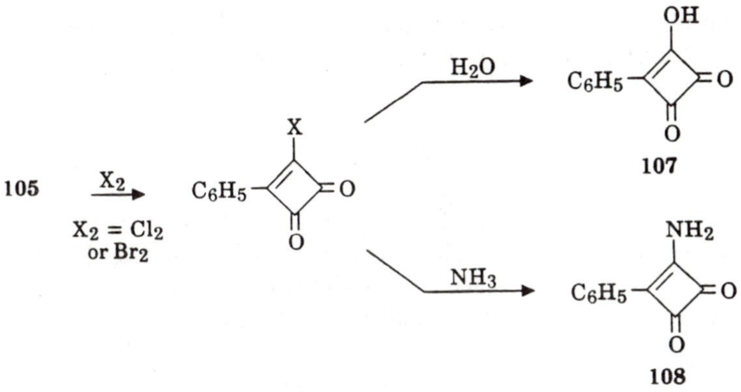

(51)

gen derivatives are reactive in the manner of acyl halides. The amino compound, **108b**, was very weakly basic and did not form a benzoyl derivative. The hydroxy compound, **108a**, was an extraordinarily strong acid with a pK_a of 0.4, which makes it the strongest known neutral organic acid of carbon, hydrogen, and oxygen.

This series of compounds finally did yield a cyclobutadiene derivative,[96,97] **109** (equation 52), but, like other cyclobutadienes, it promptly dimerized to **110** (equation 53). It came from two different

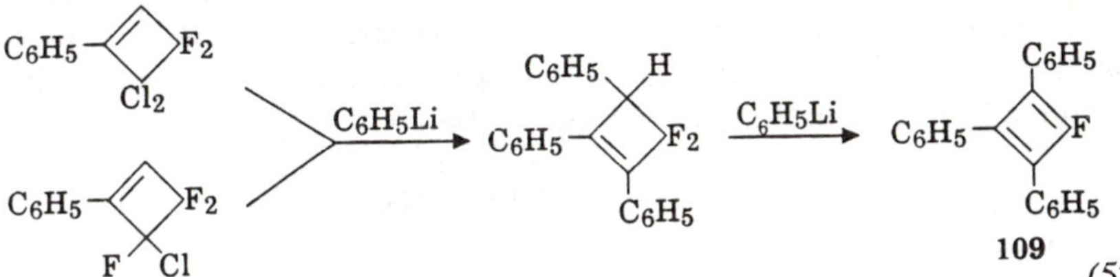

(52)

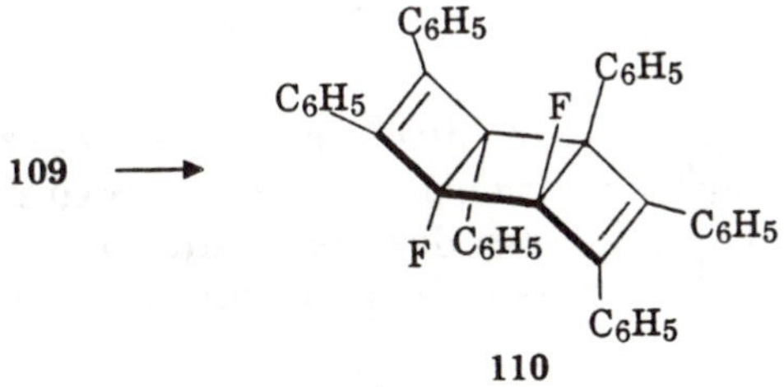

(53)

starting materials on treatment with phenyllithium. The structure, **110**, was indicated by strong double-bond absorption in the Raman spectrum, but not in the infrared, which requires $C_6H_5C{=}CC_6H_5$ and not $C_6H_5C{=}CF$ groupings. Of the possible dimers with this general arrangement, **110** is the only one in accord with the dipole moment associated with the C–F bonds. The structure, **110**, was confirmed by an extraordinary X-ray diffraction study[98] by E. W. Hughes and Charles Fritchie, Jr., at Caltech. When completed, this study was the most complex, complete structure determination by the Patterson projection procedure without benefit of a heavy atom to assist in getting the phases.

Benzyne

I still have a vivid memory of being told, while at Penn State in 1941, that Henry Gilman of Iowa State University of Science and Technology found chlorobenzene to react with sodium amide in liquid ammonia to give aminobenzene (aniline), equation 54. I hardly believed that report,

$$C_6H_5{-}Cl + NaNH_2 \xrightarrow{NH_3} C_6H_5{-}NH_2 + NaCl \qquad (54)$$

at the time being imbued with the lore that halobenzenes did not undergo S_N2 reactions. However, there were already reports of such reactions in the literature going back as far as 1895, and there was also the work of F. W. Bergstrom at Stanford University, published in 1936.[99] I don't recall ever meeting Bergstrom. He did not seem to be part of one of the "in" groups of organic chemists, centered around Illinois and Roger Adams, or Harvard and Conant, but he did do some interesting and important work. Being a Gilman follower, I was really struck by his observation[100] that 2-halo-1-methoxybenzenes react with lithium diethylamide to form 3-amino-1-methoxybenzene (equation 55). Gilman

$$2\text{-}X\text{-}C_6H_4OCH_3 \xrightarrow{LiN(C_2H_5)_2} 3\text{-}(C_2H_5)_2N\text{-}C_6H_4OCH_3 + LiX \qquad (55)$$

$X = Cl, Br, I$

published this rearrangement just after a much-less-noticed report by Urner and Bergstrom[101] of similar processes occurring in the naphthalene series. Bergstrom also discovered the surprising catalysis of halobenzene-substitution reactions by potassium amide (equation 56).

$$C_6H_5Cl + \overset{\oplus}{K}\ \overset{\ominus}{:}C(C_6H_5)_3 \xrightarrow[\text{fast; } \overset{\oplus}{K}\ \overset{\ominus}{:}NH_2]{\text{very slow}} C_6H_5{-}C(C_6H_5)_3 \qquad (56)$$

This was really amazing chemistry, and I was fortunate to enlist Edward M. Kosower, then an undergraduate at MIT, to work on related reactions for his senior thesis. At that point in his career, Ed Kosower was a bit of a "flake". His grade record at MIT was hardly consistent with the brilliance of his intellect. Quite undisciplined, erratic in his enthusiasms, and not an especially neat laboratory worker, he was hardly an obvious choice to solve the mechanistic problem posed by this collection of rather simple-appearing reactions. In fact, he didn't solve the problem, but he did lay the groundwork we needed. Ed's work was important in telling us about some areas where we should not concentrate our attention, because the structure proofs were difficult and difficult mixtures of products were obtained. Ed worked with the chloromethylbenzenes and lithium diethylamide in ether, a system that had been investigated in a rather indecisive way by Bergstrom and Horning.[102]

At this time, there was no vapor-phase chromatography, no NMR, and no routine infrared, so the product analyses of the materials formed from the chloromethylbenzenes were not easy. Ed tried to identify the products by making picrates and differentiating possible 3-methyl-1-*N,N*-diethylaminobenzenes from the 2- and 4-isomers by nitrosation (equation 57). What Ed found was that the 2- and 3-chloro-1-

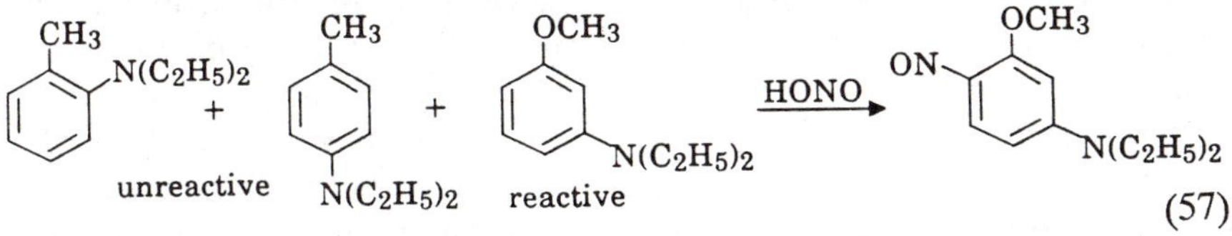

CH_3 $N(C_2H_5)_2$ + CH_3 $N(C_2H_5)_2$ + OCH_3 $N(C_2H_5)_2$ $\xrightarrow{HONO}$ ON OCH_3 $N(C_2H_5)_2$

unreactive reactive

(57)

methylbenzenes gave mixtures of 2- and 3-methyl-1-*N,N*-diethylaminobenzenes, although the 4-isomer gave a mixture of the 3- and 4-methyl-1-*N,N*-diethylaminobenzenes. But he was not able to quantify the results. From 3-chloro-1-methylbenzene, he isolated products that he believed to be 111–113 or isomers thereof. To account for the rearrangements, Ed wrote mechanisms such as equation 58, here using 2-

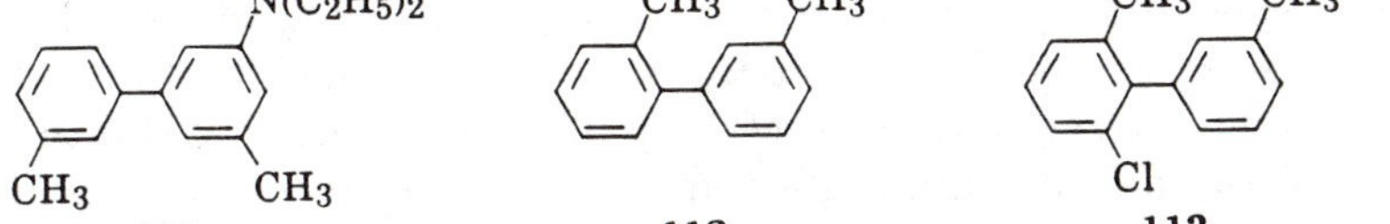

$N(C_2H_5)_2$ CH_3 CH_3 **111**

CH_3 CH_3 **112**

CH_3 CH_3 Cl **113**

CH_3 Cl $\underset{Li^{\oplus}}{\rightleftharpoons}$ CH_3 Cl Li $\xrightarrow{\ominus :N(C_2H_5)_2}$ CH_3 $N(C_2H_5)_2$ + CH_3 $N(C_2H_5)_2$

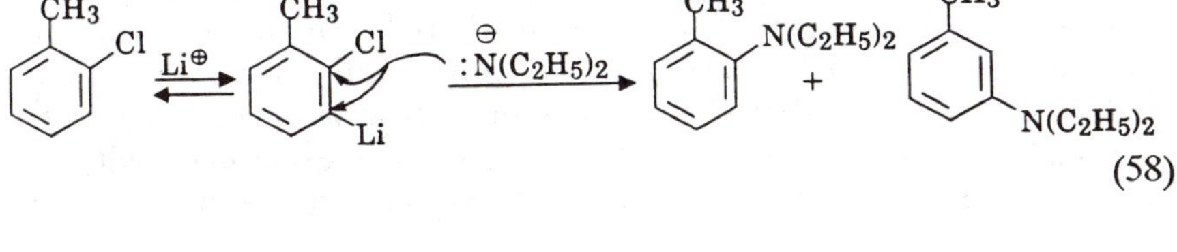

(58)

chloro-2-methylbenzene as an example.[103] These mechanisms do not stand up well when examined in detail, but they do reflect our thinking, at the time, that metalation is likely to play some kind of a role in the overall amination reaction. Obviously, metalation is vital to an understanding of how compounds like **111–113** might be formed.

Ed wanted to go to graduate school. I thought he should go to UCLA, because I thought I could get UCLA to take him despite his academic record. I did, but it was not easy. Amusingly, another MIT undergraduate named Kent was admitted to graduate work at UCLA at the same time. Kent had a superb record in courses and spectacular success in his senior thesis research with Gardner Swain. To everyone's surprise, Kent dropped out of UCLA after just a few months. Although Kosower's somewhat erratic behavior caused some early problems for him, he settled down and did a first-rate thesis with Winstein. Subsequently, he has had a distinguished career in physical organic chemistry at Stony Brook and in Israel.

The amination problem lay fallow for a couple of years in our laboratory after Kosower left, but in 1950 I was able to interest a senior thesis student, C. Wheaton Vaughan, in working on the problem. He ultimately produced what was surely one of the most significant senior theses ever at MIT.[104] Wheaton seemed to have come from an aristocratic background. He was not really much in tune with the other MIT undergraduates. He was smart as a whip, but overly patrician, if not arrogant, and not very tolerant. Other than in intellect, he was the antithesis of Ed Kosower.

By the time Wheaton started, I was more and more convinced that metalation was involved in the process. To understand better what was going on, we decided to look at orientation patterns, choosing sodium amide in liquid ammonia as the reaction medium and the CF_3– group as our first orienting substituent. We checked out Benkeser's results[105] with 2- and 3-chlorobenzotrifluorides, both of which gave the 3-amino product. However, with the 4-chloro compound, we found nearly equal amounts of the 3- and 4-aminobenzotrifluorides. The analyses were easier by this time because we finally had routine infrared spectroscopy available to us at MIT. Now it appeared that the electrical property of the orienting substituent was ineffective, whereas it was highly effective with the 2- or 3-chloro starting materials.

Then, going back to the chloromethylbenzenes that Kosower had studied with lithium diethylamide, Wheaton found the pattern of amination-product percentages shown in Chart 1. There was a new element here, formation of the 4-amino compound from the 3-chloro derivative, something that simply does not occur with methoxy or trifluoromethyl starting materials. Drawing on my experience with the metalations of benzotrifluoride and methoxybenzene, it did not take

Chart 1. Amination products from the chlorotoluenes

	CH_3, NH_2 (ortho) %	CH_3, NH_2 (meta) %	CH_3, NH_2 (para) %
CH_3, Cl (ortho)	45	55	none
CH_3, Cl (meta)	40	52	8
CH_3, Cl (para)	none	62	38

long to envision a scheme such as equation 59 for the course of the reaction of the 2- and 3-chlorobenzotrifluorides with sodium amide. We could clearly rate **114** as being less stable than **115**, because the metal

CF_3, Cl, H $\xrightarrow[-NH_3]{\oplus\ \ominus\ NaNH_2}$ CF_3, Cl, Na (**114**) $\underset{\leftarrow}{\xrightarrow{fast}}$ CF_3, Na, Cl (**115**) $\xleftarrow[-NH_3]{\oplus\ \ominus\ NaNH_2}$ CF_3, H, Cl (59)

position between two electron-attracting groups is particularly favorable with **115**. Now, simple displacement of the chlorine of **115** with $-NH_2$ will give the 3-amino product.

Similarly, with the 4-chloro compound, we can construct equation 60, where **116** and **117** are sure to be more nearly equal in energy

CF_3, Cl $\xrightarrow[NH_3]{NaNH_2}$ CF_3, Na, Cl (**116**) $\overset{fast}{\rightleftarrows}$ CF_3, Cl, Na (**117**) (60)

than **114** and **115** because of the remoteness of the CF_3- group. The result could well be formation of a nearly 50:50 product distribution. The application of the same kind of mechanism to the products from the chloromethylbenzenes should be obvious.

This formulation of the reaction explained everything up to that point, even the amide catalysis of tetraphenylmethane formation (equation 56), by having $NaC(C_6H_5)_3$ reacting to displace the chlorine from **118** (equation 61). Although this satisfied me about the orientation pat-

$$C_6H_5Cl \xrightarrow[-NH_3]{NaNH_2} o\text{-}ClC_6H_4Na \quad (\mathbf{118}) \qquad (61)$$

terns and catalysis, in a session with Wheaton on January 9, 1951, I said: "But why are we hanging NaCl on the benzene ring? Won't it be easier to just form a triple bond? Hey! That's benzyne!"

Why this (equation 62) had not occurred to me much earlier is not clear. I guess it became clear only when I was forced to see that we

$$\mathbf{118} \xrightarrow{-NaCl} C_6H_4 \quad (\mathbf{119}\text{, benzyne}) \qquad (62)$$

Georg Wittig and John D. Roberts in Pasadena, about 1960. Wittig received the Nobel prize in 1979, for his many contributions to synthetic chemistry, notably the Wittig reagents.

had to have equilibrations such as **114** with **115** and **116** with **117** to account for the reaction products. Then I had to ask myself what the mechanism might be for equilibrations involving interchanges of sodium and chlorine on adjacent carbons. It was even less understandable for me to fail to recognize the possibility of benzyne because there was highly suggestive work by Wittig in the literature on couplings of phenyllithium with halobenzene derivatives. It almost got Wittig himself to propose benzyne as an intermediate.[106,107] He came very close indeed when he suggested a C_6H_4 dipolar intermediate, **120** (equation 63), which was formulated with nonequivalent carbon atoms and was

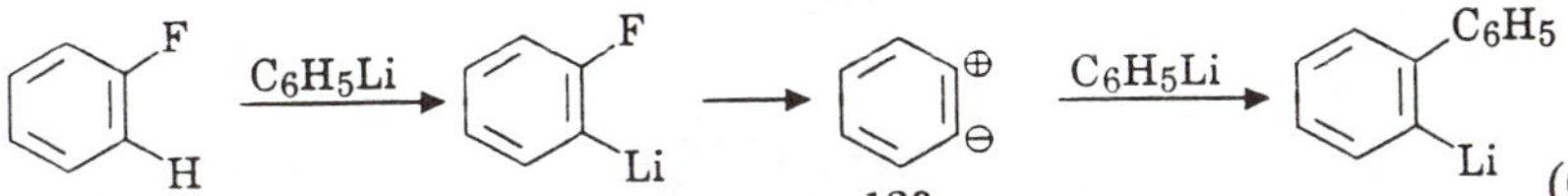

F, H; C_6H_5Li → F, Li → ⊕, ⊖ **120**; C_6H_5Li → C_6H_5, Li (63)

not proposed to account for rearrangement processes. I had read Wittig's 1940 paper much earlier, but it certainly was not in the forefront of my consciousness when we were struggling to understand rearrangements in the amination of halobenzenes.

With this flash of understanding, pieces began to fall together, and an amazingly concordant picture of an enormous amount of chemistry became clear. If benzyne were the intermediate, and highly reactive, we could account for almost everything we knew, such as the formation of diphenyl- and triphenylamines as side products and some of the kinds of odd products found by Kosower. Orientation became an exercise in predicting, first, which way $NaNH_2$ would react to eliminate HX when there were alternative possibilities. The favored pathway would be expected to involve the more acidic hydrogen adjacent to the halogen. Thus, formation of **121** would be favored over **122** (equation 64) because the 2-hydrogen should be the most easily removed by

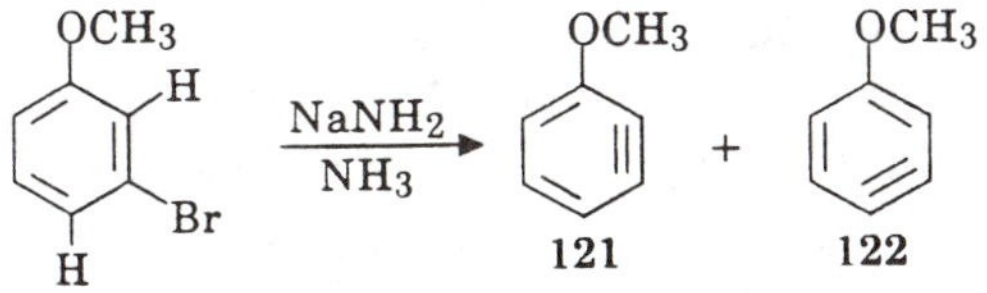

OCH_3, H, Br, H; $NaNH_2$ / NH_3 → OCH_3 **121** + OCH_3 **122** (64)

amide ion. The second consideration was how $NaNH_2$ would add to a particularly substituted benzyne, and the criterion was whether the sodium could wind up next to a reasonable electron-attracting substituent. Thus, formation of **123** would be favored over **124** (equation 65), although there would be much less preference for **125** over **126** (equation 66).[110]

The telling experiment seemed to be to try to aminate something that did not have a hydrogen next to the halogen. The first thing that came to mind was bromomesitylene, **127**, which obliged us beautifully

OCH_3 121 $\xrightarrow{NaNH_2}$ OCH_3, Na, NH_2 123 + OCH_3, NH_2, Na 124 (65)

OCH_3 122 $\xrightarrow{NaNH_2}$ OCH_3, Na, NH_2 125 + OCH_3, NH_2, Na 126 (66)

by not reacting at all (equation 67), even though many nonrearranging reactions, such as with 3-chloro-1-methoxybenzene, were known. With this very simple experiment, we were off and running. Wheaton immediately looked hard in the neutral residue of the reaction of chlorobenzene with sodium amide in liquid ammonia for evidence of formation of biphenylene (equation 68), but we found none—that,

CH_3, CH_3, Br, CH_3 127 $\xrightarrow[NH_3]{NaNH_2}$ no reaction (67)

2 ? (68)

indeed, would have been great. But, of course, we did not know then about forbidden $2\pi + 2\pi$ cycloadditions and the Woodward–Hoffmann rules. Wheaton was winding up his senior thesis in April, and at the Cleveland meeting of the American Chemical Society on April 8, 1951, Benkeser and Buting[109] reported the transformations of equations 69 and 70.

CH_3, Br, OCH_3 $\xrightarrow[NH_3]{NaNH_2}$ CH_3, NH_2, OCH_3 CH_3, OCH_3, Br $\xrightarrow[NH_3]{NaNH_2}$ CH_3, OCH_3, NH_2 (69)

CF3 / OCH3 / Br —NaNH2, NH3→ CF3 / H2N / OCH3 but CH3 / Br / OCH3 —NaNH2, NH3→ no reaction (70)

It was fantastic. Everything fitted with what we predicted from benzyne as the intermediate. I gave an informal seminar about our results at UCLA during the summer of 1951. Saul Winstein did his best to disbelieve, but finally agreed that the mechanism was a good one. Of course, by then, Wheaton was on his way to Berkeley as a graduate student, flush with success, and apparently the attitude that had seemed only patrician was beginning to border on arrogance. I heard that he was not terribly popular with his peers at Berkeley, but he did a strong thesis with W. G. Dauben. Then, in 1954, he went to Du Pont for several years as a research chemist. He switched to business in 1958, earned an MBA at Harvard, and was director of planning at Scott Paper Company and later vice president for business planning at Honeywell Corporation. At this point, he disappeared off the radar screen of *American Men and Women of Science.*

The heat was on us to publish benzyne or possibly get scooped, but I wanted to do the clinching experiment of showing that, even with chlorobenzene itself, isotope-position rearrangement accompanied substitution. Thus, if one started with chlorobenzene-1-^{14}C, equal amounts of aminobenzenes with the ^{14}C at C-1 and C-2 would be formed (equation 71). This experiment, which turned out to be a real classic, was carried out by L. A. Carlsmith, an M.S. student, and Howard E. Simmons, Jr. Howard, who later became a great success in the Central Research Department at Du Pont and is currently vice president and the department's director, had done an excellent senior thesis with me.

14 Cl —NaNH2→ 14 —?→ 14 NH2 + 14 NH2 (71)

Although it was not Cope's policy that MIT undergraduates be allowed to continue at MIT for their Ph.D. work, some notable exceptions made in my time included Robert H. Mazur, G. Bruce Kline (returned from military service), E. J. Corey, and Simmons. I initially recruited Howard to follow up on some of Bruce Kline's cyclobutenone chemistry. He did some interesting work on that, while, at the same time, beginning to see if there was isotope-position rearrangement in

Howard E. Simmons in his office at the Du Pont Central Research Department about 1956. A dyed-in-the-wool Virginian, Simmons is culminating a great industrial research career as vice-president and director of research of his department. He has made many important basic research contributions, especially to the chemistry and properties of macrobicyclic amines and, in 1989, coauthored with R. E. Merrifield, a striking book, Topological Methods in Chemistry *(Wiley).*

the reaction of cyclohexyl-1-^{14}C-amine with nitrous acid. Because the degradative schemes for the cyclohexylamine products could be transferred over directly to the products of equation 71, it was not hard to persuade Howard to take on THE benzyne experiment instead. It was a hot item, and I really liked the idea of getting a "known quantity" to work on the problem, because I was shortly going off on a Guggenheim Fellowship to Caltech and, later, to Europe.

The work was divided; Carlsmith optimized the procedure for converting chlorobenzene to aminobenzene and Simmons ran the degradation reactions. The published values for the degree of rearrangement, which supported our contention of the intermediacy of benzyne, were obtained in early March 1953, and the received date on the communication to the *Journal of the American Chemical Society* was March 12, 1953.[110] By coincidence, our communication was published in the same issue of the *Journal* as an article by Benkeser and Schroll[111] suggesting an addition–elimination mechanism.

The arguments we gave for benzyne, besides the isotope-position rearrangement observed for chlorobenzene-1-^{14}C, can profitably be summarized here. First, the reactions are very rapid, even with chlorobenzene, in liquid ammonia at –33 °C. Second, the entering amino group has never been found more than one carbon away from the position occupied by the leaving group. Third, the starting halides and resulting amino compounds are not isomerized under the reaction conditions. Fourth, no reactions occur with halides, such as bromomesitylene, bromodurene, and 2-bromo-3-methyl-1-methoxybenzene, where no hydrogen is attached to the position adjacent to that occupied by the halogen.

We then undertook to determine whether there was a kinetic deuterium-isotope effect in these reactions. One can expect, for an E2 elimination reaction on chlorobenzene, wherein breaking of the C–H and C–Cl bonds is essentially simultaneous, that there should be about a sevenfold rate difference if deuterium is substituted for hydrogen in the C–H bond (equation 72). Of course, there is an alternative—

$$C_6H_4DCl + {:}\overset{\ominus}{N}H_2 \xrightarrow{E2} C_6H_4 + \overset{\ominus}{Cl} + NH_3 \tag{72}$$

operation of an E1 mechanism where D–H exchange would be expected. The isotope effect was especially important with respect to other mechanisms that could be written to account for many of the other facts about the reaction. The key to these mechanisms is formation of **128** or **129** (equation 73). Operation of mechanisms like those of equation 73 would hardly be expected to show a kinetic isotope effect because the loss of the proton on the carbon adjacent to the chlorine occurs well after the loss of that chlorine and, indeed, also after the making of the C–NH_2 bond.

Howard did several experiments to establish the magnitude of the kinetic isotope effect on the basis that elimination might occur as

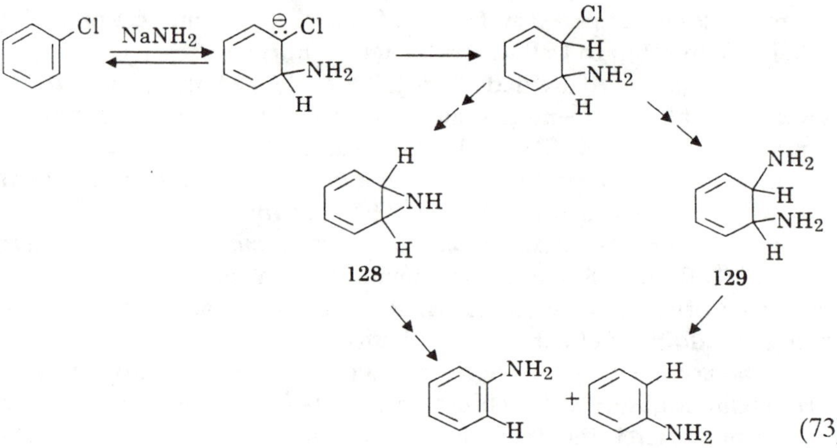

$NaNH_2$

128

129

(73)

shown in equation 73. However, at this point in Howard's superb thesis work, I decided to go to Caltech. Howard, who was already two years along toward his Ph.D., decided it was better for him to stay at MIT and shift to Art Cope as his thesis adviser. I could understand, even though I hated to lose him. You never win them all!

The kinetic-isotope effects were put on a firm basis by Dorothy Semenow in her thesis work at Caltech. She found that with sodium amide in liquid ammonia, the measured isotope effect is rather small with chlorobenzene because of deuterium–hydrogen exchange. In one run, it actually looked like k_H/k_D was less than 1, because the recovered starting material contained *less* deuterium than was present initially—a sure sign of exchange.

In contrast, bromobenzene, which has a more easily broken C–halogen bond, showed a more substantial kinetic-isotope effect, $k_H/k_D = 5.5$, with sodium amide in liquid ammonia. This finding helps to explain why the order of reactivity of the halobenzenes toward sodium amide in liquid ammonia is Br > I > Cl >> F,[112] whereas the usual order of reactivity of organic halides is I > Br > Cl >> F. All of the evidence showed that fluorine is the most potent of the halogens for activating an adjacent hydrogen by amide ion, but it is the poorest leaving group. On the other hand, iodine should be the best leaving group, but it is surely the poorest hydrogen activator. Bromine appears to offer the best balance between activating the hydrogen and being a good leaving group. Furthermore, Dorothy determined k_H/k_D to be about 5.8 and 5.6, respectively, for 2-deuterio-1-bromobenzene and 2-deuterio-1-chlorobenzene with lithium diethylamide in ether.

In early 1955, Erwin Jenny[113] showed that Wittig's reaction of phenyllithium with fluorobenzene to give 2-lithiobiphenyl occurred with the benzyne-expected degree of isotope-position rearrangement when run with fluorobenzene-1-^{14}C (equation 74). By this time, Rolf Huisgen in Munich was well into his elegant and comprehensive stud-

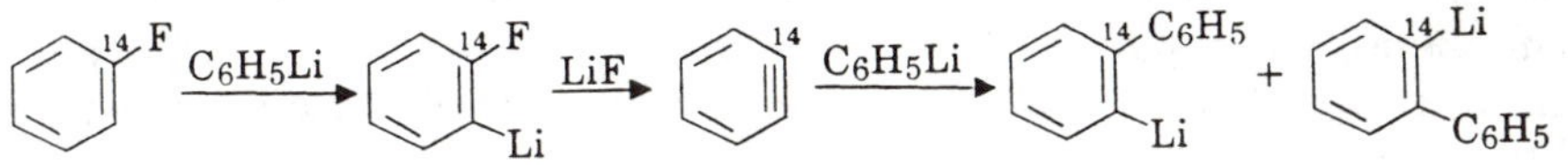

$$C_6H_5{}^{14}F \xrightarrow{C_6H_5Li} \text{o-}Li\text{-}C_6H_4{}^{14}F \xrightarrow{LiF} {}^{14}C\text{-benzyne} \xrightarrow{C_6H_5Li} \text{o-}Li\text{-}C_6H_4{}^{14}C_6H_5 + \text{o-}C_6H_5\text{-}C_6H_4{}^{14}Li \quad (74)$$

ies of arylation with lithium reagents, which established the benzyne mechanism so well in this kind of reaction.[114] Nonetheless, the idea that reactions could proceed easily and quickly by way of benzyne as an intermediate was hard for many people to swallow. I particularly remember giving a lecture in 1953 at Cambridge, in England, and being vehemently attacked by a heterocyclic chemist named Frederick G. Mann, to the great amusement of the lead professor, Alexander Todd, and his crew of marvelous lieutenants.

About this time, I made a psychologically interesting discovery. I had been drawing benzyne as **130a**, assiduously avoiding (as you have may have noticed earlier in this section) drawing its perfectly legitimate resonance structure, **130b**. In my lectures on benzyne, I explained that **130b** looked just too much like a misprint. Then, one day, I drew the structure as **131a** and **131b**, with the extra bond *outside* rather than *inside*

130a ⟷ **130b**

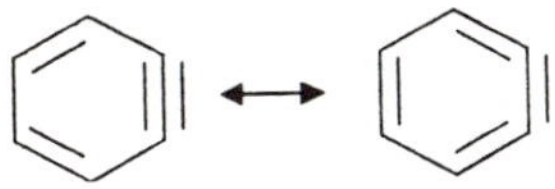

131a ⟷ **131b**

the benzene ring. I was quite surprised how much more palatable this representation seemed to be, and I adopted it forthwith.

A couple of very nice experiments on amination in liquid ammonia were carried out by Manuel Panar, a Canadian Ph.D. student recommended to work at Caltech by Ruben B. Sandin of Alberta, who had sent Chuck Lee to MIT earlier. Manny, later research manager of exploratory polymer research at the Central Research Department of Du Pont, showed, first, that the degree of isotope-position rearrangement observed with 2,4,6-trideuterioiodobenzene-1-^{14}C is the same as for iodobenzene-1-^{14}C and, second, that the *nonrearranging* amination of 2,4,6-trideuterio-3-bromo-1-methoxybenzene shows a kinetic-isotope effect.

These results show decisively that the rearrangement is not the result of a fortuitous mix of a simple nonrearranging displacement and some other rearranging reaction, which gives other than the 50:50 isotope-position distribution characteristic of benzyne.[115] When you take into account the kinetic-isotope effect, rearrangement, activation, and so on, you might write, as an alternative to benzyne, the symmetrical intermediates **132** and **133**. However, neither of these structures is

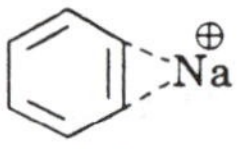

132 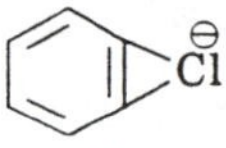 133

reasonable with respect to the very elegant benzyne-trapping experiment carried out by Wittig and Pohmer,[116] equation 75, which is strong evidence for a rather bare benzyne.

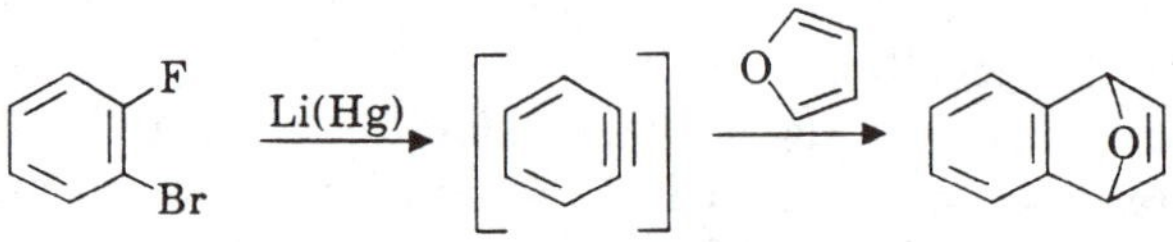

Li(Hg) (75)

Wittig entered the lists on the naming front with a valiant effort to establish the name "dehydrobenzene" for benzyne. Of course, I resisted, although many authors, particularly in Germany, used dehydrobenzene in deference to Wittig; at least "arynes" came to be rather common usage. A serious problem with dehydrobenzene is that you need to specify from which carbons the hydrogens are lost. Besides 1,2-dehydrobenzene, there are also the 1,3 and 1,4 isomers, **134** and **135**,

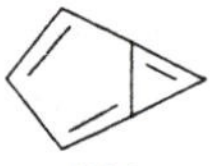

134  135

respectively. Ed Smutny tried very hard to synthesize these isomers at Caltech by vacuum pyrolysis of 1,3- and 1,4-lithiohalobenzenes, as well as by a number of other reactions, with no visible success.

On a related front, graduate student Al Bottini, later at the University of California at Davis, showed that the high-temperature hydrolysis of halobenzenes, a classical and sometimes commercially viable procedure for synthesis of hydroxyarenes, usually involved a mix of rearranging and nonrearranging substitution reactions. When hydrolysis was induced with concentrated strong base at high temperatures, much rearrangement occurred, as in equation 76. But with fewer, or weaker, bases and lower temperatures, the nonrearranging displacement was favored.[117]

The other major development in this general field of displacement reactions was achieved by graduate students Franco Scardiglia and

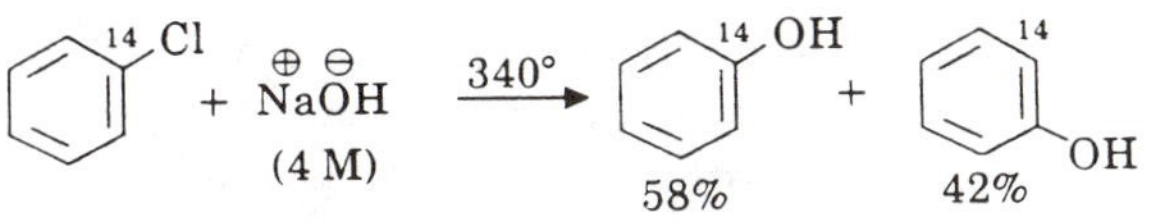

$$C_6H_5{}^{14}Cl + \overset{\oplus\ \ominus}{NaOH}\ (4\ M) \xrightarrow{340^\circ} C_6H_5{}^{14}OH\ (58\%) + {}^{14}C\text{-}C_6H_4OH\ (42\%) \quad (76)$$

L. K. Montgomery.[118,119] They did a superb job of showing, through isotope-position rearrangements, that 1-chlorohexene and 1-chlorocyclopentene appear to react with phenyllithium by way of elimination–addition, forming the corresponding cycloalkynes as intermediates (equations 77 and 78). And so we finally abandoned further

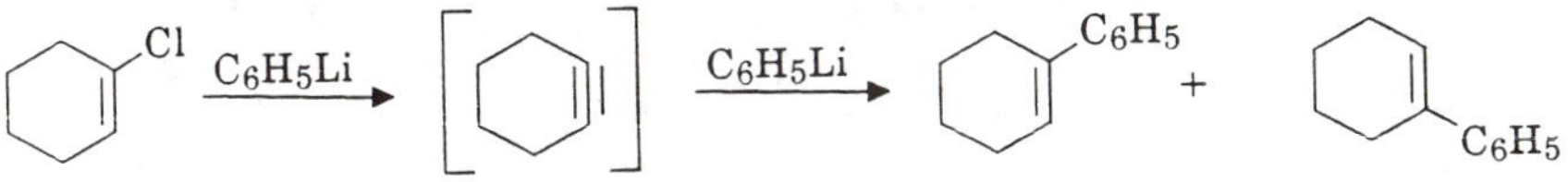

$$\text{1-chlorocyclohexene} \xrightarrow{C_6H_5Li} [\text{cyclohexyne}] \xrightarrow{C_6H_5Li} \text{1-}C_6H_5\text{-cyclohexene} + \text{2-}C_6H_5\text{-cyclohexene} \quad (77)$$

$$\text{1-chlorocyclopentene} \xrightarrow{C_6H_5Li} [\text{cyclopentyne}] \xrightarrow{C_6H_5Li} \text{1-}C_6H_5\text{-cyclopentene} + \text{2-}C_6H_5\text{-cyclopentene} \quad (78)$$

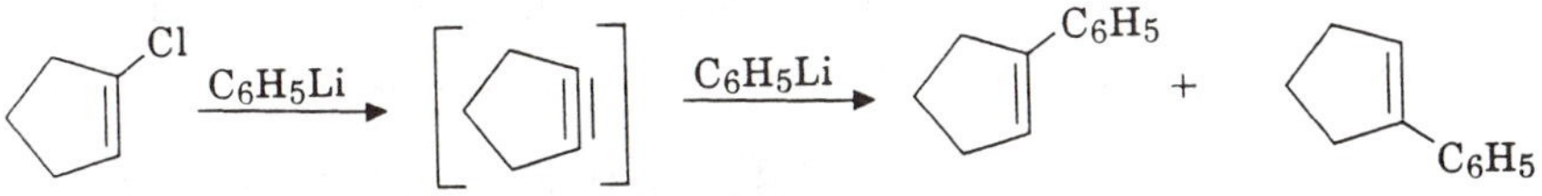

work on benzyne and related reactions. It was a wonderful field to be in, and it has been heartwarming to see the many subsequent developments.[120]

Teaching at MIT

Much of my early teaching at MIT involved running a quiz section for elementary organic chemistry, a section of the organic chemistry laboratory. I was really pleased that Art Cope asked me to develop a new physical organic chemistry course for seniors and beginning graduate students. This was my first really serious teaching responsibility, and I didn't do well in the first go-round. The major problem was over-preparation. I spent 6 to 8 hours getting ready for each lecture and was loaded for bear on every outing. One reason for the extensive preparation was that it was a large class of 75 or so. That was bad enough, but on top of it, most of the organic chemistry postdoctoral students came, as well as Professors Cope, Sheehan, Swain, and Ashdown.

In today's terms, the thrust of the course seems pretty conventional. An introduction with heavy emphasis on Lewis electronic structures, resonance effects on bond lengths, dipole moments, spectra, acid strengths, and chemical reactivity, followed by a rather thorough survey of S_N1, S_N2, E1, and E2 reactions of organic halides. Finally, a detailed exposition of the Hammett equation. It was a thorough job, but it sailed away over the heads of those in the class. Then there was the true–false test fiasco in which I asked the students to classify statements as true or false, with six points for correct, minus six for incorrect, and zero for no answers. Some of the statements were somewhat ambiguous—an example was "The first-order elimination reaction of alkyl halides in aqueous ethanol is actually E2 with solvent, because the ratio of elimination *vs.* substitution varies with temperature." True or false? I don't know now what I had in mind. A histogram of the student responses showed that the average overall grade on these questions was very close to zero!

Some of the students complained to the dean about being overwhelmed by content and assignments, the dean complained to the department, and Cope and Sheehan administered a severe dressing down on a day when I was almost *hors de combat* with the flu. Actually, it really wasn't that bad a course. I learned a lot that I didn't know or needed to know better. But, I was clearly ahead of my time in giving that kind of material to undergraduate and beginning graduate students.

The next year I was more relaxed, the course content was about the same, the exams were comparably tough, but there were no complaints. I enjoyed giving the course, and it loped along until the summer of 1951, when a big change took place. For several years, there was a growing consciousness of molecular orbital theory. Mulliken, Coulson, Wheland, and others were publishing what seemed like rather esoteric papers, some of which clearly were relevant to physical organic chemistry.

Resonance theory was well entrenched, generally highly satisfactory in a qualitative way, even if you could not claim to have the faintest understanding of the quantum mechanics involved and had no real idea of what to make of a statement like "Compound X resonates between structures A and B." To be sure, there were books such as Eyring, Walter, and Kimball's *Quantum Chemistry*[121] or Pauling and Wilson's *Introduction to Quantum Mechanics*[122] that covered matters such as these, but precious little of it was either understood by or useful to organic chemists. Most of us were resigned to explaining resonance to students by some variation on Wheland's rather tortuous qualitative descriptions of pendulums and similar analogs in *The Theory of Resonance*,[123] with the hope that, in the repetition, familiarity could mimic understanding.

I was uncomfortable with this approach but, even today, quantum mechanics is bound to be a bit uncomfortable if you require things to be intuitively obvious. Nonetheless, when one gets beyond the description of the general nature of resonance theory, there is little question of the enormous usefulness of resonance theory in correlating, predicting, and providing a kind of understanding of an immense range of organic phenomena. Its use in such connections was very ably documented by Wheland, a consummate scholar, in his 1944 and later books.[124]

Despite the widespread acceptance of resonance theory, there were some cracks in the façade. One was the question of cyclobutadiene, **136**, for which two equivalent resonance structures can be written, and which, by qualitative resonance theory, should be expected to be reasonably stabilized in something like the same way as benzene, **137**. Of course, the bond angles of **136** would be far from the "normal" 120° values that benzene would possess.

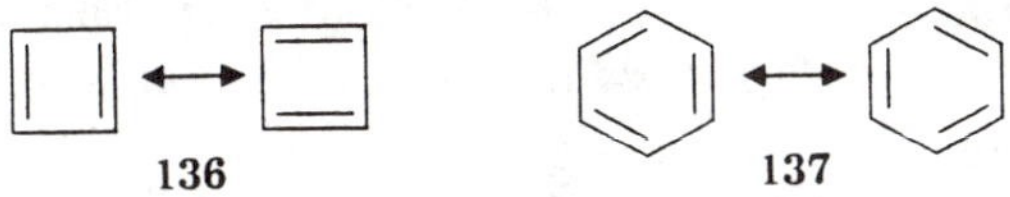

136 **137**

Linus Pauling told me around 1950 that he believed that cyclobutadiene would be stabilized by resonance, but that the stabilization might not be enough to overcome the angle strain in the small ring, and that cyclobutadiene could well decompose into two molecules of acetylene. I did not challenge him, but I should have, because straightforward calculation from known thermochemical data shows that, even if there were no resonance energy and no activation energy at all for scission of cyclobutadiene to acetylene, the extra *angle strain* required for scission would have to be about 60 kcal/mol over and above that of cyclobutane, which is about 19 kcal/mol.

It is hard to explain by resonance theory why 1,3-cyclopentadiene, **138**, is a quite strong acid for a carbon–hydrogen compound (equation 79) and yet 1,3,5-cycloheptatriene, **139**, is not (equation 80). Here, the resonance structures look very comparable. With respect

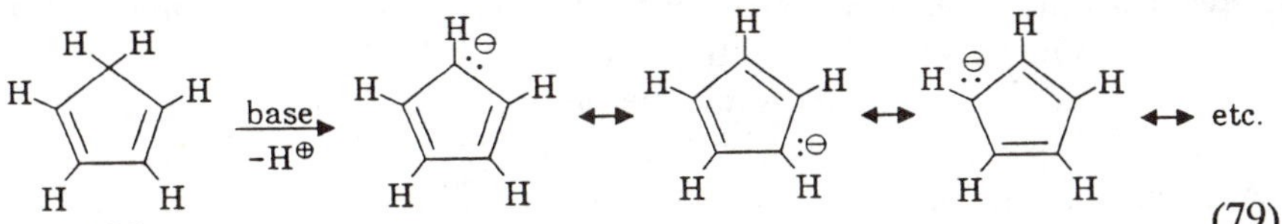

138 base, $-H^{\oplus}$ ↔ ↔ ↔ etc. (79)

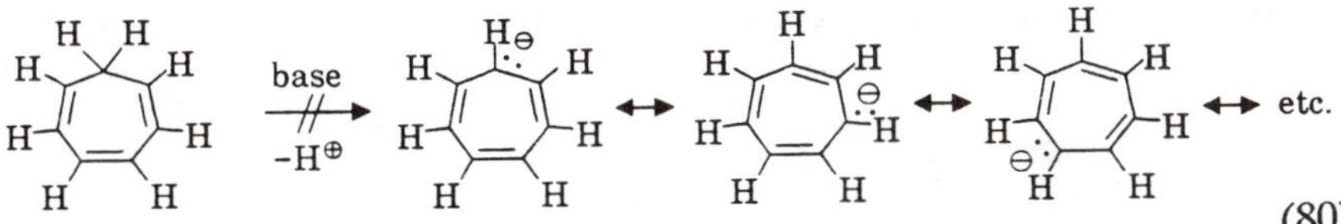

139 base (no reaction), $-H^{\oplus}$ ↔ ↔ ↔ etc. (80)

to the nonacidity of 1,3,5-cycloheptatriene, Wheland, in *The Theory of Resonance*[125] (1944), suggests as one possibility that the resonance structures of the corresponding anion might not be all planar.

Way back in 1934, Wheland had shown that the relatively low acidity of 1,3,5-cycloheptatriene is in accord with predictions of the Hückel molecular orbital theory.[126] Nonetheless, in 1944, he indicated strongly that the molecular orbital results are not to be trusted because they take improper account of interelectronic repulsion. Just how many organic chemists were reading the *Journal of Chemical Physics* in 1934 is unclear (I know I wasn't!). Any who were seemed unable to understand or use the clear message of the Wheland paper, that much could be predicted by molecular orbital theory for compounds in which they had an interest. A bit of a breakthrough for popularization of molecular orbital theory came in 1947, when C. A. Coulson published an article in the first issue of *Quarterly Review* that made it easier to understand how to formulate atomic orbital models like those for cyclopropane and carbocations.[127] The winds of change had started to blow. In 1950 I got a copy of Michael Dewar's *The Electronic Theory of Organic Chemistry*[56] and was tantalized by the claims, in typical Dewarian style, that resonance was, in effect, old hat and that we should hitch our chariots to molecular orbital theory. Perhaps so, but oddly enough, as one reviewer of Michael's book pointed out, the latter part of it explained almost everything in terms of resonance!

Anyway, I listened to the siren song and, inflamed, told my class in the first lecture of the usual Fall semester that this year was going to be different—resonance would be abandoned and the course would be taught *solely* on the basis of molecular orbital theory. In preparing for lectures, I began to think about the allyl radical, **140**. The overlapping p–π orbitals and their σ-bonding frameworks did seem elegant, convincing, and ready to press on the students. However, a little bird chirped, "You've drawn the atomic orbitals for the allyl radical, which resonance theory formulates as **142a–142c**, but what does molecular orbital theory tell you about the way the electrons are distributed?"

Dewar[128] gives a diagram for the allyl radical in which the three p–π orbitals are represented by **141**. Then Dewar says: ". . . the three (π) electrons can be fitted into two orbitals of low energy covering all the atoms; the lower orbital is filled, but the higher orbital is not, so the compound is still unsaturated; . . . the resulting radical is completely symmetrical." That may seem clear, but it doesn't really tell you how

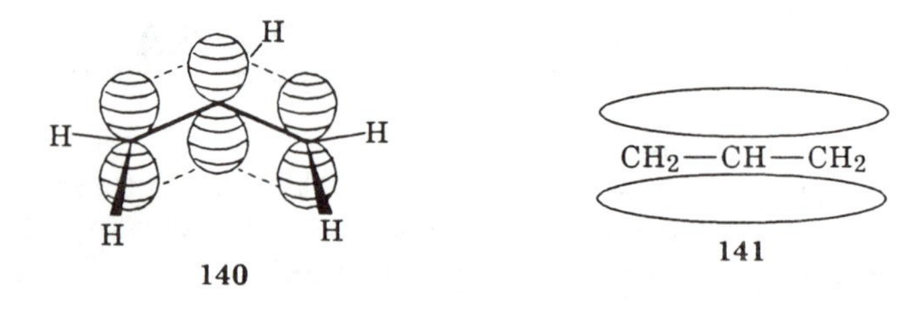

140 **141**

the odd electron is distributed through the molecular orbital(s). The odd electron acts as if it were on C-1 and C-3 but not on C-2, a result that falls out immediately from looking at **142a** and **142b** (equation 81).

$$\underset{\textbf{142a}}{CH_2{=}CH{-}\dot{C}H_2} \longleftrightarrow \underset{\textbf{142b}}{\dot{C}H_2{-}CH{=}CH_2} \sim \underset{\textbf{142c}}{\overset{1/2\,\bullet}{CH_2}{\,\overset{\cdots}{-}\,}CH{\,\overset{\cdots}{-}\,}\overset{1/2\,\bullet}{CH_2}} \quad (81)$$

As for the cation and anion—how does the Hückel molecular orbital (HMO) theory guide us in predicting the charge distribution in these entities? At this point, I began to have a sinking feeling in my stomach, began to curse Dewar and others for not letting me in on the secret, and began to curse myself for being so foolhardy as to think I could teach this subject to undergraduates. Panic set in. How was I going to understand how those electrons get distributed?

I rushed to the library and tried to see what those great works on quantum chemistry had to say about these problems. They said absolutely nothing in words that I could understand. Remember, I had taken no graduate courses in chemistry. I had never heard anything at all about quantum theory, in either my undergraduate chemistry or physics courses. The books all started from the Schrödinger equation, replete with wave functions; Hamiltonian operators; overlap, resonance, and coulomb integrals; and grisly looking summations. It was nothing I could learn in the 2 weeks or so before I had to explain to my class where those electrons are, or *admit defeat.*

Then I remembered that my old friend and physical chemistry lab partner, Bill McMillan (pp 19–22) was visiting professor at Harvard. I was sure Bill knew all about this stuff and might be able to get me going, so I called him up and told him my problem. He was faintly amused, but hardly helpful. "Just get a copy of Eyring, Walter, and Kimball's *Quantum Chemistry*[121] and work it through until you find what you want." I told him I had tried that, but what I wanted seemed to be in Chapter 19, and he should know I'd never get there in a couple of weeks. Then I asked him if he really did know how you found out about these things. He said, "Of course; it's easy." "Okay, I'm coming tomorrow afternoon," I said, "and you've got to show me!"

Tomorrow afternoon came, and I met Bill at Harvard. It was pretty clear that he hadn't done any preparation, but when I said I had to know where the odd electron is predicted to be in the allyl radical, he said, "That's only three carbons; I guess we can work that out. Hand me the Eyring book. I want to look at the character tables." I had not the faintest idea of what character tables were for, so he explained they were used in group theory and could make the mathematics simpler. Then he started doing something mysterious with the character tables, wrote down and solved the simple algebraic equation, $x^2 - 2 = 0$, and finally drew out the orbitals in proportion to their coefficients in the equations. In summary,

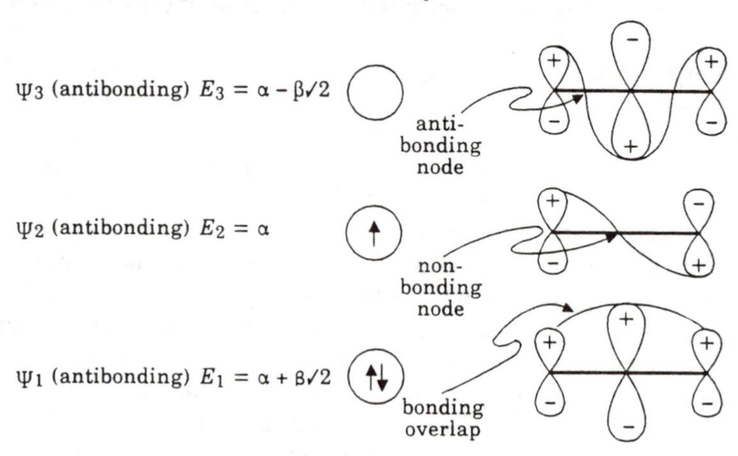

For the allyl radical, two electrons with paired spins (↑↓) could go in the lowest, most stable molecular orbital, which I could now see was made up of a linear combination of atomic orbitals (LCAO). The odd electron (↑) could go into the second most stable orbital, which, from the coefficients of the atomic orbitals involved, clearly and unequivocally showed the odd electron being confined to the *end* carbons, just as orbitals predicted by the resonance method!

I was dazzled by the simplicity and elegance of this *quantitative* treatment of π-electron systems and set to work to understand what α, the coulomb integral, and β, the resonance integral, stood for and, in general, what kind of approximations were involved. The overlap integrals, degenerate orbitals, character tables, and $E = \int \Psi H \Psi d\tau / \int \Psi^2 d\tau$ took a little longer. I often had to make frantic calls to Bill to get on to the right track, but I was learning how to go to a variety of sources and look for anything at all I could understand, then try to transport that back to solving the problem that really interested me. Several wonderful papers of C. A. Coulson and H. C. Longuet-Higgins[129] were an enormous help.

I never did go back and slog my way through the first 18 chapters of Eyring, Walter, and Kimball, or otherwise try to understand those impressive basic molecular orbital equations such as

$$E\Psi = -\frac{h^2}{8\pi^2 m}\nabla^2\Psi + \left(\frac{e^2}{r_a} - \frac{e^2}{r_b}\right)\Psi \qquad (82)$$

for the simple entity H_2^+. When the copy of *Quantum Chemistry* that I used is closed, it has a thin, use-darkened zone on the page tops at

Robert A. Mulliken and Charles A. Coulson in Coulson's office at Oxford University in the spring of 1953. Mulliken, a professor of physics at the University of Chicago, was very important in the development of molecular orbital theory, especially in its applications to spectroscopy, for which he received the Nobel prize in 1966. In my few encounters with Mulliken, I found him to be a good listener but slow to offer opinions. At best, he would have something to say in answer to a question the next day. Coulson, trained as a mathematician, published a very influential and clear book entitled Valence *(Oxford, 1952), which was extremely helpful for understanding the basics of molecular orbital theory but gave little hint about how the calculations are actually done.*

Chapter 19, but the rest of the page tops are pretty virgin-white. Bill McMillan does things differently. He works through books like that from page 1, and use-darkened page tops show just where he stopped, if he did.

Andrew J. Streitwieser, Jr., had come from a very successful graduate career with W. v. E. Doering at Columbia, on an independent postdoctoral fellowship, to work with me, and he got excited about HMO calculations also. We started to calculate everything we could think of that might be of interest to organic chemists and did not offer too great computational difficulties. First, we applied group theory to the extent possible, then broke down the so-called "secular determinant", and finally used a Marchant calculator to get, by successive approximations, the roots of the resulting high-order algebraic equation.

None of what we did could be regarded as high-order theoretical chemistry, but we were able to publish most of our results in the *Journal of the American Chemical Society*.[130] That paper had some impact in drawing attention to the possible large stabilization of the cyclopropenyl cation, **143**, and in suggesting that π-electron systems of **144–147** are predicted by the HMO method to have interesting properties.

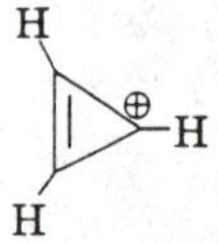

H H H CH_2 CH_2 CH_2

143 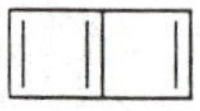 **144** 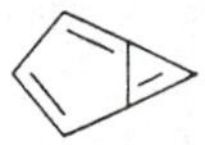 **145** 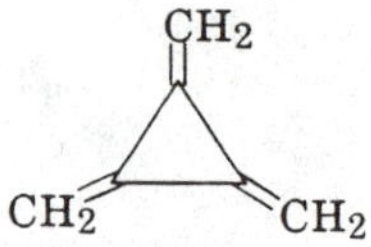 **146** 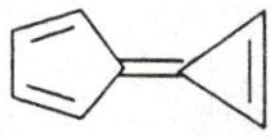 **147**

Later, both Andy and I got heated up enough to write books for organic chemists for the purpose of introducing them to HMO theory. Andy's was far more complete and scholarly[131a]; mine was quite simple-minded and designed to help people get over the activation barrier.[131b] Students seemed to like the idea of knowing what HMO theory can help correlate. My book actually went through 13 printings!

I also gave quite a few short courses for industrial chemists at company sites so that those who had not had an opportunity to find out about HMO theory in school could get involved. One of the most enjoyable occasions for this sort of thing was 3 weeks at the end of May 1962, in Munich. Professor Dr. Rolf Huisgen was in charge. He had planned for a small informal seminar on HMO calculations, but too many people showed interest, and the lectures had to be given in one of the large lecture halls.

I decided to operate American style, with audience participation. This style was discomforting for some of the faculty and greatly enjoyed by the students. The faculty would sit in the front row, and when I wanted to show how to get the individual matrix elements in a secular determinant, I would go to the front row with the question: "Okay, now what is H_{14}?" When one professor couldn't answer, I would go to the next, and the students would roar. Actually, Rolf Huisgen was pretty good at all of this, but Professor Klaus Hafner was really the star performer. During all of my lectures, Professor F. Wille, a theoretical chemist, kept clucking about the lack of rigor and lack of attention to detail. Later, he translated my book into German and allowed himself a substantial clarifying appendix.

Included here is a photograph taken during my lectures, in which I was explaining how HMO theory might be applied to 1,3-dipolar additions, a very hot topic in Munich at the time and for many years to come! The little drawing, to the upper right of the blackboard, showed a graveyard of organic chemists who tried to scale the heights

John D. Roberts lecturing on Hückel molecular orbital theory in Munich in 1962.

John D. Roberts receiving an honorary degree of Dr.rer.Nat.h.c. from Rolf Huisgen at the University of Munich in 1962. The costume worn by Professor Huisgen signified his position at the time as Decanus *of the university.*

of HMO theory up the sheer wall of the Schrödinger equation. In contrast, I was to offer them an easy back route to the top.

Having had at least superficial success in using and helping to popularize HMO theory for organic chemists, I became quite interested in the next level of such calculations. Unfortunately, HMO theory neglects repulsion between electrons in the molecular orbitals. This deficiency is in sharp contrast to the valence-bond theory, in which calculations in the simplest form place a premium on interelectronic repulsion. With neutral molecules, this repulsion does not allow more than one electron in each p–π orbital.

One, allegedly simple, way to take interelectronic repulsions into account was being espoused by Parr and Pariser.[132] These worthies ignored many types of the more messy overlap integrals and treated interelectronic repulsion in a quite straightforward way. The problem with their approach was that they used a number of semiempirical parameters for which they gave "best" values. My interest in their work was sparked by getting to know R. A. Pariser at Du Pont's Jackson Laboratory while acting as a Du Pont consultant. Rudy and I started on different tracks with Du Pont at about the same time, and one of his first jobs for the company was to see how well the properties of dyes could be calculated by molecular-orbital theory. The Parr and Pariser procedure involved calculating the energy of various HMO-electronic configurations, including interelectronic-repulsion terms, of a given molecule with a π-orbital system. When this step was complete, the energy was minimized by mixing together those states of the same symmetry. The mixing procedure, called *configuration interaction,* is somewhat analogous to valence-bond calculations, except that the off-diagonal matrix elements of the secular determinant are simpler.

At this time, I had an MIT graduate student, W. F. Gorham, who was trying to synthesize pentalene (p 60) and, as I mentioned earlier, there was some ambiguity about whether or not HMO theory could predict, even qualitatively, its energy and reactivity. Pentalene seemed an excellent candidate for a high-level calculation, so, in the fall of 1952, when I was a Guggenheim Fellow at Caltech, my project was to use the Parr and Pariser procedure for that purpose. It was a far from trivial task. Once you start to take into account interelectronic repulsion, the computational difficulties increase enormously and the arithmetic is tedious when worked out on a Marchant electric calculator, whirring and clanking away. Pariser was better equipped and carried out his calculations on some sort of an early digital computer at IBM. Neither his program nor the computer was available to me. But I really didn't care much, because I wanted to work through the calculations to see if I could understand what was involved. It was hard work, especially because of my inexperience, but the pile of results on the various configurations was complete and ready for the configuration-interaction calculation in early December, when I visited Rudy Pariser at the Jackson Laboratory, full of pride in my achievement. In a very few minutes, my spirits dropped to my shoe soles when Rudy told me (with his own great pride) how he and Bob Parr had concocted a whole new and, of course, far better set of semiempirical parameters, and I should use them in place of the older ones. This would mean at least a month of work, and I sensed that finishing up on pentalene would be like writing on water; each new iteration of parameters would invalidate or supersede what was done before. Because HMO theory had no comparable semiempirical parameters, this was not a problem with it. It was a bitter lesson and made me resolve to avoid such projects in the future, but that resolve was not permanent, as later history will show.

I also had a fling at simple valence-bond theory, which uses the same atomic orbital models as HMO theory. It was not a very successful endeavor because I simply never got the hang of calculating the matrix elements of the secular determinant. Pauling had developed a supposedly simple, but actually quite esoteric, routine for this task that involves analysis of geometrical figures made by superimposition of diagrams reflecting bonding patterns of the various resonance structures.[133] This procedure made me uncomfortable because I did not, and could not, understand its basis. Along with this, there are difficulties with the numerical calculations without a digital computer (not then available), because of the form of the valence-bond secular determinant.

In later years, after Raoul Hoffmann had popularized the extended Hückel theory (EHT), which was HMO theory applied to all of the valence-shell electrons, not just the π electrons, I spent a quarter at Caltech trying to teach the students how that was done. Unfortunately, although I learned a lot, the students did not find EHT very interesting, no matter how useful it was in practical situations. One problem was that the calculations could not be done on the back of an envelope. A sizable computer was involved, and the computer program we used had more "black box" characteristics than the one that I had written for HMO calculations in my earlier courses.

Life in Cambridge

Life for us in Cambridge was very full and very rewarding, even in our cramped quarters in Holden Green. We had excellent relationships with the Harvard organic chemistry faculty. In fact, Gardner Swain and I played an important role in helping to bridge the chasm that had existed for some years between chemistry at Harvard and at MIT. We had several wonderful parties in our apartment, a few with more than 20 people in our 12- × 18-foot, non-air-conditioned apartment on hot and humid evenings.

One occasion I remember particularly well was a party for Dave Curtin, who had come up from Columbia. Several of the group had overflowed into the large closet that passed for a kitchen, with Bob Woodward holding court and pontificating. Then about 30, Bob, by citing well-chosen and very particular examples, was obdurate on the theme that chemists are clearly over the hill at age 35. Some 15 years later, at the height of his fame, I remember asking him how he felt about the situation now. His reply was, "It's still true, but, indeed, experience counts for a lot."

Prelog, Plattner, Robinson, Criegee, Arndt, Todd, Jones, and others visited Cambridge after the 1951 75th anniversary ACS meeting. I had been interested in photography and had done a lot of developing, printing, and enlarging in my college days, but had given it up during the war. About 1948, I began to take portraits of my chemical acquaintances and, of course, of our visitors. The best picture of fall 1951 is of Sir Robert Robinson, taken in the large lecture hall at MIT after his lecture on the synthesis of steroid rings by the Robinson annelation reaction. The funniest was a not-too-well-focused, off-the-cuff shot of Robinson and Woodward together, trying very much to look as though they did not know each other. Of course they did, but the Woodward correction of Robinson's strychnine structure, stresses over the chemistry of penicillin, and a 30-year age gap kept them from being very

Sir Robert Robinson and Robert Burns Woodward after a seminar at MIT in 1951 by Robinson on the synthesis of the steroid nucleus by the Robinson annelation reaction.

friendly. They did not want to have their picture taken together, but the pleas of those surrounding them were not to be resisted, and they stood in juxtaposition for the shortest possible time.

One of the truly influential events in the development of physical organic chemistry was the founding of the biennial Reaction Mechanisms Conferences by Paul Bartlett and Charles C. Price. The first conference was held at the University of Notre Dame in early September 1946 with about 90 attendees, many of them straight organic chemists wanting to see what all the fuss was about. Among the torrid topics of the time were S_N reactions, peroxide decompositions, front- and back-strain as influences on the acidities of ammonium salts, and mechanisms of aromatic substitution. Unfortunately, I can't recall as much as I would like about the details of the conference. But I do remember Art Cope's pleasure in being accosted by a priest on the campus (it was before the school year started) and being told, "It is good of you football players to come early and start practicing for the fall season." Another highlight was Louis Hammett patiently lecturing Herb Brown on stage about the necessity of considering the role of entropy when trying to evaluate steric effects. Another personal milestone at this conference was my first opportunity to meet Frank

Participants in the Second Organic Reaction Mechanisms Conference at Colby College, New Hampshire, 1948. See if you can identify the following who are especially prominent in the picture: Saul Winstein, C. Gardner Swain, Stanley J. Cristol, W. v. E. Doering, Frank H. Westheimer, Louis P. Hammett, Cheves Walling, and Arnold Weissburger.

Westheimer, fresh from his wartime research success in unraveling the mechanism of the oxynitration of benzene to picric acid.

The 1948 Reaction Mechanism Conference was held at Colby College (now called Colby–Sawyer College) in New Hampshire as part of

Paul Bartlett indulges a wry smile as John D. Roberts scores a wicket with his deadly mallet at Colby College, 1948.

the Gordon Research Conference series. It was clear that the Gordon Conferences would have been glad to take over for future Reaction Mechanism Conferences, but the attending group voted to preserve their autonomy and formalized a system whereby each organizing committee picked a successor group that would be given carte blanche to arrange another conference or even, circumstances warranting, to discontinue the series. At this time, Cheves Walling, Robert Burwell, and I were chosen as the organizers for the 1950 conference. The 1948 session was heavy on free-radical chemistry; the featured speaker was the dean of radical chemistry, Morris Kharasch of the University of Chicago. Kharasch, an extraordinary chemist, discovered many new reactions and became especially well known for his classic work with Frank Mayo, which established that the anti-Markownikoff addition of hydrogen bromide to alkenes is a free-radical chain reaction that turns out to be faster than the normal polar Markownikoff addition. This classic of research has become an integral part of elementary organic chemistry.

Kharasch had ties to the pharmaceutical industry and tended to play his research cards close to his chest. He did not seem to regard himself as a physical organic chemist and was clearly a bit put out by the invasion of newcomers into his research field. He started his lecture with a story that illustrated how he felt. Told with a thick accent, it was about two jewelers in the Bronx. One had a fine watch, which he sold to the other at a bargain price. But then he had second thoughts and got the watch back, at a higher price. Then the other jeweler bought it back at a still higher price and so on. After several cycles, one of the jewelers sold the watch to a third party. The other jeweler was infuriated. "Why did you do that?" he demanded. "We had such a nice business going."

The featured speaker at the 1950 Reaction Mechanisms Conference at Northwestern University was to be none other than C. K. Ingold of University College, London. He was surely the best known physical organic chemist of that particular time because of his path-breaking work with E. D. Hughes on the workhorse synthetic reactions of nucleophilic substitution and elimination. It is clear from his papers that Ingold did not suffer fools (or indeed anyone who disagreed with

Sir Christopher K. Ingold at Northwestern University in 1950, at the time of the Third Organic Reaction Mechanisms Conference.

him) gladly. Robert Robinson and Ingold were mortal enemies.[134] Robinson was absolutely and irrevocably convinced that Ingold had tried to steal his electrochemical theory of 1927 away from him. With this background, we got ourselves prepared to deal with an acerbic, growly Briton. Ingold came to the United States to go to Cornell as Baker Lecturer, and, as I remember, I first met him with a relatively small group in New York. None of us was prepared for the tall, lean, gentlemanly, and charming Ingold, who was completely different from what we expected.

When we were introduced, he immediately asked, "Columbia Roberts or UCLA Roberts?" This Ingold was a very good listener and offered excellent suggestions for others' work. His lecture at Northwestern was jammed to the doors. The subject was nitration of aromatics, and he had both a wealth of data and an excellent manner of presentation. The audience was enthralled and immensely impressed by a box he passed around that contained samples of crystalline nitronium, NO_2^+, salts. One of these, nitronium perchlorate, was actually quite dangerously explosive. No matter, it didn't explode. It was a momentous day for chemistry, and I took an excellent picture of Ingold on that occasion. Amusingly, the Ingold we had expected was lurking behind the façade of courtesy and charm. His book, *Structure and Mechanism in Organic Chemistry*,[135] was vintage acerbic Ingold.

Our Cambridge period was not wholly bucolic. Besides the stresses associated with internal chemistry problems, there was the ominous rise of McCarthyism and its threat to the freedom of our academic institutions. After being brought up on the staunch republicanism of the *Los Angeles Times* of the 1930s, I was appalled to learn from Senator McCarthy that the State Department was a hotbed of communism. It did not take long before I was converted to skepticism of such claims.

Particularly disturbing was the news that Iz Amdur, a chemical physicist at MIT, was, with others, a member of some sort of alleged communist cell. Somehow, Iz didn't fit my notion of a bearded, dark-cloaked subversive. Iz was a mild man, almost, if not actually, a saintly man, clearly with principles, but hardly a danger so far as I could see. I was always very fond of him and proud, for myself and MIT, to be associated with someone who was achieving a breakthrough with atomic beams and the collisions of such beams with atoms and molecules.

I could see the Amdur results making a substantial impact on calculations of interaction energies of nonbonded atoms and, in fact, on ones I used in a semiquantitative estimation of the energy of interconversion of three possible conformers of 1,5-dichlorocyclooctadiene, **148**.[136] We showed, by dipole-moment measurements, that **148a** was the most stable of these, but we were uncertain as to whether the preferred conformation was **148a** or whether, when **148** was formed by dimerization of $CH_2{=}CHCCl{=}CH_2$, **148a** was the initial product and was unable to be converted to **148b** or **148c** (equation 83). In working through this

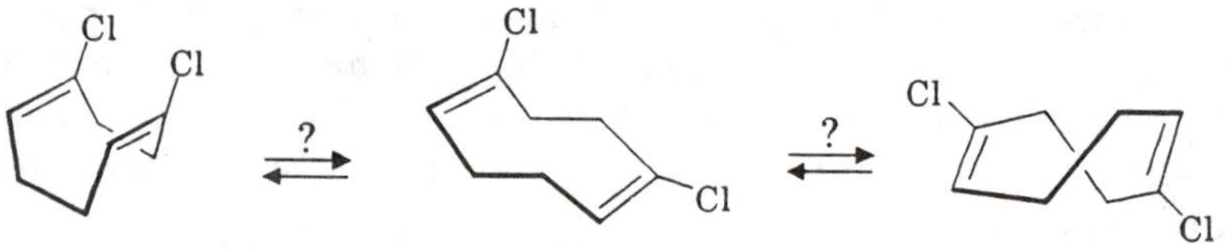

Cl Cl ? Cl ? Cl Cl Cl

148a **148b** **148c** (83)

problem, I got invaluable advice from Derek H. R. Barton, who was then a visitor at Harvard. Derek suggested paths of interconversion that I had not thought of and impressed me enormously with his incisive intellect, as well as the way in which he was applying it to the mutarotation of the bromine-addition products of cholesterol.[137]

But let us return to Iz Amdur and his association with a communist cell at MIT. Somehow, along with a rather broader vision of the world that I had been exposed to in Cambridge, I began to see things in a different way. I recognized that if this kindly man could be crucified for following some sincere belief, however misguided, any of us could be vulnerable if it suited the purpose of a demagogue of the left, the right, or the center. I came to see in McCarthy the same shrill appeal to fanaticism that we had experienced, from a distance, with Hitler and Mussolini.

Amdur surely suffered, but so far as I know, no action was taken against him. Not all were so fortunate. A strong effort was made to tar Linus Pauling with the same brush, but this effort was not very successful because Linus fought back. Because of his prominence and long association with liberal causes, Pauling presented an inviting target for the McCarthyites. But he was far from a pushover, and many were heartened by the example of his resistance. The situation was not an easy one. Many universities, especially the University of California, were in the throes of a push for binding loyalty oaths, and defiance of these were grounds for suspension or dismissal. Matters were not made easier by the Klaus Fuchs case, the prosecution of the Rosenbergs, then of Alger Hiss and, still later, the trials of J. Robert Oppenheimer.

Linus was the subject of an inquisition by the House Committee on UnAmerican Activities, which so pleased him that he had the transcript of his session reprinted under the title "Old Sour Wine in New Bottles". The title was a play on words of the name of the chief investigator of the committee. Nonetheless, Linus came under vigorous attack from some of the Caltech trustees, who wanted him fired forthwith. Fortunately, Robert Bacher, for many years Caltech's provost, worked with the then Caltech chairman of the trustees, James R. Page, to establish with the board of trustees that this would be a violation of academic freedom and the institute's commitment to its tenure policy. The line was held, but some conservative trustees resigned as a result, most notably Herbert Hoover, Jr.

Several of the remaining trustees were far from being Pauling fans and were vocal in their belief that he had cost the institute millions in donations. I am less sure. It seems to me that, in the long run, you do better to be known as a bastion of integrity than as a weather vane, responsive only to the directions whence the money winds blow.

By the end of 1951, I had achieved at least a modest level of notoriety through publications and seminar talks, as well as meeting presentations. Chemistry departments were still very much in an expansion phase, and I began to be sounded out for positions elsewhere. The first real proposal was from Florida State College, which had started to build up in a very serious way and had a winner in physical organic chemistry, Ernest Grunwald, Saul Winstein's best graduate student.

Somewhat later, in February 1952, Columbia weighed in. Louis Hammett, then the executive officer of the department, invited me to discuss the possibility of a full professorship at Columbia. In his letter, he said that

> . . . the man who takes this place will determine the future of organic chemistry in this institution. He will not be hampered by seniors in his field who are resistant to change. He will not be limited by any commitments to younger men except the Department's guarantee of a chance to advance by proving merit. His recommendations will largely determine what appointments are made and how available funds are assigned.

Needless to say, that was a very strong statement to a 33-year-old who was chafing a bit at the situation at MIT. On my visit, however, it was clear the Columbia situation was not the most attractive in the world. At that time, the laboratories were old, the offices were broom-closet-sized, and apartments in New York were expensive. Still, I liked Hammett very much, Columbia was a great university, and New York was exciting, even if expensive. Hammett was not so encouraging when he told me that "the salary would be $9000 a year, higher than most universities to compensate for the cost of living in New York, but you have to understand that it would essentially be a life-long salary. Columbia does not have a practice of raising the salaries of full professors after they are hired." Hammett also asked me later, by letter, if I thought Gilbert Stork would be an appropriate choice for a second organic chemistry appointment, and I was very enthusiastic about this. Gilbert was unquestionably a superb first choice. However, right in the middle of these negotiations, I received word that my Guggenheim Fellowship would be forthcoming, and I lost interest in making a change in the fall of 1952.

Back to Caltech

The Guggenheim application specified that I would study nonclassical

Howard J. Lucas just before his retirement at Caltech in 1952. Lucas was hired by Caltech to teach organic chemistry in 1916, when his only published papers represented his master's thesis at Ohio State University and his work at an agricultural experiment station in Puerto Rico, where he studied the milk of Puerto Rican cows and the determination of nitrobenzene in peanut oil. Why nitrobenzene in peanut oil? It turns out that it was a common cheap adulterant in Puerto Rico to provide peanut oil with an attractive odor and taste! Legend has it that Howard was brought to Caltech to teach and not make waves in research; whether true or not, he made waves in both teaching and research through being in the right place at the right time with high standards and an open and inquiring mind.

ions, but I was actually absorbed in trying to understand the higher flights of molecular orbital calculations. I chose to do this at Caltech for

Edwin R. Buchman, at a mountain picnic near Pasadena in 1953. Buchman did his Ph.D. work with the famous German organic chemist, von Braun. Subsequently, Buchman participated in the synthetic work that led to the first commercial process for preparing vitamin B_1 *with Robert R. Williams. Royalties from patents on the synthesis allowed Buchman to do research at his pleasure as a senior research associate at Caltech. His consuming interest was the preparation of cyclobutadiene, which he pursued with painstaking care. Saul Winstein once observed that Edwin was the only person he knew so free of pressure that he could put a draft of a paper in his desk drawer and not look at it for a year, just to see if it would stand the test of time. Buchman was legendary for his friendships, generosity, and guidance for talented people who had difficulty in making social compromises.*

several reasons. It was not too far from the homes of our parents, and our son, Don, was due to arrive on the scene shortly. Also, if we lived with our parents, UCLA, with Bill Young, Saul Winstein, and Don Cram, was easily accessible—a big plus. To be sure, I knew that Howard Lucas, Caltech's original and foremost organic chemist, the mentor of both Bill Young and Saul Winstein, would be retiring in 1953. If, indeed, they might be interested in my succeeding Howard, it would not be bad to get mutually acquainted before the decision was made, at least in case it had not already been made.

As a Guggenheim Fellow at Caltech, I was given a writing desk in a back bay of Edwin Buchman's laboratory and access to a Marchant calculator, downstairs in a pleasant room next to Pauling's office, which contained the IBM punched-card tabulators used in X-ray diffraction calculations. Pauling's office was something to behold. A converted laboratory, it still had a hood and lab bench, but was otherwise jammed with molecular models, papers, and books. The blackboard was Pauling's memo board, and it was crowded with reminders, messages,

Linus Pauling at a Caltech Chemistry and Chemical Engineering Department picnic in 1953.

and things to follow up on. Pauling's usual habit in those days was to come early to the laboratory, take care of the Division's business or research matters, and then go home about one o'clock for lunch. Unless there was a seminar, he would spend the rest of the day working at home. When in town, he never missed Wednesday seminar, and almost always asked the first question. He strongly believed in seminars as an educational experience for both students and staff.

In late December, Edwin Buchman indicated to me in his usual oblique way that I was a candidate to succeed Lucas. A few days later Pauling called me into his office, told me that Caltech would be offering me a professorship, and asked if I would be interested. The answer was definitely yes.

The indicators for Caltech were almost all positive, although Bill Young expressed concern that, as division chairman, Pauling might run over me. Still, it was a fair question to ask why I should leave MIT (and Harvard) when support and students were so good, and where Cope had given me a wonderful helping hand. The major difficulty at MIT was internal. With few notable exceptions—Stockmayer, Amdur, Stephenson, and Ashdown—the pre-Cope professors hardly welcomed the Cope crew with open arms. My relationships with some of the older staff members were cool and with a very few they were anta-gonistic. It was not comfortable.

One other aspect of MIT concerned me, and that undergraduate–faculty relations in the area of written examina[illegible] MIT had no honor system. Examinations for large classes were gi[illegible] the large drafting rooms, one student to a table, with a vigilant [illegible] teaching assistants and faculty proctors. The students acc[illegible] challenge, and there was a lot of cheating or attempts a[illegible] Later, after moving to Caltech, the honor system came to [illegible] culture shock. At first, I could hardly believe that it work[illegible] settled into it with a sigh of relief. What a wonderful [illegible] human relations!

When it came time to give the final decision, [illegible] to change my mind. He had Jim Killian, MIT's p[illegible] But Jim realized that I was going to be obdurat[illegible] was short. After the decision was final, I offere[illegible] of the Guggenheim fellowship and teach for [illegible] session, but Art generously told me to go ahe[illegible]

Visit to Eur[illegible]

That was a wonderful experience. E[illegible] before. We sailed the Atlantic Oce[illegible]

on the *Queen Mary*, in the cheapest cabin class. The *Queen Mary* was a bit starchy and its decor was the English version of art deco, but the trip was infinitely interesting. We enjoyed our multiple nightly turns around the boat deck, with the strong winds and sometimes stinging spray.

My favored seminar talk during my English university tour was on benzyne, and I gave it many times. In London, C. K. Ingold got so excited that he jumped on top of the lecture desk from his front-row seat to get more rapidly to the other side and start his commentary. In Oxford, Sir Robert Robinson (OM and Nobel prize) was most complimentary and affable, but only after he had again expressed to me in his office his reservations about Ingold. After that was over, his mood changed, and he wanted me to tell him what was new. I was astonished at first. I expected that, like many oldsters (as I am), he would be pushing his past accomplishments, defending himself against the intrusion of Prelog and Woodward into strychnine. But no, he wanted to hear about new things and he absorbed them like a sponge—nonclassical ions, benzyne, electrical effects, and so on. Leslie Sutton, a very well-known dipole-moment chemist, took me to dinner with the tutors and students of Oxford's Maudlin College, first in the great hall and later to have port with the tutors in their private room, where the port bottle was passed around on a kind of track.

We experienced no chemistry in France, where a principal goal was to visit the grave at Omaha Beach of Edith's brother, Don, a paratrooper who was killed in the invasion of Normandy. There was plenty of chemistry in Switzerland. In Basel we visited Cyril Grob, whom we h met when he was a postdoctoral fellow with Don Cram at UCLA. I s period, Cyril was fretting about his position in the university. gh he had an independent research program, he felt very much the thumb of the senior organic chemistry professor, Thaddeus in, Nobel prize winner for his work on steroid hormones. n invited me to lunch, and we met in his office. This was in pring of 1953, long before the output of chemical research aging torrent. In Reichstein's elegant office there was a large stacked high with journals. Reichstein was cordial, but ned toward the table, he smiled ruefully and said it was his ading, and this seemed to be a real drag on his spirit. e got out of the office and to the restaurant, his mood was delightful.

ad a wonderful visit with Prelog at the ETH (Eidgenos- Hochschule, Switzerland's MIT). Vlado introduced us a, another Nobel prize winner and, at the time, the uzicka took us on a personal tour of his wonderful Dutch masters and to a fine luncheon at the

Sir Robert Robinson in 1951. Sir Robert received the Nobel prize in chemistry in 1947 and was a member of the Order of Merit. A forthright man on many issues, he and John D. Roberts had several excellent discussions of chemical interest between 1953 and Robinson's death in 1975. Robinson wrote a generally very favorable review of the first edition of Roberts and Caserio, Basic Principles of Organic Chemistry. *He must have read it closely because he noted that a hydroxyl group had been left off the structure of quercitin.*

Kronehalle, one of Zurich's great restaurants. After leaving Zurich, we went to Zermatt to see the Matterhorn. Then we made the trip back across the Atlantic in cabin class on the *America*, a relaxed, enjoyable boat. It was nearly July when we got back to Cambridge. Before long we packed up, said our farewells, and headed off to California.

California Institute of Technology

When we arrived in Pasadena, the Buchmans took us into their wonderful, large house in Altadena and moved their family to Big Bear Lake until we could find a place to live. We were a bit desperate because our second son, John Paul, was on the way, but after a bit of

Edith Roberts with (left to right) Don, Anne, John Paul, and Allen at the Altadena home in 1956. In 1989, Don was an orthopedic surgeon in private practice in Vancouver, Washington; Anne was an assistant professor of radiology at the University of California—San Diego; John Paul was an assistant professor of surgery and associate director of the Liver Transplant Service at the University of California at San Francisco; and Allen was an electrical engineer in charge of advanced chip development at the fabulously successful MIPS company in San Jose.

searching, Edith found a nice small home to rent in northwest Altadena, and we set up housekeeping there. This two-bedroom house served us rather well until Allen was born in 1956 and the children's activities began to expand. In 1958, we were able to purchase a rather old (by southern California standards), five-bedroom home on almost an acre of land, which Edith gradually had rebuilt over the next 20 years. With an enormous, rather unkempt, fenced yard, the children had plenty of room for pets, outdoor games, tricycles, and creating mammoth mudholes by hydraulic mining.

When I decided to leave MIT to come to Caltech, the question arose immediately about which students would want to move and what the arrangements would be for them. My research group had swollen to 19, a terribly unmanageable number. Fortunately, quite a few were close to finishing, quite a few were just beginning, and a few more were not interested in making a switch, most likely because Caltech had a reputation for being strong in physical chemistry and they were worried about how they might fit in. Four students decided to come, and that brought with it a new problem. One of them was Dorothy Semenow, a very talented young woman from Mt. Holyoke College.

Dorothy A. Semenow in 1955, the first woman to be admitted as a graduate student at Caltech. She had an excellent thesis on the intermediacy of benzyne in aromatic substitution reactions (p 61).

Caltech, unlike MIT, did not admit women as students, although there were a few female postdoctoral fellows. I talked to Linus about Dorothy and her strong desire to come to Caltech. To my surprise, he showed immediate interest. He told me that the question of admitting women had been raised not many years earlier, and that the faculty had voted not to change. Furthermore, he said that the Institute's trustees had taken note of the faculty action and had endorsed it. But he said he wanted to try again with a specific case, and asked that Dorothy submit an application as soon as possible.

I wasn't on hand and had no idea what happened at Caltech during the decision-making process. It was certainly to the credit of both Caltech and Linus that the matter was settled by the end of the academic year, including approval by the trustees. There were stories that I had said I would not come if Dorothy was not admitted. That was not true. I only presented the case and others carried the ball, but it was wonderful to become associated with an institution that could act so quickly to change a very strong tradition.

Howard Lucas was wonderful about making room for me. He made a clean sweep out of his office and laboratories and went off to Hawaii on a visiting professorship. Besides Edwin Buchman, who had the title of research associate, the organic chemistry staff had two full professors, Laszlo Zechmeister and Carl Niemann, both very interesting people.

Laszlo Zechmeister was born in Gyor, Hungary, in 1889. He had his higher education in Zurich and received the Dr. Ing. degree in 1913. After holding positions in Germany and Denmark, he was a professor in the medical school of the University of Pecs in Hungary, finally escaping to the West in 1940. With the aid of liquid–solid chromatography, Laszlo had done extraordinary work on the coloring matter of plants, for which he had received the Pasteur Medal and the Grand Prize of the Hungarian Academy of Science. I do not know the sequence of events whereby Pauling got Laszlo to Caltech in 1940, but I do know that Linus was interested in the work that Laszlo had done on substances such as lycopene and carotene with substantial chains of conjugated double bonds.

Laszlo was tall; he had black eyes, rugged features, and a high, bald head. He was extremely courteous and kind, worked extensively with his own hands in the laboratory, and was famous for his skill at manually separating the colored adsorption bands on extruded solid substrates from large-scale liquid–solid chromatography by the classic Tswett procedure. I was very fond of Laszlo and we got along very well, but there was a culture, generation, and chemical-interest gap that kept us from being very close.

Carl Niemann was wholly different. He was 45 years old in 1953, had obtained his Ph.D. with Karl Paul Link at the University of Wisconsin, and had been a postdoctoral fellow with Max Bergman at the Rockefeller Institute, where he became interested in proteins and enzymes. After a year at the University College Medical School in London as a Rockefeller Foundation fellow, Carl came to Caltech in 1937 as part of Pauling's growing interest in chemical biology. Even in those days, Linus could foresee the future importance of bio-organic chemistry. Indeed, strong efforts were made in the early 1930s to hire James B. Conant because of his work on chlorophyll and, later, Alexander Todd of nucleotide and vitamin B_{12} fame. Carl's office was adjacent to mine.

After the war, Carl decided to work on the mechanism of enzyme action, and he chose for this purpose the proteolytic enzyme α-chymotrypsin. He had thought about the problem with great care, and his decision was based on availability, stability, general significance, and

Carl Niemann in his office at Caltech in 1954. Niemann was a pioneer in the systematic study of proteolytic enzymes through the techniques of organic chemistry. He was a talented synthetic chemist and served for many years on the editorial board of Organic Reactions.

lack of high specificity. The last consideration was especially important because he wanted to make the problem a problem in organic chemistry, not biochemistry. He planned to use his synthetic capabilities to make a very systematic investigation of the effect of structure, including stereochemistry, on the rates of cleavage of peptide and ester bonds. The idea was that this might allow him to map out the active site of the enzyme. His sudden death of a heart attack in April 1964, at the prime of his life and while riding into Chicago in a taxicab, was a terrible shock to the division.

Carl introduced me to Heathkits, wonderfully well-documented kits from the Heath Company for building electronic instruments, computers, and audio-visual equipment. He liked to work with his hands, but we both found that laboratory work was extremely hard to do because it could not usually brook interruptions. Assembling electronic gear was ideal from the standpoint of interruptions, offered an excellent opportunity to learn something about electronics, the products were useful, and it was good mental relaxation. Over the next few years, I must have assembled more than 50 of these kits, some for home use, the others for the laboratory. It was an interesting period during which to be involved, because the early kits were all based on vacuum tubes set in steel chassis. Later came single transistors, then printed circuit boards with individual transistors, and finally, the transition from simple, integrated circuits with just a few transistors to the modern multipin, multifunction electronic chips. Building these kits provided invaluable experience when I got into NMR instrumentation, because it gave me some confidence that I could actually make minor electronic repairs and understand better how the instrumentation worked. This is just another example of being at the right place at the right time—with no very good reason and not even knowing, except in retrospect.

An enormously hard worker, Carl Niemann spent several nights a week at the laboratory, almost always writing. He had beautiful handwriting, and his papers were meticulously crafted by the cut-and-paste procedure through many revisions. While he participated heavily in divisional affairs, most notably as the chairman of the chemistry graduate committee, it was clear that these duties and most of his other nonresearch activities were not what he wanted to be remembered for by posterity. When he answered a letter, the original and the carbon were both dropped into the wastebasket. After his death, I was asked to clean up his desk and files. His desk drawer had an inch or so of current business. There were four or five cabinets—not just full, but bulging—containing not only the final form, but each revision of his manuscripts. Such was the measure of what this man regarded as important.

Carl Niemann was a strong influence in the Institute as a whole. He made every effort to make me feel welcome, as well as to introduce me to a wide range of colleagues outside of chemistry. Nearly every day he would gather me up at noon, and we would go to lunch at Caltech's unique and beautiful Athenaeum, not quite a faculty club, but wonderfully fulfilling that function. Within a few months I knew most of the people and, after the initial culture shock, I came to regard these varied lunchtime periods as one of the enormous assets and advantages of Caltech.

One of the great opportunities I had in the relatively early part of my lunchtime career at the Athenaeum was to become acquainted with the famous astronomer and astrophysicist Fritz Zwicky. Fritz, born and raised in Switzerland, came to this country and to the Caltech faculty in the 1930s. He was especially well-known for his work on supernovae, postulation of neutron stars, and rocketry. Fritz was strongly built, and his almost permanent scowl masked a real, but acerbic, sense of humor. Blunt, outspoken, and controversial, he was clearly not everybody's cup of tea. He used to come to lunch relatively early and would sit at one of the rear faculty tables facing the entrance to the dining area. When others who knew him came in, you could see them consciously decide whether this was a day they were ready for lunch with Zwicky. I always was and would make a beeline to sit beside him, if that space was vacant. He was fascinating on any subject and was interested in almost everything, even including organic chemistry. T. Reichstein in Basel was one of his good friends, and I believe that Reichstein worked with Fritz in some way on developing equipment for the Swiss ski troops.

Besides Fritz and my immediate colleagues, I immensely enjoy my Athenaeum interaction with Francis Clauser, a great and vivid personality and a highly respected professor of engineering, who played an important role in the development of intercontinental ballistic missiles. Francis loves to give seminars at lunch. Early on, they covered his passion for improving the internal combustion engine to reduce hydrocarbon and nitrogen oxide emissions; then, later, he included his massive and imaginative participation in the design of America's Cup yachts. The latter was a long, slowly unfolding story, described in detail in published articles, which a number of faithful Athenaeum lunch-goers followed for more than a year in vicarious excitement over the technical challenge, the politics of yacht-racing syndicates, and the arcane rules that govern the America's Cup races. When Francis gives lunchtime lectures, he writes in large letters, so all can see, using the backs of the Athenaeum paper place mats. When developments are hot, or he has a lot of explaining to do, all eight place mats will be commandeered and covered with engineering sketches, performance charts, and equations. These developments—as well as hearing Francis argue on such subjects as the need for teaching tensors in elementary physics, aircraft design, what General Motors needs to do to improve their gasoline engines, and investing in the stock and bond markets—constitute what the Athenaeum is all about for fostering faculty togetherness and intellectual breadth.

My Entrance into Nuclear Magnetic Resonance Spectroscopy

My entrance into NMR spectroscopy was hardly the result of simple revelation from a first hearing as to what it was all about. Indeed, the first hearing resulted in no connection whatsoever. It involved Richard Ogg, professor of chemistry at Stanford, one of the birthplaces of chemical NMR through the efforts of Felix Bloch and his co-workers. Ogg, a physical chemist, was a very broad-gauge person and quite in demand for meetings centered on physical organic chemistry. He loved to dis-

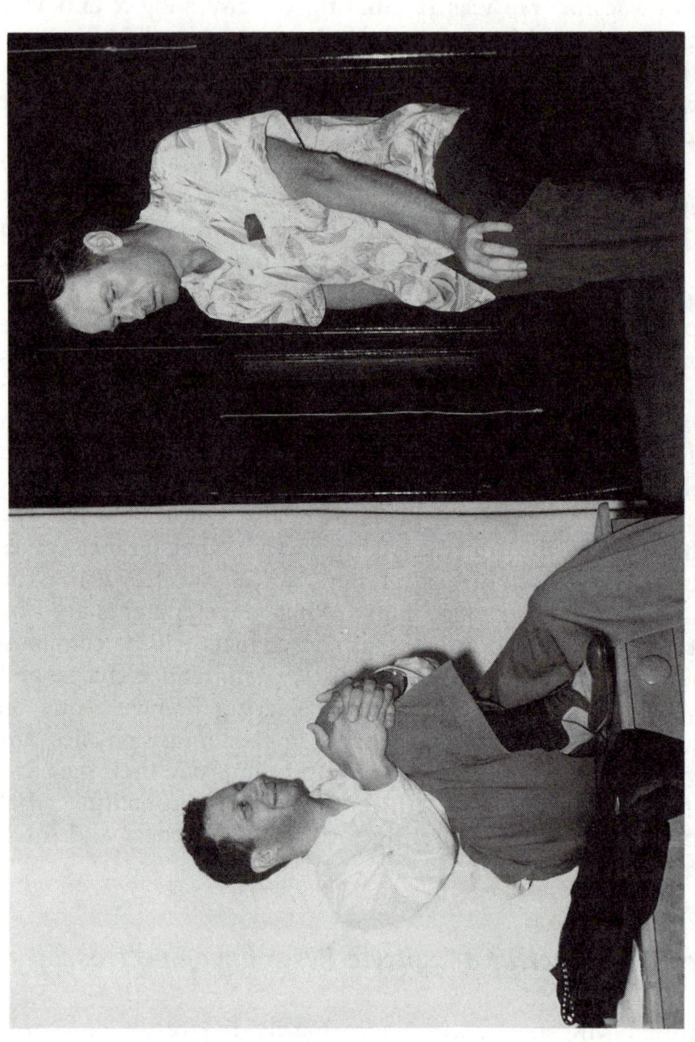

Frank Westheimer and Richard A. Ogg at a Reaction Mechanisms Conference in 1952. A brilliant and imaginative physical chemist, Ogg was initially criticized for his innovative work on solvated electrons in liquid ammonia solutions. He also published some disturbing kinetic evidence on the addition of bromine to carbon–carbon double bonds, which contradicted the common textbook view of such reactions and has largely been ignored over the years. High-strung and sensitive, he committed suicide in 1962.

cuss, although his tone was often strongly pontifical. Some regarded him as excessively garrulous, and one wag suggested that a new unit of silence be defined as the reciprocal Ogg.

I got to know Ogg in 1945, when he visited Harvard. Later, when he was a visiting professor, we got along very well, indeed, probably because I can be a good listener. One day in late 1949 or 1950 he was at MIT, and I invited him to lunch. He was really wound up and proceeded to tell me about the wonderful new magnetic resonance spectroscopy, with such promise for chemistry. I wish I could say that I could understand even 5% of what he told me, but I had too little knowledge of magnetism and absorption of radio frequency radiation—indeed, hardly any knowledge of other radiation except ultraviolet, visible, and infrared. It was clear there were applications to chemistry, even if I didn't understand what they were. That, along with Ogg's obvious enthusiasm, was what stuck in my mind.

About the same time, I was approached by the famous physicist, Francis Bitter. He had heard from some official source that I had been allocated some of the country's valuable supply of ^{13}C-enriched methyl iodide, and he wanted to use it in an experiment. I was protective, asked what for, and expressed grave concern about whether he planned to open the glass tube with its delicate breakoff seal, because methyl iodide is very volatile. He said he wanted to measure the magnetic moment of ^{13}C. No, he wouldn't open the tube; he would just wrap a wire around it, put it in a magnetic field, and measure the absorption frequency.

I made no connection between what he wanted to do and Dick Ogg's description of NMR for a few years. Anyway, I told Bitter to go ahead. Some days later, he returned the sample and said, obviously pleased, that he had pushed the accuracy out three decimal places. Chemical shifts in NMR were not then all that well-known. Of course, we know now that Bitter was measuring a combination of the magnetic moment of ^{13}C and the diamagnetic shielding provided by the special electron environment of the carbon nucleus of methyl iodide.

I continued in complete ignorance of NMR until some time in 1954 when I visited Du Pont and was scheduled to talk with William D. Phillips, a former infrared spectroscopist, who had been a graduate student with Dick Lord at MIT. Bill had to educate me from square one about NMR, and he was really generous about doing it. He showed a lot of the NMR data that they had been getting, using both hydrogen and fluorine chemical shifts and spin–spin splittings to establish structures. It was a revolutionary way to get structural data. He told me about *N,N*-dimethylformamide, with which, at room temperature, you could see in the proton spectra two methyl-group signals that coalesced to one signal above 120 °C because of increased rates of rotation about the C–N amide bond. By then I was surely hyperventilating with excitement.

The only problem that I had was when Bill described the fluorine spectrum of a dimer of 1,2,2-trifluoro-1-phenylethene, **149**. Bill was adamant that the coupling pattern proved that the dimer was **149**, but from all my experience with the chemistry of such dimerizations, I was positive that it had to be **150** (equation 84). On this, we had to agree to

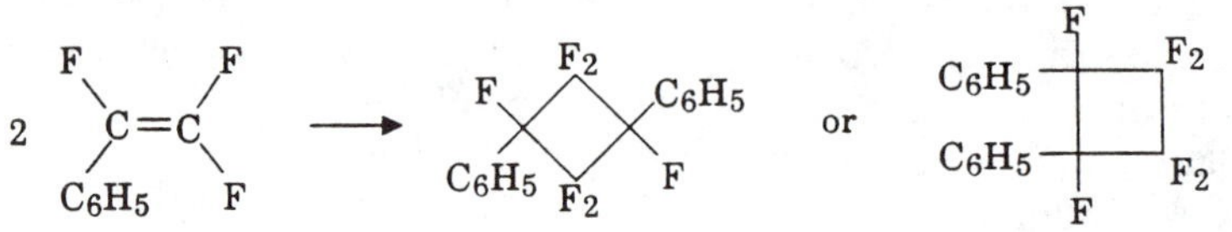

$$2\ (F)(C_6H_5)C{=}CF_2 \longrightarrow \underset{\mathbf{149}}{\text{F, } C_6H_5\text{, } F_2\text{, } C_6H_5\text{, } F\text{, } F_2} \quad \text{or} \quad \underset{\mathbf{150}}{\text{F, } C_6H_5\text{, } F_2\text{, } C_6H_5\text{, } F\text{, } F_2} \qquad (84)$$

disagree. Quite a few years later, Bartlett[137] showed that the correct and different interpretation of the NMR spectrum showed that the structure was indeed **150**.

According to Robert M. Joyce, who later was director of the Central Research Laboratory, the finding by Bloch and co-workers[138] of three different proton resonances for the three varieties of protons in ethanol convinced Du Pont of NMR's chemical utility. Apparently, Varian had some fiscal problems in producing the first commercial spectrometers and asked Du Pont for an advance of $10,000 against the purchase price of about $25,000. It was not at all a usual arrangement for Du Pont to make, but the company bit the bullet and got one of the first three instruments. The others went to Shell Development in Emeryville, California, and to Humble Oil in Baytown, Texas. In the university world, applications of NMR to chemistry involved home-built spectrometers, with Herbert Gutowsky as the leading practitioner. Important studies were also carried on by the physics group at Stanford: James Arnold, Martin E. Packard, and Weston Anderson.

The first result of my Du Pont visit was an attempt to persuade Linus Pauling to find $26,000 for a Varian NMR spectrometer. It turned out that Pauling already knew quite a bit about NMR. In fact, Don M. Yost had used it at Caltech to show that the hydrogen ion is symmetrically, $[F{-}H{-}F]^-$, rather than unsymmetrically, $F{-}H{-}{-}{-}F^-$, located between the fluorines.[139]

Anyway, when I first put pressure on Linus to mobilize an effort to obtain an NMR spectrometer, he agreed that it was an important area for Caltech to be in. However, he was quite positive that the way to do it was to hire a person trained in NMR, preferably a chemical-physics type, and have that person develop a research program. Rebuffed, I was forced to regroup and rethink the strategy. Clearly, research on NMR itself would be wonderful to have going, but the problem was that enough research had been done already to show that NMR was a superb analytical tool. I wanted to have it set up for analysis, with the organic chemists in control of its use, just as we already had an infrared spectrometer in a little room across from my office on the third floor of Crellin.

We were beginning to move into larger involvement with federal support of research. The National Science Foundation was just getting started; I had a grant from them for about $15,000 to study small-ring compounds, and the Office of Naval Research was giving us some support for work on rearrangements. Some money was available for equipment. I went around to potentially interested colleagues and tried to collect as many promises of contributions as possible to get a start on the $26,000 needed for a Varian spectrometer, then operating at 30 MHz.

The most I could come up with was about $15,000, and I went back for another round with Pauling. He was reasonably impressed with having some pledges in hand, but still skeptical about the research. As my punch line, I told him that "with NMR, we can investigate the borderline between resonance and tautomerism." It seemed to spark his interest, and I explained how we could study cycloheptatriene as a function of temperature to determine if it was a tautomeric equilibrium (equation 85) or actually what might be called a monohomobenzene, **151** (equation 86).

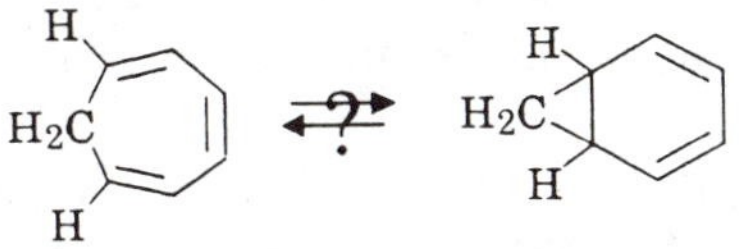

H, H_2C, H ⇄? H, H_2C, H (85)

H, H_2C, H ⟷ H, H_2C, H ~ H_2C

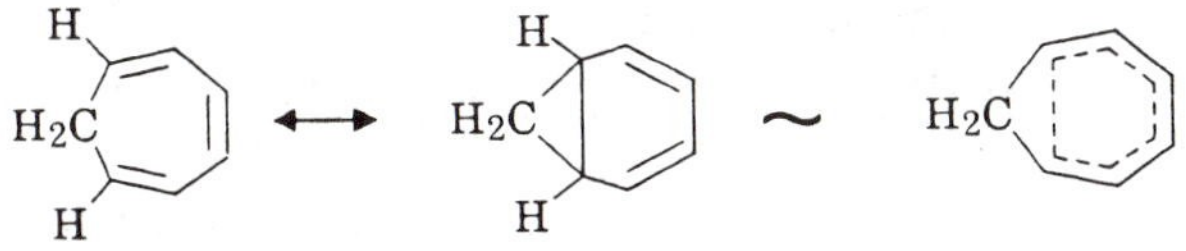

151a **151b** **151** (86)

After I finished, he said that would be nice to do, but there wasn't enough money available to buy the spectrometer. I countered with, "We have about $15,000; why can't you ask the board of trustees for the rest?" Linus said perhaps that would work, and he would try. He had Carl Niemann put in a request, and the board approved not just the balance but $26,000, the whole purchase price. This was especially fortunate, because it left money for installation and extra things we would need to keep the operation running. I was soon to find out that a lot was needed. We planned for the then-current 30-MHz spectrometer, but were soon informed by Varian that an improved 40-MHz machine would be available. To my knowledge, it was to be the first commercial NMR spectrometer to be sited in a university. If it was not the first piece of such equipment, I'm sure it was the first to be put under the jurisdiction of an organic chemist.

I had hoped that we could place the NMR spectrometer near my office, but I was told that no way could it go into an upper floor on Crellin because it was too heavy for the floors. The brand-new Norman W. Church Laboratory of Chemistry and Biology was just becoming available for occupancy. J. Holmes Sturdivant, our super laboratory director, had specified that the beams run through the center of the rooms in Church, rather than along the walls. The magnet, which weighed about 2000 lbs, could then go into 301 Church, the first room across the bridge between Crellin and Church. The NMR spectrometer thus turned out to be the first working research equipment in the chemistry end of the Church building.

When the spectrometer arrived in the summer of 1955, I'm not sure whether I had ever even seen a Varian spectrometer, and I had almost no idea of how they worked. George Pake had written two wonderful articles in the *American Journal of Physics*,[140] which were extremely clear, but more than clarity was needed for the unprepared mind. A brief "instruction manual" came with the spectrometer, but it did not really tell what to do to make the spectrometer function or what the NMR phenomenon was all about. As with molecular orbital theory, I was starting from square one.

The 40-MHz spectrometer was my first experience with large instrumentation. It was mightily impressive with the enormous electromagnet with 12-inch pole faces; the console's array of pilot lamps, dials, and meters; and the towering magnet power supply with its gigantic voltage control for the magnetic current, roaring cooling air, and the bright yellow glow from its voltage regulator tubes. The installer was James N. Shoolery, a Caltech Ph.D. (student of Norman Davidson), later to become widely known for pioneering NMR studies as head of Varian's NMR applications laboratory.

By then Jim had already gained quite a bit of practical experience in interpreting NMR spectra. When he asked for a sample to check the performance of the machine, I decided to give him some 1,3-dimethylenecyclobutane, **152**, to see if he could figure out its structure from the spectrum. The **152** was no normal everyday compound. It had, in fact, just been synthesized in a six-step synthesis by Fred Caserio, a very skillful postdoctoral fellow who had obtained his Ph.D. with Bill Young at UCLA. The last step of the synthesis was a Hoff-

The first Varian NMR 40-MHz spectrometer to be installed in a university. This picture was taken in early 1956, after addition of the Super Stabilizer (p 88), the box that sits on top of the magnet to the right. The magnet for this instrument went through many changes in mission and was finally discarded in 1989.

mann degradation, as shown in equation 87. The spectrometer was not yet stable enough to allow a meaningful strip-chart recording, and Jim

$$CH_2{=}\square{-}CH_2\overset{\oplus}{N}(CH_3)_3\,\overset{\ominus}{O}H \xrightarrow{\text{heat}} \underset{\mathbf{152}}{CH_2{=}\square{=}CH_2} + H_2O + N(CH_3)_3 \qquad (87)$$

had to rely on the cathode-ray tube (CRT) tracing. The CRT had a special, long-persistence, bright-yellow trace. If the magnetic relaxation of the protons was reasonably rapid, the spectrum could be run every second or two.

I put the sample of **152** in the usual 5-mm tube and Jim started displaying the spectra. I had little idea of what he was doing, twirling all of those knobs, but did my best to feign informed interest. At the same time I was shaking in my boots in fear of what would happen when he would go home and leave me with this complex mass of electronic equipment with the most meager of instruction books. While I was in the midst of worrying about that, Jim announced that squiggles on the screen showed the compound had a methyl group and a single hydrogen on a double bond, plus other groups. I was flabbergasted and dismayed, but it took only a little while to realize that **152** could have rearranged in the presence of base to **153** (equation 88), which would show the peaks Jim was talking about.

$$152 \xrightarrow{\text{base}} CH_2{=}\text{(cyclobutene)}{-}CH_3 \quad (153) \qquad (88)$$

This episode redoubled my respect for NMR as a structural tool for organic compounds and made even more imperative my need to understand both how the machine worked and what NMR was all about. So, basically, I gave up everything else and lived at the machine. It was a slow process to get recorded spectra that were interpretable.

The magnetic field of about 6000 G was supplied by a water-cooled electromagnet with 12-inch pole faces and a gap of about 2 inches. The strength of the field was held ostensibly constant by regulation of the current flowing in the coils. The power requirements were considerable and there was a substantial bank of high-power rectifier tubes in the power supply cabinet. Above, there were eight air-blast-cooled 304TL regulator tubes, each about 4 inches in diameter and 10 inches high. On top of this cabinet, there was another box with a second array of 304TL tubes for final current regulation.

Our early version had a chopper amplifier in which the voltage of the power supply (suitably stepped down) was compared with that of a mercury battery. Differences in voltage could then be amplified as alternating current and used to change the current in the 304TL tubes. The NMR signals were located by making small changes in the magnetic field in the neighborhood of the sample with field-sweep coils located in the probe. Appropriate radio frequencies were generated and detected in individual rack-mounted units. The oscillators were driven by thermostatted quartz crystals. The samples were sealed in 5-mm glass tubes (you selected the glass by hand for uniformity and straightness) and placed in the aluminum probe, where they could be spun by an air turbine to help average out inhomogeneities in the magnetic field of the magnet.

To get the apparatus going, you switched on the power and, after a warm-up period, turned up the voltage to cycle the magnet. Cycling was required because if you simply brought up the current to the desired value, the field would be correct in the center but would fall off seriously toward the outer part of the magnet gap. When cycling, you brought the field up to perhaps 10% higher than what you actually wanted, waited for an empirically determined interval, then turned the current down to the desired value. This produced a final "dished" field, a bit too high at the outer part, but reasonably flat in the center.

With a standard sample, almost invariably ethanol containing a trace of acid, you turned up the rf, adjusted the detection system, started a field sweep going, and tried to locate a spot in the field that would produce first a signal and later useful resolution. The sample was located by moving the probe around, vertically or horizontally, with hand-driven cranks. One soon learned that changing the probe location when there were gradients in the magnetic field caused the resonance to appear at different places in the field sweep. It took a long time, however, before the potential benefit of such observations in measurement and medical imaging was recognized. In the early days, such gradients were, at best, an annoying nuisance.

Instability was the problem. Changes in line voltage, air temperature or cooling-water temperature, movement of magnetic materials, and so on, caused the spectra to be poorly reproducible, even over seconds. Substantial precision was necessary, because we were interested in phenomena that required accuracy and resolution to 1–2 Hz. At 40 MHz, this meant *stability* of at least two parts in ten million over seven seconds in both field and frequency. It also meant a *homogeneity* of the magnetic field to the same degree of precision over a cross-sectional area of perhaps 2 cm^2. With vacuum-tube technology and no positive feedback to control the field, such precision was hard to achieve.

It was often almost as useful, if not more useful, to observe the results of repetitive sweeps on the CRT rather than to record spectra. However, when we did record, we used a relatively high-speed Sanborn recorder, which wrote the output with a hot wire on a narrow strip chart. The normal procedure was to run off a series of seven-second sweeps. We could usually get 15–20 recordings before a major change occurred that would make the spectra unrecognizable. The strip chart was then cut up, and comparisons were made to see if two or more spectra could be found that were the same. If that was not possible, then the probe would be relocated, perhaps the magnet recycled, and the taking of the spectra repeated. A typical early "high-quality" spectrum of ethanol is shown in Figure 3.

These rapid-fire spectra are characterized by relaxation wiggles, whereby the change in magnetism induced at resonance causes a beat frequency with the oscillator, which decays away at a rate reflecting the T_2 relaxation time of the protons causing the resonance and/or homogeneities in the magnetic field. The decay of the wiggles from the left-most proton resonance of acidulated ethanol, which corresponds to the OH proton, was most often used to find the "sweet spot" in the field to use for taking recorded spectra. If the spot was good, the wig-

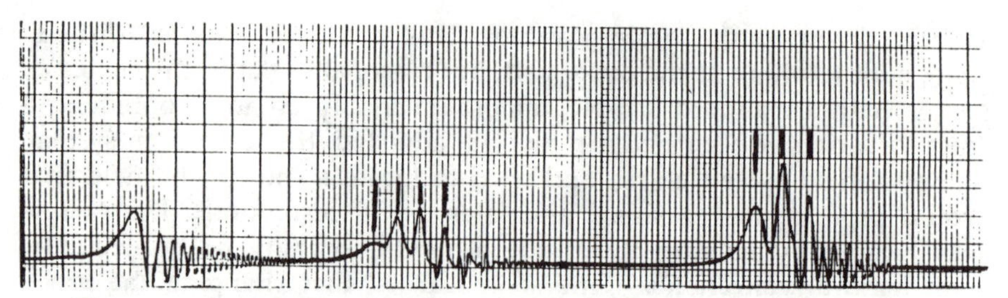

Figure 3. A 40-MHz spectrum of acidified ethyl alcohol taken over 7 s in 1955 with a Varian NMR spectrometer. In early 1955, spectra had to be run off rather quickly because of instabilities in the magnetic field and oscillator frequency. The positions of the resonances corresponding to the CH_2 quartet and CH_3 triplet are marked by vertical bars.

gles from this resonance died away with a simple logarithmic envelope. Of course, the CH_2 and CH_3 were multiplets and would not give a smooth decay.

I had a terrible time a bit later, when one of the magnet coils clogged up with scale from the cooling water. The scale was more or less soluble in sodium hydroxide solution. I spent a solid week, 16 or more hours per day, evacuating the coil, then allowing sodium hydroxide solution to be sucked back to the plug. This was followed by air pressure to facilitate penetration of solution into the plug, and repetition of the cycle. Finally, applied pressure blew about a liter of sodium hydroxide solution out of the third-floor window onto the parked cars below. After that episode, we built an elaborate closed system for the cooling water, which would prevent this from happening again.

My first NMR research was planned to elucidate the structure of Feist's acid, a nagging problem with a rather long history.[141] This substance was thought to have either structure **154a** or **154b**. We were able to show that **154a** was the correct structure;[142] after all, the problem of Feist's acid is the problem of **152** or **153** in a slightly different guise.

HO_2C H, $=CH_2$, HO_2C H — or — HO_2C, $-CH_3$, HO_2C H

154a **154b**

This was not, in fact, my first NMR publication. As part of an orienting study of amines and their salts, I had become interested in the temperature and exchange effects of ^{14}N quadrupole relaxation on the resonances of hydrogens attached to nitrogen. I was enormously impressed by Jim Shoolery's finding that the NH resonance of pyrrole was broadened to unrecognizability by such relaxation. I was able to show that this was not the result of proton exchange, because raising the temperature caused the ^{14}N splitting of the hydrogen resonances to become more, rather than less, apparent for pyrrole, amides, and amine salts.[143] It was not a very high-powered result, but it was a start in a field of which I still had little real knowledge and understanding.

In December 1955, I was scheduled to give a seminar at the University of Rochester and traveled to the East by train. It was a rather bitter winter, and one of those classic heavy snowstorms hit Rochester just as I arrived. I had decided, and was committed to trying, to talk about NMR and what it could do for chemistry, but an important detail of how the spectrometer worked still evaded me. I worried and struggled with it all the way across the country. Then, while I was tossing and turning, unable to sleep, the night before the lecture, somehow the pieces all came together. The seminar was a substantial success, and I returned to Pasadena with renewed confidence.

It is now hard to believe that our initial instrumentation was so primitive. For example, consider temperature control. We had a small, clear Pyrex Dewar flask with a receiver coil wound around it that would fit in the probe. We could heat or cool the sample in a bath, then insert it in the probe and take a spectrum. It had to be taken fast, before the temperature changed. Some cooling could be maintained with dry ice or low-boiling fluorocarbons, but it was not satisfactory to use hydrogen-containing liquids because of the strong signals they produced. One of my first projects was to make a probe insert that had some prospect of maintaining a temperature different from ambient and would allow sample spinning. I worked out a device for the purpose with Jim Shoolery.[144]

Our earliest real triumph had to do with nitrogen inversion. Al Bottini, a fine graduate student, was to verify the Ettlinger structure[145] of **155** by its NMR spectrum. The two protons on C-3 turned out unexpectedly to have the same chemical shift, although in either **155a** or **155b** (equation 89), the hydrogens would be expected to have noniden-

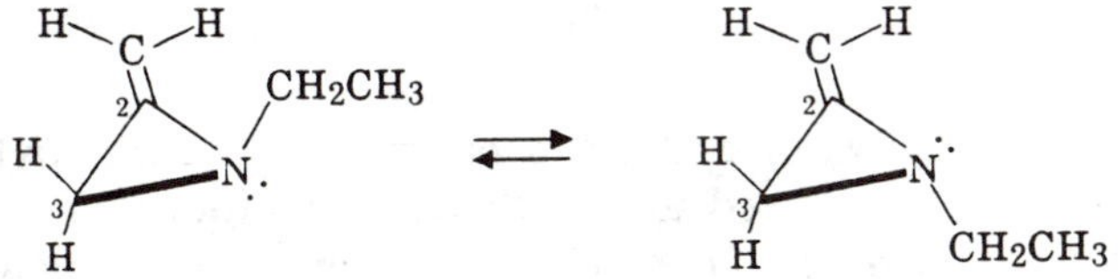

155a ⇄ **155b** (89)

tical chemical shifts because the $-CH_2CH_3$ would be on the same side of the ring as one of the hydrogens and opposite from the other. There was no way of knowing whether the shift difference would be large or small, but some difference seemed reasonable. The fact that none was observed led at once to the notion that **155a** and **155b** were in rapid equilibrium.

The situation with **155** at room temperature resembles that with **156** above 120 °C. The pyramidal nitrogen is undergoing rapid stereochemical inversion, and the CH_3CH_2– group flops back and forth from one side to the other. By the Heisenberg uncertainty principle, this flipping, if fast enough, makes the ring hydrogens at C-3 seem to have no chemical-shift difference. It was an enormous thrill to cool **155** to –80 °C and have the inversion become slow enough to allow observation of a substantial chemical-shift difference between the hydrogens at C-3.

The double bond at C-2 of **155** greatly facilitates the inversion process. When we made **156**, the hydrogens at C-2 showed a chemical-shift difference at room temperature (equation 90). Complete averaging

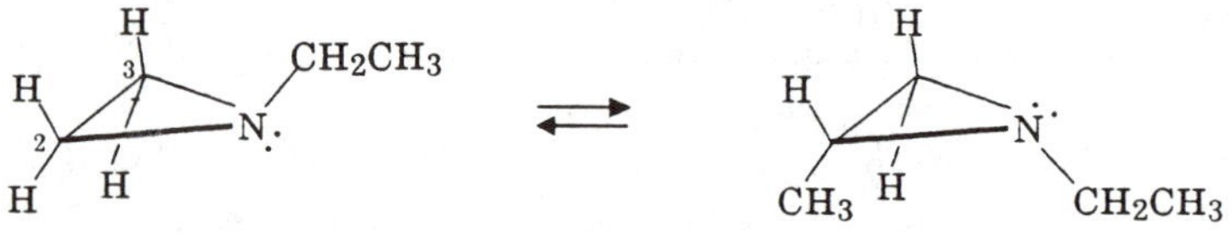

156a ⇌ **156b** (90)

did not occur until the temperature was raised to 120–130 °C. The likely reason for this is resonance involving an interaction between the double bond of **155** and the nitrogen as symbolized by **157a** ↔ **157b** ~ **157** (equation 91). Starting here and continuing over quite a few years, we studied a variety of imine inversion processes by NMR.[146]

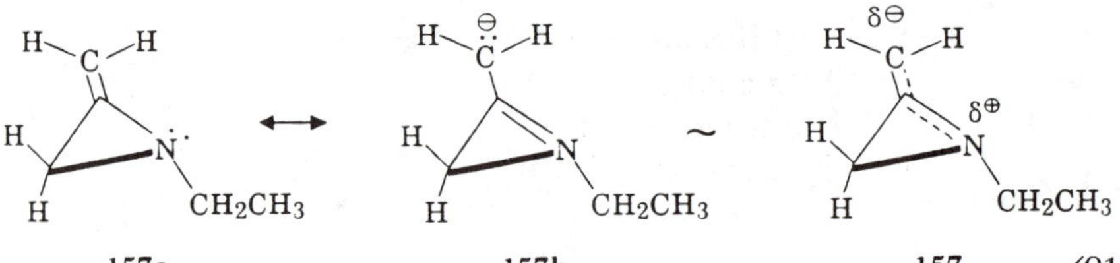

157a ↔ **157b** ~ **157** (91)

The NMR spectra of cycloadducts from tetrafluoroethylene with phenylacetylene, **158**, showed unexpectedly large, four-bond, cross-ring H–F spin–spin splittings. It was then a generally accepted rule of thumb that splittings between nuclei separated by one, two, or three chemical bonds are variable but significant. When four or more bonds are involved, the splittings could be neglected according to this rule.

We were apparently the first to demonstrate four-bond couplings involving protons of magnitudes comparable to or greater than those involving fewer than four bonds, as with **159**. This result sparked our interest in multibond coupling and the stereochemical relationships that maximize it.[147]

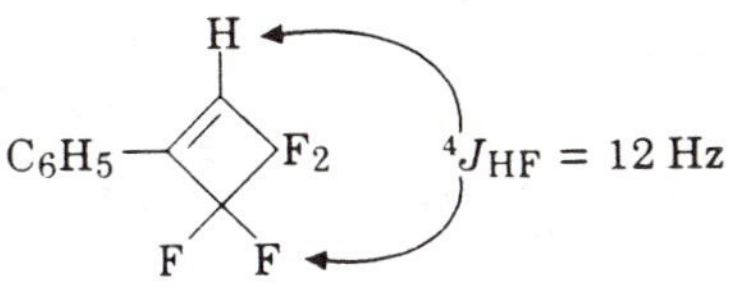

C_6H_5, F_2, F, F, H — $^4J_{HF} = 12$ Hz

158

C_6H_5, F_2, F, F, H — $^3J_{HF} < 1\text{-}2$ Hz

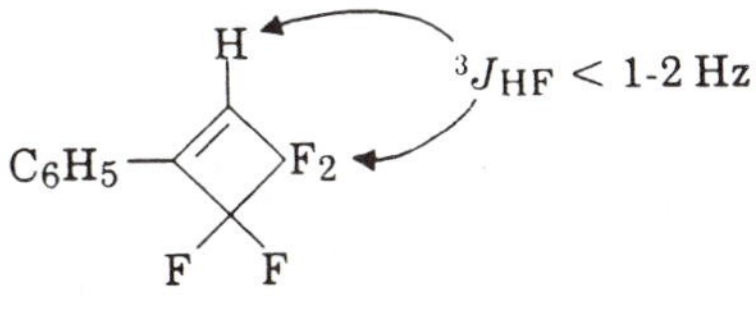

159

The record in our laboratory for the number of bonds between interacting nuclei with an observable splitting was 0.4 Hz over nine bonds for **160**.[148] For unsaturated compounds such as **160**, qualitative understanding of the role played by the conjugated system in transmission of the coupling can be gained by consideration of resonance structures **160a** and **160b** (equation 92), obviously of a type not possible for a

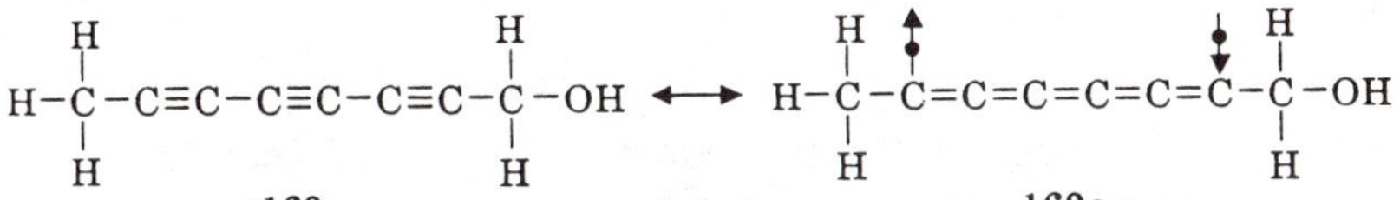

$$H{-}CH_2{-}C{\equiv}C{-}C{\equiv}C{-}C{\equiv}C{-}CH_2{-}OH \longleftrightarrow H{-}CH_2{-}\dot{C}{=}C{=}C{=}C{=}C{=}\dot{C}{-}CH_2{-}OH$$

160 **160a**

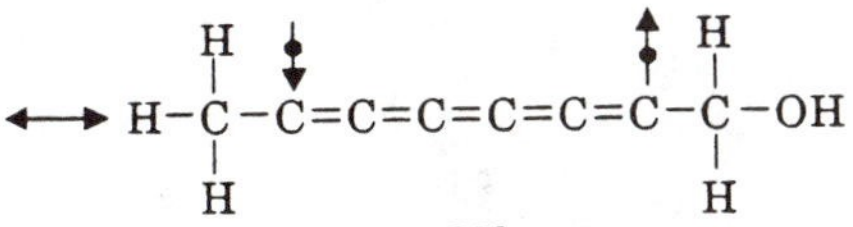

$$\longleftrightarrow H{-}CH_2{-}\dot{C}{=}C{=}C{=}C{=}C{=}\dot{C}{-}CH_2{-}OH \qquad (92)$$

160b

saturated chain with six $-CH_2-$ groups separating the CH_3- and $-CH_2OH$ groups. Interactions between the magnetic dipoles of protons of the CH_3- and $-CH_2OH$ groups and those of the electrons of the multiple bonds provide a modus for producing different magnetic-energy states.

Another interesting facet of long-range couplings pertains to those in systems in which either arguments based on resonance such as **160a,b** cannot be valid or else the couplings seem too large for that to be the exclusive mechanism. Indeed, Sederholm[149] proposed that some sort of proximity or through-space mechanism can be involved, although opinion is not unanimous on that score. In this connection, graduate student Ken Servis found a 31-Hz coupling between the five-bond, separated but close together, fluorines of **161**. However, the corresponding butadiene, **162**, has a 36-Hz, five-bond coupling.

Infrared, Raman, microwave, and dipole-moment studies suggest that the butadiene molecule does not exist to more than a few percent in the *cisoid* conformation, **163** (equation 93).[150] On the other hand, Frank

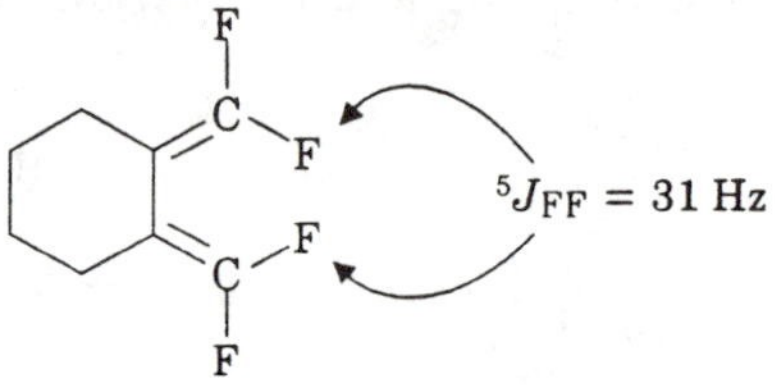

$^5J_{FF}$ = 31 Hz

161

$^{15}J_{FF}$ = 36 Hz

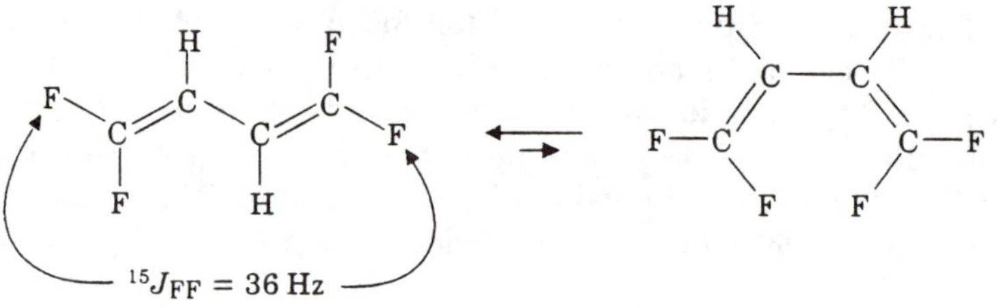

162 **163** (93)

Weigert (of whom I will say much more later) observed a 75-Hz, five-bond coupling in the *gauche* conformation of **164** at low temperature,

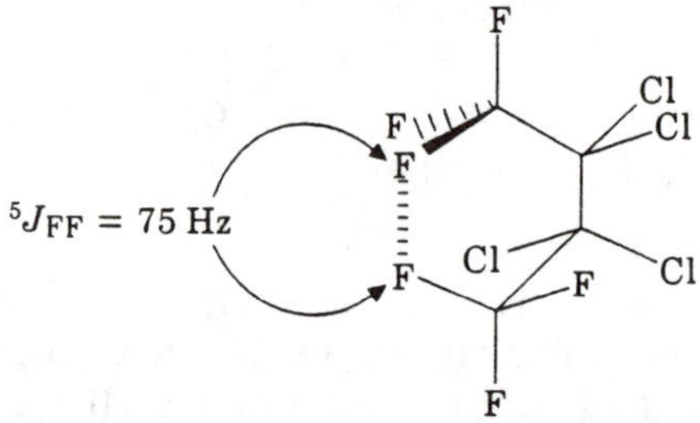

$^5J_{FF}$ = 75 Hz

164

which would seem to arise from the indicated, quite unique, proximate pair of fluorines.[151] No other $^5J_{FF}$ couplings in **164** were large enough to be measured under the conditions used to obtain the spectrum. (J is the coupling constant, the superscript 5 denotes the number of bonds between the nuclei involved, and the subscript FF denotes coupling between fluorines.) My former student, Frank Mallory of Bryn Mawr College, has done some very nice work on this problem.[152]

NMR: Rotation about Single Bonds and Magnetic Nonequivalences

Probably the most important early work we did on NMR was in connection with rotation about single bonds and magnetic nonequivalences arising from internal molecular asymmetry. Our investigation was sparked by a report by Jack Drysdale (who did his senior thesis with me at MIT[69]) and Bill Phillips of Du Pont on the ^{19}F NMR spectra of substances such as **165**.[153] The fluorine spectrum was quite complex, and it

$$C_6H_5CHBr-CF_2Br$$

165

was suggested that (1) only a single conformation, most likely **166**, was populated and (2) rotation about the central C–C bond was slow, even up to 200 °C, because it was believed that rapid rotation would make the spectrum simpler than observed.

When interpreted at face value, the Du Pont results immediately indicated to me the possibility that suitably substituted fluorohaloethanes could exist as stable optical enantiomers at room temperature. We set out to check this possibility by using several compounds, most notably **166**, which is expected to have three rotational conformations (equation 94). Of these, **166b** and **166c** are nonidentical mirror

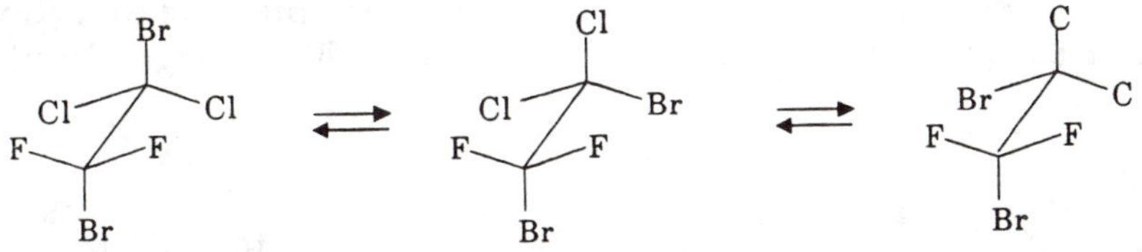

Br, Cl, Cl, F, F, Br ⇌ Cl, Cl, Br, F, F, Br ⇌ C, Br, C, F, F, Br

166a **166b** **166c** (94)

images; if isolated separately, they would show optical activity. Inspection of these individual conformations leads to the expectation that their fluorines will be nonequivalent, one being between two chlorines, the other being between a bromine and a chlorine. Thus, if rotation is slow and there is any significant amount of **166b** and **166c** present, then their nonequivalent fluorines should show up as a chemical-shift difference. Of course, the fluorines of **166a** are equivalent, each being located between bromine and chlorine, and, therefore, would show no chemical-shift difference.

When we took the spectra, no chemical-shift difference was observed, which meant either that only **166a** was present or that rotation was rapid and the NMR "camera" was only seeing the average of the different ^{19}F resonances of **166b** and **166c**. One has no idea when entering uncharted waters whether it would be possible to slow down interconversion of **166b** and **166c** or whether **166a** was the only substance present. Still, the obvious experiment was to lower the temperature and hope something interesting would happen. And, indeed, it did. Below –30 °C the ^{19}F resonance lines began to broaden. At –80 °C the spectrum was that of a "frozen" set of the conformations **166a** and the **166b–166c** pair in the ratio of about 1.4 to 1. A similar test on **167** showed that, at –80 °C, the enantiomeric pair with nonequivalent fluorines dominated to the extent of 90%!

$$C_6H_5CBr_2CF_2Br$$

167

These experiments appear to be the first example of the use of NMR to demonstrate slow rotation about single bonds in ethane.[154] We made a number of subsequent investigations[155] of rotational barriers and populations of rotational conformations. Among the most spectacular of these investigations was the work of Frank Weigert on **164**.[151]

The other important facet of our first communication on rates of rotation was the recognition that, contrary to Drysdale and Phillips, slow rotation was not a necessary criterion to obtain complex spectra. Thus, the NMR spectrum of any compound that has two magnetic nuclei attached to an atom carrying a group X and also three different groups, R_1, R_2, and R_3, in some proximity to the magnetic nuclei can be expected to show a chemical-shift difference for the magnetic nuclei. For a simple ethane derivative, there will be three conformations, **168a–168c** (equation 95), and in no one of these will the magnetic nuclei

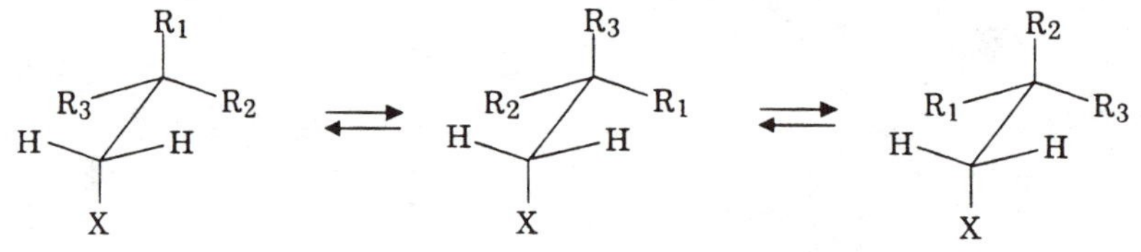

R1 R3 R2 H H X <=> R2 R1 H H X <=> R1 R3 H H X

168a **168b** **168c** (95)

(here taken to be protons) be equivalent, that is, have identical chemical shifts.

The NMR spectra will depend on the relative populations of the conformations, and it seems reasonable that, if all of the conformations have exactly the same energy, the chemical-shift difference between the protons will be negligibly small. Our key example was comparison of **169** and **170**, each of which can be presumed to prefer the conformations shown with bromines in the *trans* relationship. Both substances can be presumed to be undergoing rapid rotation at room temperature, and **169** shows clearly nonequivalent protons, although **170** does not.

Arrangements such as R_1, R_2, and R_3, as in **168**, correspond to optical enantiomers, and my first thought was that nonequivalence might be used as a test for chirality. A little reflection and a quick perusal of possible candidate compounds in the organic stockroom gave me the example of **171**, which is an achiral substance but still has nonequivalent $–CH_2–$ protons. The point is that, even though the molecule

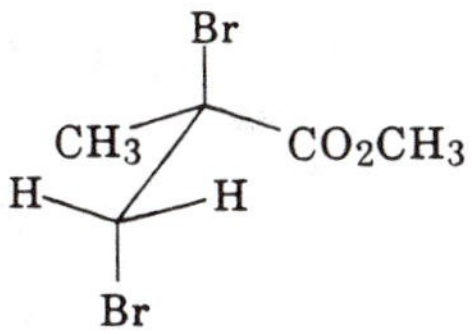

169 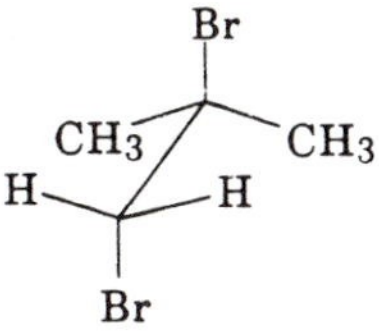 **170**

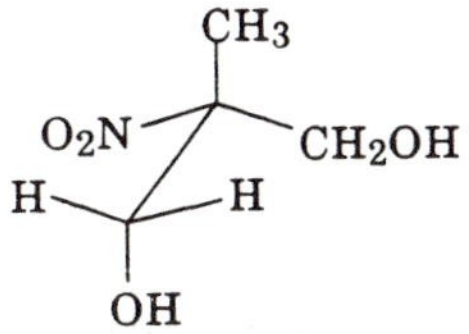

171

is achiral, each $-CH_2-$ group is in proximity to a R_1, R_2, and R_3 arrangement.

In later work, we were able to use magnetic nonequivalence to measure the rate of an amine inversion. Thus, with **172**, the $-CH_2-$ protons show no chemical-shift difference until the temperature is lowered sufficiently so that nitrogen inversion becomes slow (equation 96).[156] Another and, in fact, earlier evidence that an electron pair could

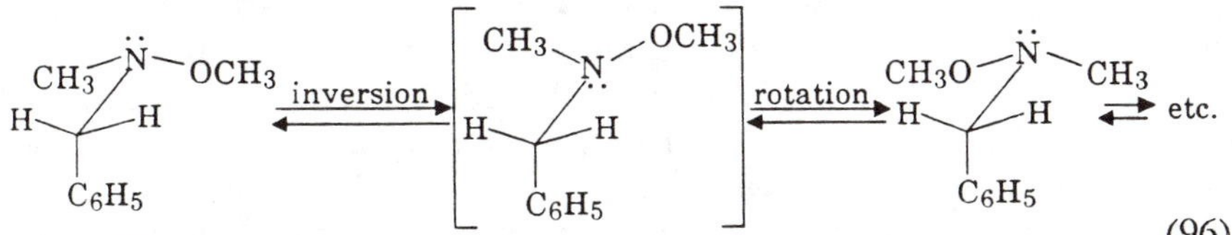

inversion rotation etc.

172

(96)

be one of the R_1, R_2, and R_3 groups was the $-CH_2-$ magnetic nonequivalence demonstrated by diethyl sulfite, **173**.[157] The sulfite group

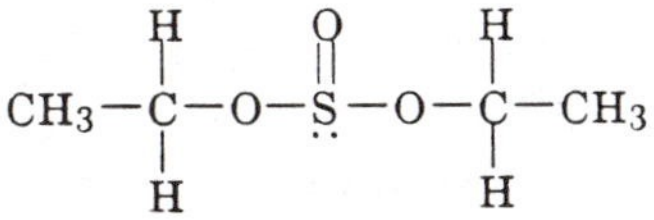

$CH_3-CH_2-O-S(=O)-O-CH_2-CH_3$

173

has long been known to have characteristics that can lead to molecular dissymmetry. It should not come as a surprise that **173** should be like **171** in having chemical-shift differences between its methylene protons. In our study of **173** we were able to show that the $^3J_{HH}$ of the ethyl groups has an opposite sign to the $^2J_{HH}$.[157]

Another marvelous application was measurement of how fast the double bond in a deuterated *trans*-cyclodecene, **174**, turns over within the relatively tight chain of methylene groups connected to the ends. This rotation interconverts the optical enantiomers **174a** and **174b** (equation 97), much as rotation about the C–C single bond of **166** can convert **166b** to **166c**.

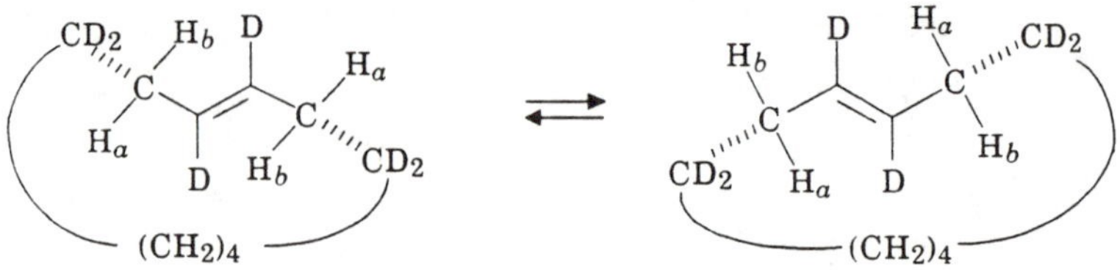

174a $\rightleftarrows$ **174b** (97)

The molecule was deuterated in the particular positions shown to reduce the proton–proton couplings at the $-CH_2-$ protons adjacent to the double bond and marked *a* and *b*. It will be seen that H_a and H_b are not equivalently located in **174a** or in **174b**. Turning over the double bond reverses the relationship of H_a and H_b. If the rotation is fast, then H_a and H_b will appear to be equivalent in NMR spectra.

The synthesis of the deuterated *trans*-cyclodecene was carried out with great skill by Gerhard Binsch, a postdoctoral fellow from Rolf Huisgen's laboratory in Munich and later a most powerful NMR and quantum theoretician. Gerhard found that, although rotation appears to be very fast at room temperature, the resonance arising from averaging the shifts of H_a and H_b began to broaden at –40 °C and appeared essentially "frozen" at –75 °C. The free-energy barrier to turning over the double bond was shown to be about 11 kcal.[158] This barrier can profitably be compared to the 35.6-kcal barrier measured by Cope and co-workers[159] for *trans*-cyclooctene, which is **174** with the $-(CH_2)_4-$ replaced by $-(CH_2)_2-$ and, of course, has a much tighter ring of atoms to slow rotation of the *trans* double bond.

One other facet of the magnetic nonequivalence that might well be mentioned was investigated by George M. Whitesides, one of those graduate students (like Bob Mazur, Frank Weigert, and Keiko Kanamori) whom you hope to have working for you at least once in your lifetime. I will say more about George later. One of the problems he investigated was how far the dissymmetry element could be from the groups displaying nonequivalence, in an open-chain compound, before the chemical-shift difference became undetectable. It is not good to use methylene hydrogens for this problem because, at small separations of the resonance lines, the resolution is complicated by the $^3J_{HH}$ coupling between hydrogens. In this situation, it was better to use isopropyl methyl groups. Four examples of such groups are provided by **175**–**178**, where, in correspondence to **168a**–**168c**, $R_1 = C_6H_5-$, $R_2 = CH_3-$, and $R_3 = H-$. The chemical-shift separations in hertz between the geminal methyl groups were measured at 60 MHz in benzene solution.[160a]

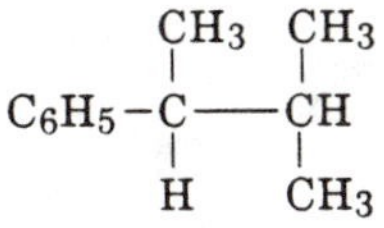

$C_6H_5-C(H)(CH_3)-CH(CH_3)_2$

175, 8 Hz

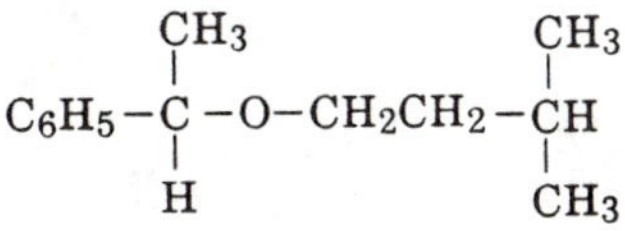

$C_6H_5-C(H)(CH_3)-O-CH_2CH_2-CH(CH_3)_2$

176, 1.8 Hz

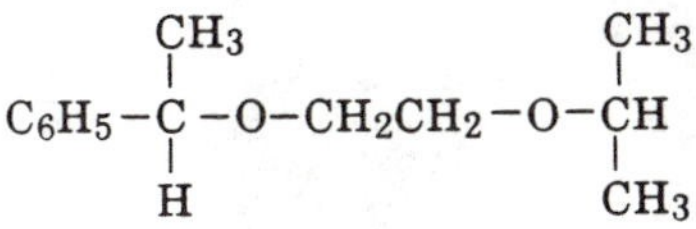

$C_6H_5-C(H)(CH_3)-O-CH_2CH_2-O-CH(CH_3)_2$

177, 0.8 Hz

$C_6H_5-C(H)(CH_3)-O-CH_2CH_2-O-CH_2-CH(CH_3)_2$

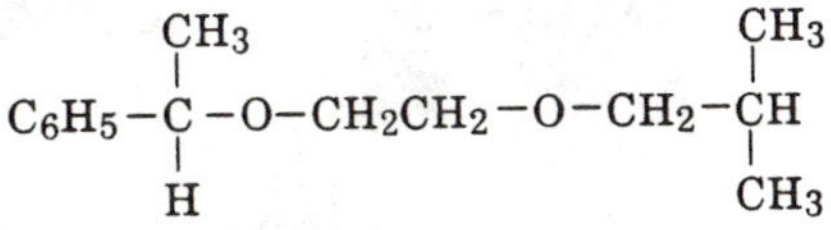

178, 0.0 Hz

Whitesides and David Holtz, an undergraduate research student, studied the effects of structure and also of solvent.[160b]

These were the fruits of our earliest discoveries in the NMR area. As I have indicated, the initial work was made terribly difficult by instabilities and the constantly changing environment typical of most research laboratories. Cranking the probe around, recycling, and recycling the magnet again, along with taking seven-second spectra to try to get useful resolutions, made NMR not nearly so good as it could be for structural analysis of large molecules or measurements of reaction rates.

NMR: Improved Instrumentation

After the first few months of frustration, I visited Varian in late November 1955 to talk to Jim Shoolery about the temperature-controlled probe insert. After a few minutes, he said, "I'd like to show you something, but for now, it is secret." I was willing, and he took me into the applications area for a display of spectra such as I had never seen. These were not seven-second spectra. These were 1- to 2-minute spectra, and the relaxation wiggles were small shoulders on the sharp peaks, rather than running from one widely separated group of peaks over to another, as in Figure 3. For the first time, the full 16 possible resonances from the three nonequivalent hydrogens of a vinyl group were there, beautifully resolved. I had begun to hyperventilate when finding out from Bill Phillips about specific applications of NMR to organic chemistry, but these spectra caused an even more severe reaction because of my continuing difficulties in obtaining reasonably reproducible spectra and then trying to analyze them when the relaxation wiggles were running riot over the peaks.

The device that produced the magical results was the "Super Stabilizer", which, I have been told, was suggested to Varian by Mel Klein, a most imaginative and inventive worker in the NMR field. Today we

An A–60 NMR spectrometer with the development team of Varian Associates that made it possible. This spectrometer, one of the first shipped in 1961, was retired after 20 years of useful life and returned to Varian in 1982, when this picture was taken. Seated at the console is Edward L. Ginzton, chairman of the board of Varian Associates when the A–60 was announced. Standing from left to right: Tom Kingston, electronics engineer; Wes Anderson, a distinguished physicist who was involved in the early development of NMR with Felix Bloch at Stanford; Andy Baker, electronics technician; George Schulke, electronics engineer; John Moran, electronics engineer and project leader; Jim Shoolery, an applications chemist who helped define the concepts of what the A–60 was to do for its users and who installed our first spectrometer (pp 154–156); and Bob Gang, magnet engineer.

would do it differently, but the solution derived at the time was a very useful one. The basic idea was to sense changes in the magnetic field and to provide a compensating current that would counteract the changes. The sensors of the field changes were coils within the pole-face covers. A field excursion, no matter what source, generated a dc voltage that was sent to a Leeds and Northrup galvanometer where a narrow light source is reflected from the galvanometer mirror to a scale. Varian installed two phototubes, one on each side of the zero of the scale. These detected which way the mirror was deflected and, hence, the direction in which the field was changing. The compensating current was sent to buckout coils surrounding the pole faces. The field stability was greatly improved thereby. Slow drift could still occur, but most of the short-term fluctuations disappeared.

Needless to say, I put in an order as soon as possible. To have what one now would call a β-test site not too far from Palo Alto, Varian sent us a Super Stabilizer with serial number 1. With various updates, it worked until we finally turned over our original system to Robert Vaughan in chemical engineering in 1974. It was sitting on top of the magnet, and it crashed six feet to the concrete floor in the 1971 earthquake. No one expected it to have survived, but we bent various parts back into shape, resuspended the galvanometer, changed the broken light-source lamp, and away it went, like new!

Our next big change in NMR instrumentation came in February 1961, when, on one of my visits, Jim Shoolery told me they had developed a more analytical instrument and asked if I would like to guess what properties it ought to have. Actually, I did pretty well in describing the features I hoped would be in what turned out to be the very popular and widely used model A–60, except that I did not foresee stabilization of the field by locking onto the proton signal from a doped water sample placed inside the probe. Again, this was a vacuum-tube instrument, full of 12AX7 and similar tubes. It was made for easy servicing; at the first sign of trouble, a tube-changing frenzy was the first line of defense. This machine was so suited to hands-on use by anyone interested in structural and quantitative analysis that it was not hard to arrange for its purchase.

I took personal charge of the A–60 and its maintenance. In order to make the set-up procedure more straightforward for the users, I generated an elaborate series of spectra that showed what happened with changes of the field-homogeneity settings, rf power levels, spinning rates, and so on. At this point, NMR became routine for organic chemists. Later, at the end of 1965, we purchased an A–56/60 spectrometer from Varian, because we had set up a very active program for studying conformational analysis and equilibration using ^{19}F NMR. This instrument could easily be switched between 60 MHz for protons and 56 MHz for fluorine, with no change in magnetic field. Further, the A–56/60 had good capabilities for going to low temperatures, which was very important to our studies of conformational equilibration.

NMR: Educational Activities

Before describing some of our other NMR research, it may be well to describe some of my NMR educational activities. In the spring of 1956, I worked up a graduate course at Caltech, in which I lectured as I learned. The spectra illustrating the various points I wanted to make were generated as needed (still using the Sanborn recorder) and affixed to my lecture notes with Scotch tape. I was invited to give a lecture at the Reaction Mechanisms Conference in September 1958 and decided that this was *the* opportunity to really get organic chemists interested in NMR.

We spent several weeks of the latter part of August at Edwin Buchman's cottage at Big Bear Lake. To the dismay of my family, who thought this was supposed to be a vacation, I spent almost all of the time hunched over a drafting board, drawing slides. The original drawings were done in India ink, about 11 × 14 inches, and were colored with water colors. About 50 drawings in all, they were photographed with Kodachrome in the mountain sunlight. I wanted to have the slides mounted in glass, but time got very short and I had to do it myself. I finally finished in New York, just the night before the conference started. On balance, I would say that this was surely the most influential lecture I have ever given. The timing was right, the audience was eager to find out what NMR was about, and the slides were sensational. The aftermath included some 40 invitations to give essentially the same talk about NMR at universities and companies around the country and abroad. One of the great experiences of this period on the lecture circuit was a two-week visit as Karl Folkers Lecturer to the University of Wisconsin, where I became a co-worker of William S. Johnson, the great synthetic chemist, later to move to Stanford and to elevate the department there to the very top rung. The story of our dynamic NMR encounter has been published.[161] I was a stellar salesman for Varian in those days, but never received any compensation from them.

These activities all happened in the days when the publishing of chemical texts and monographs was flourishing. One day William A. Benjamin, a young editor from McGraw-Hill, dropped in and asked if I would consider writing a book about NMR. Bill was enthusiastic about the prospects for sales. He was also enthusiastic about modeling the art work after my slides, even wishing to see if the illustrations could be done in color. The latter surely would be a first at that level of monographs, and Bill's juices flowed at the prospect. He liked to be first.

We were still living in northwest Altadena at the time, about a 20-minute drive from Caltech in my old Chevrolet sedan. I had bought one of the early portable dictaphones, which used magnetic tapes, and I dictated the first draft of the book while driving back and forth to work. Because I was so familiar with the material from giving lectures, I needed no notes or references. The appendix was quite another matter. It dealt with the Bloch equations, how they could help in understanding absorption- and dispersion-mode spectra, and how one could be obtained, essentially independently of the other, by adjustment of the leakage current in the probe. I struggled with the Bloch and Pake articles for quite a while before I got it straight and, I hoped, made it understandable.

To my knowledge, this book was the first to include the now-ubiquitous spectroscopic problems that, in the simple way I used them, had only a proton spectrum and a molecular formula. In principle, the problems could be solved from the chemical shifts and coupling patterns. Fairly simple, but interesting, examples were necessary. I scrounged through the stockroom and our research samples looking for suitable candidates.

In the process I discovered that the proton spectrum of neohexyl chloride, **179**, showed a curiously complex pattern for the $-CH_2Cl$ protons. It was quite different from the simple 1:2:1 triplets expected from

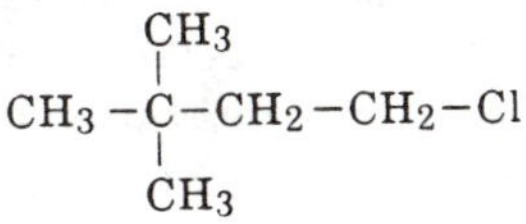

$$CH_3-C(CH_3)_2-CH_2-CH_2-Cl$$

179

the simple qualitative rules of spin–spin splitting for protons adjacent to a $-CH_2-$ group and uncomplicated by other nuclei. Such triplets are observed for the $-CH_2Cl$ protons of $CH_3CH_2CH_2Cl$ and $CH_3CH_2CH_2CH_2Cl$.

Clearly, something strange was going on, and it stuck in my mind. However, then I did not connect it with the earlier, very disturbing report by Harden McConnell[162], then of the Shell Development Corporation, that 1,1-difluoroethylene, **180**, gave a very complex proton or

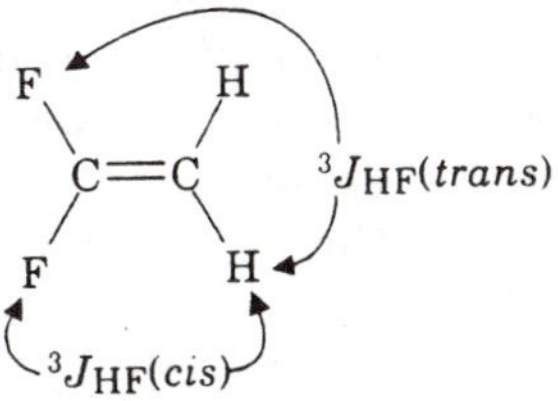

$F_2C=CH_2$

$^3J_{HF}(trans)$

$^3J_{HF}(cis)$

180

fluorine NMR spectrum rather than simple triplets. McConnell's published spectrum seemed so crude that I thought it might be impure. I got some **180** and, sure enough, it gave a beautifully resolved, very complex spectrum.[163]

The problem is that there are two different $^3J_{HF}$ couplings corresponding to a *cis* and a *trans* H–F magnetic interaction, **180**. By the simple qualitative rule, I expected four equally intense lines for the resonances of the hydrogens or the fluorines on the reasonable assumption that $^3J_{HF}$ (*cis*) is different from $^3J_{HF}$ (*trans*). The hydrogen resonances would be split into a doublet by the *trans* coupling and each of the doublet lines split into doublets by the *cis* coupling. But the fluorine spectrum showed 10 lines in all, some very strong, some very weak. So again, I had something to learn, but not for this NMR book. I hoped the problem posed by McConnell was not going to be a prevalent one. That hope was not to be realized, and I had to deal with it later.

Bill Benjamin was eager to get my book out because it would be complementary to Pople, Schneider, and Bernstein,[164] which was also being published by McGraw-Hill. When it went to press in the fall of 1958, I was visiting professor at Harvard and went down to New York to watch the first print run. The color printing was a fascinating process. The book[163] turned out to be about 125 pages long and, despite a rather high price of $7.50, sold about 8000 copies.

My next excursion into book writing came in 1960, after I had helped Bill Benjamin to found the publishing firm of W. A. Benjamin, Inc. He was desperate to get a book out. Although he had Marjorie Caserio and me under contract for an organic chemistry text, that was not going to be finished for quite some time. Fortunately, he caught me at a time when I had an extra-high level of missionary spirit about disseminating an understanding of spin–spin splitting in NMR.

After our initial and rather shallow sweep through applications of NMR, I decided to concentrate for a period on spin–spin interactions and reaction-rate problems. If you expect to try to investigate spin–spin J values as a function of structure, it certainly helps to be able to analyze the spectra and understand the ambiguities involved. K. B. Wiberg,[165] A. A. Bothner-By,[166] and J. D. Swalen[167] were devising computer programs to simulate complex spectra and also to compare calculated and experimental line positions and iterate to find the best possible shift and coupling parameters.

My trouble was the usual one—I didn't understand how spin–spin splittings came about once you got into situations where the simple qualitative rules did not apply. Even for two nuclei, there were problems. When you observe the proton spectrum of a substance such as $HCCl_2–CCl_2F$ (where the chlorines act as nonmagnetic nuclei), it is not hard to see how there could be two distinct species producing spectra, one with fluorine having a magnetic quantum number of +½ and the other with −½. One of these ^{19}F spin states would augment the magnetic field at the proton and the other would decrement it so that there would be two equally intense proton resonance lines, one corresponding to each species. The same reasoning could be used to predict two equally intense lines in the ^{19}F NMR spectrum.

It was harder to understand why no splitting was observed with $HCCl_2–CCl_2H$, and why, with substances such as $HCCl_2–CBr_2H$, the predicted splitting of the chemically shifted $HCCl_2–$ and $–CBr_2H$ protons into two doublets was observed, but the resonance lines were not equally intense. The inner lines were always stronger than the outer lines, although the intensities began to become more nearly equal if the shift difference, $\Delta\nu$, could be increased (i.e., $\Delta\nu >> J$). In the crude qualitative theory, one used catechisms such as "nuclei with the same chemical shift do not split one another" or "unless the shift differences are very large, the number of lines, their intensities, and their positions may not be just what you would expect." This is what made the McConnell case of $CF_2{=}CH_2$ so troublesome, because here $\Delta\nu >>> J$, and yet there were "unexpected" resonance lines and intensities. I knew I had to understand as much as I could of this.

For spin–spin splitting, as for molecular orbital calculations, the theory was already well worked out. The problem was that it was not that easily accessible to the mathematically unwashed. Quantum mechanics was involved, of a type rather akin to resonance theory, and account must be taken of spin states. In the simple Hückel molecular orbital theory, you derived the orbital energies as linear combinations of atomic orbitals, but the spin states of the electrons did not affect the electron energies. However, when you calculate interatomic repulsions, as in the Parr and Pariser theory, then the electron spin states have to be taken into account. Consequently, I did have some prior experience with such matters.

The necessary formulas for actually calculating the energies of various simple nuclear spin configurations of molecules with magnetic nuclei had been derived and listed by a number of workers,[164] but I could not understand the formalism that they used. I came to a screeching halt whenever I tried to do more than substitute in their formulas. Frustrated, I finally got Harden McConnell, by then a professorial colleague at Caltech, to sit down, on July 4, 1956, and explain what was involved. He was an excellent teacher, and I got the drift of the thing well enough to try to explain it to others who were interested.

It worked out rather well. I told Bill Benjamin that, if he wanted a book, I would write *An Introduction to the Analysis of Spin–Spin Splitting in High-Resolution Nuclear Magnetic Resonance Spectra*. It was a long title for what was to be a short book. Indeed, it was not a lot more than a fuller explanation of a relatively few pages in Pople, Schneider, and Bernstein.[164] He said, "Fine, get to work."

A wonderful opportunity to do just that came when Paul Bartlett, George Hammond, Cheves Walling, and I were invited to give a two-week refresher course in physical organic chemistry for college teachers, sponsored by the National Science Foundation, at Fort Lewis College in Durango, Colorado. The lectures were given in the mornings and evenings, with afternoons free for hiking or touring the countryside. We drove to Durango with the children and, while everyone else was off on afternoon trips, I stayed at the college, in complete isolation before a picture window, writing furiously until a fresh idea or phrase was needed. Then I watched the afternoon cumulonimbi gather over the Colorado mountains until inspiration came. The first draft was just finished in the two available weeks.

Bill Benjamin wanted this book to be a showpiece to convince potential authors that he could do a very-high-quality publishing job in half or less of the time normally required. The heat was on, and this was a text with a lot of mathematical equations, graphics, and spectra. The spectra were planned for white lines on a gray background. Because each had to illustrate exactly what I wanted to illustrate, I ran them all myself, again scrounging through our stockroom and research chemicals in search of appropriate examples. The design of the book was quite unconventional and indeed strikingly beautiful for a monograph. The final added fillip was to have my portrait on the back of the dust jacket, bled all the way to the sides.[168]

The book sold several thousand copies, but I had remarkably little feedback on its utility. Perhaps there weren't many people who were that interested in understanding the basis of what is involved in spin–spin splitting. I have always been intrigued by the amazing precision to which even an exceedingly complex NMR spectrum could be calculated by knowing only chemical-shift differences, coupling constants, and relaxation times. The reverse process, that of obtaining the shifts and so on from complex spectra, was originally much more difficult, but has lately become almost routine with the 2D NMR procedures.[169]

Expanding Computer Horizons

In this connection, I cannot forget trying to make a precise fit to the rather complicated proton spectrum of vinylacetylene, **181**, with four chemical shifts and six couplings.[170] I could use the IBM 704 to calcu-

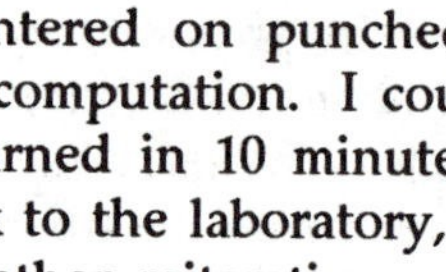

H–C≡C–C(H)=CH–H

181

late the line positions, but trial and error was deathly slow, with each new set of parameters entered on punched cards at the computing center and submitted for computation. I could never be sure whether the results would be returned in 10 minutes or 3 hours. The uncertainty meant walking back to the laboratory, walking back to the computing center, starting another reiteration, and so on. It did not take me long to realize that I needed to get away from this style of computing. I began to think about how to get my own computing machinery. Perhaps a "personal" computer would not do as much as the 704 and would be more difficult to program, but waiting for output at the computing center was not the way I wanted to spend my time.

Not much in the way of personal computers was available in

those days. My first try, in the early 1960s, was with a Monroe programmable calculator. It could do arithmetic quite well, but it had only about 50 program steps. If any steps were entered incorrectly, you had to start all over again. I got that on trial, and only a brief trial was needed. A much better solution was offered by an Olivetti, which had greater precision, easier programming, and ways to save programs on magnetic cards, but it sounded like a machine gun when it was printing. That machine served me well for a while, relieving the tedium of many calculations, although it was limited in programming ability.

By this time, I wanted the convenience of some kind of a computer at home so I could program in a less yeasty environment than the laboratory or the computing center. The opportunity to do that finally came in 1968, when Hewlett–Packard offered their 9100 computer. It was programmable to about 110 steps; used reverse Polish; and had a small green CRT display, an internal printer that wrote on special paper with a spark discharge (and produced considerable radio interference), and a first-rate pen plotter. Programs were stored on magnetic cards, and the internal memory was nonvolatile, ferrite core. You could turn off the computer, start it up a week later, and be at exactly the point you left it. The programming was tedious, computation was slow, plotting even slower, and it had no alphanumeric capability, but the results were superb. The machine had enough steps to allow me to run a numerical integration of Bloch's NMR equations (which are extremely well behaved in such calculations), and I was able to simulate and better understand a variety of changes of spectral parameters. This machine served me very well, especially for theoretical calculations of pseudocontact interactions resulting from taking NMR spectra in the presence of shift reagents, a project on which I had a lot of help from my sons, Don and Allen, who worked with me as summer research assistants at Caltech.[171]

But as one's computer horizons expand, there is a need for more memory. Hewlett–Packard did have a remedy for this problem. For about $1000, you could buy a 1024-bit ferrite memory that allowed another 150 or so program steps. This memory came in a box (about 3 × 12 × 16 inches) that sat on the floor. Today such a box could easily contain 10^9 bits or more of memory, although it would be volatile to loss of electrical power.

The HP–9100 finally reached its limits. I transferred it to Harry Gray in 1973 when Hewlett–Packard announced the HP–9830, a desktop machine with a BASIC operating system in read-only memory, a variety of outputs and inputs, a plotter, digitizer, cassette tape storage, full ASCII capability, and a red LED display for input. The BASIC was the forerunner of HP's Rocky Mountain BASIC—very powerful, easy to program, and easy to debug. The original purchase provided for about 4 Kbytes of memory, but I was ultimately able to expand it to 32 Kbytes.

The 9830 allowed enormous expansion of my programming activities, and I cut the umbilical cord to the computing center forthwith. Unfortunately, because the system was at home, my students still had to struggle with FORTRAN and the IBM machines at the computing center.

After we got underway in determining coupling constants, our other priority was determination of reaction rates by NMR. Quantitative determinations were difficult. The original Gutowsky and Holm[172] procedure for determining reaction rates by measuring the separation between resonance lines that were being averaged by rapid reactions led to such variable results that, for quite a time, measurement of rates by NMR got a rather bad name. Activation energies differing by as much as 15 kcal were reported for the *same* process.

The problem was that chemical shifts are a function of temperature, just as rates are a function of temperature. Most investigators were finding the *same* temperature for the point where two resonance lines just coalesced to a single resonance line, which meant that each was getting about the same free energy of activation. The discrepancies were with procedures used to evaluate the change in rates with temperature away from the coalescence point. My problem was to try to understand how to do more than measure peak separations or line widths, to determine the whole line shape as a function of reaction rate. This determination was not too difficult when there were two uncoupled nuclei. The challenge was to be able to calculate line shapes for two coupled nuclei under conditions where the chemical-shift difference, $\Delta\nu$, is approximately equal to J, in which circumstance the lines are not all equally intense. The basic theory for such calculations was available in papers by Alexander[173] and Kaplan,[174] but if molecular orbital and spin–spin splitting could be classed as difficult, the papers by these worthies were out of sight.

Fortunately, at this time, I had a rather talented undergraduate, then a bit overweight and sloppy, named Jesse L. (Jack) Beauchamp, who was *smart* and could penetrate the fog of the Alexander and Kaplan papers. He wrote, to my knowledge (Martin Saunders at Yale was a close second), the first computer program that would calculate the line shapes of the two-nucleus case with $\Delta\nu \sim J$, a program that we put into immediate use in many connections. Jack Beauchamp received the George Green Prize at Caltech for his outstanding undergraduate research and went on to develop ion cyclotron resonance with John D. Baldeschwieler when he was a graduate student at Harvard and Stanford.[175] Subsequently he became an extraordinary professor of chemical physics at Caltech. Today, lean and hard, he has accumulated a wonderful record of work on reactions of ions with neutral molecules in the gas phase, photoelectron spectroscopy of organic radicals, and surface reactions.

Jack Beauchamp, as an undergraduate, was a contemporary of George Whitesides when he was a graduate student, and the two together made waves of considerable proportions. George was memorable. An undergraduate from Harvard, he came to Caltech for reasons unknown to me. I talked to him about research problems and, some time later, he appeared, wanting to do research. At that particular time (1961), all my space was taken by more or less eager workers. I casually told him that I was full up, and perhaps he could find someone else to work for. A few days later he was back, insistent. Remembering my own problems with Whitmore (p 24), I decided to take him on, even though he would have to share a lab bench for the first year. Being at the right place at the right time, it was nice to make the right decision as well.

Grignard Reagents

George was a phenomenon and made many things work. I have already mentioned his research on magnetic nonequivalence because of molecular asymmetry. Perhaps more important was his work on Grignard reagents, especially my old thesis problem of the structure of butenyl and related Grignard reagents. George did not actually initiate this particular line of research. I recognized its possibilities rather early, and the first to get into it was Al Bottini, who studied hydrolysis of aromatic halides (p 118) and imine inversions (pp 159–160).

Here, because of my stupidity, we missed out on a really wonderful discovery. One day in the spring of 1956, Al was making Grignard reagents and looking at their NMR spectra. He asked me, "What do you do if a proton signal is downward, rather than upward?" "Run it again", I said, "and see what happens." Al ran it again, and it was upright. So we missed perhaps the first observation of *chemically induced dynamic polarization* (CIDNP), which has become an extraordinary means of detecting free-radical reactions by NMR.[176] You can't win them all!

Al Bottini was followed by Eric Nordlander, a fine graduate student, later at Case Western Reserve, and a fairly recent cancer fatality. Eric started work on allylic Grignard reagents in 1956 and found that allylmagnesium bromide, **182**, gives an extraordinarily simple proton spectrum, a four-proton doublet and a single-proton quintet.[177] The conventional formulation of **182** would be as the covalent structure **182a**, with the rotational angle of the C–C single bond taken to have the orbitals of the C–Mg bond so oriented as to allow at least some contribution of the hyperconjugative resonance structure **182b** (equation 98). The proton NMR spectrum is simply not compatible with **182a**, or with **182a** being in rapid equilibrium with the conformation resulting from rotation around the C–C single bond, **182c** (equation 99). This rotation averages the chemical shift of the $-CH_2MgBr$ protons, but not those of the terminal $CH_2=$ protons of the double bond.

$CH_2=CH-CH_2MgBr$
182

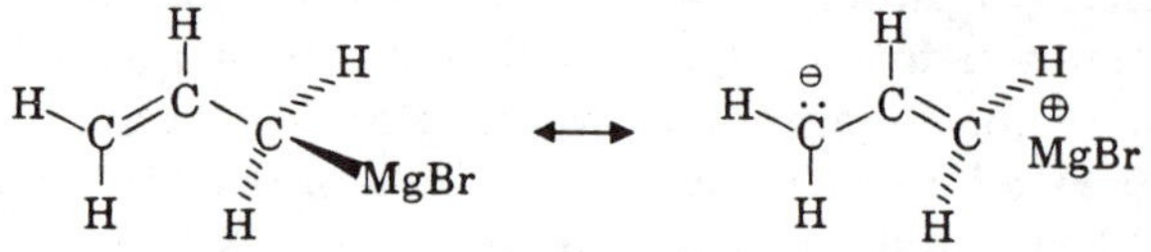

182a ⟷ **182b** (98)

182a ⇄ **182c** (99)

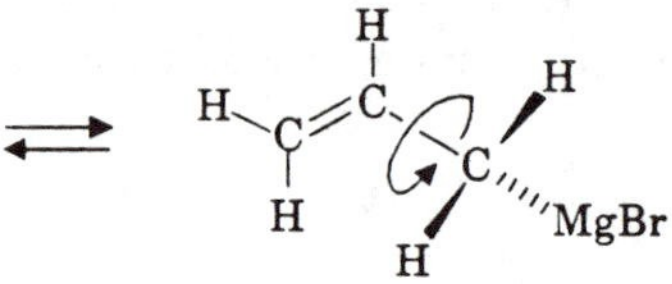

With **182a** and **182c** in equilibrium, one expects a substantial difference in shift between the $CH_2=$ and $-CH_2Br$ protons. One way out of the dilemma is to have an ionic structure, **183**. But such a struc-

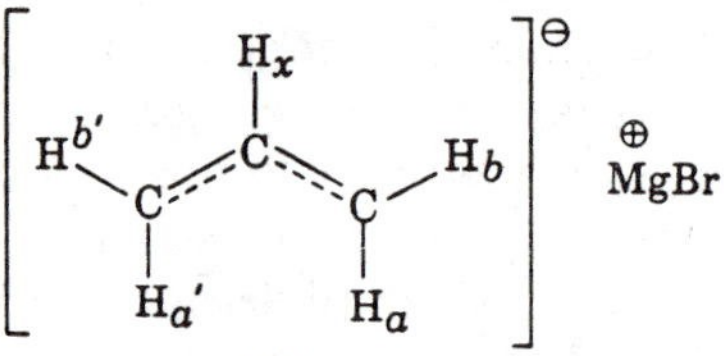

183

ture should have different chemical shifts for the *cis* hydrogens (*a* and *a'*) and the *trans* hydrogens (*b* and *b'*) unless rotation is fast enough about the partial double bonds to average these shifts and the associated $^3J_{H_{ax}}$ and $^3J_{H_{bx}}$. If so, rotation is fast at room temperature but might be expected to be slower at low temperatures. We did not observe this slow rotation.

The most reasonable explanation for the observed behavior is that there is a rapid equilibration between the eight structures designated as **184** (equation 100). It is certainly possible that **183** is an intermediate in the –MgBr shifts postulated to take place from C-1 to C-3. The sequence of **184** allows for averaging the shifts and couplings of the four terminal hydrogens and fits the observed spectrum precisely.

On the basis of my experience with allylic Grignard reagents, I was flabbergasted at this development. I would never have predicted the **184** processes to be so fast. Later we found that even diallylmag-

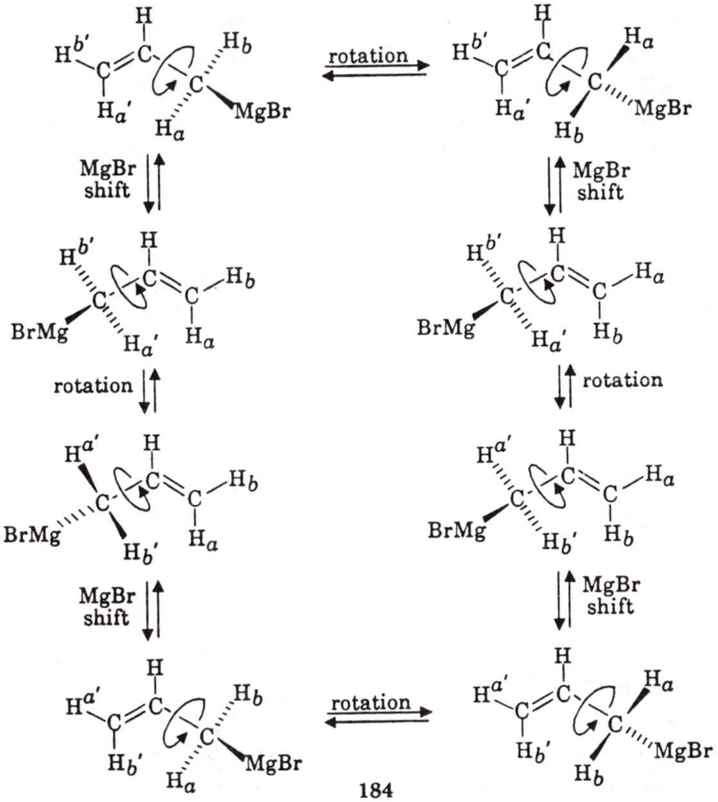

184 (100)

nesium, prepared from diallylmercury and thus not containing magnesium halides, underwent fast equilibration even at −120 °C.[178]

I had thus a bit of anxiety about what we would find with the NMR of the butenyl Grignard reagent. Indeed, it was fascinatingly different from the spectrum of the allyl Grignard in being rather consistent with what would be expected for the structure **185** I had proposed in 1944 (equation 101). But we knew that the allyl Grignard equilibration

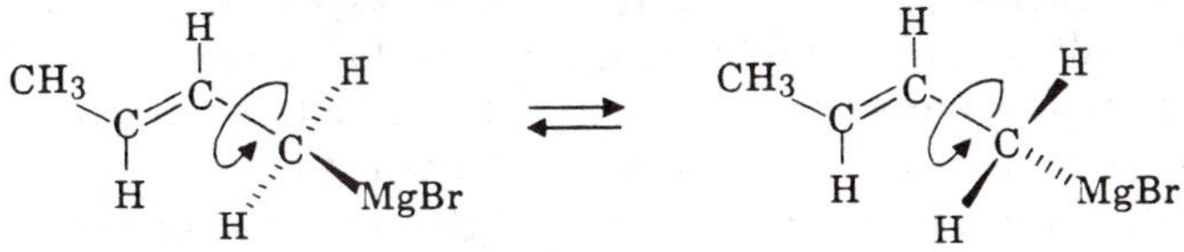

185a **185b** (101)

was fast. Without writing down all of the different rotational conformations, we can foresee the possibility of equilibration of three principal structures, **187**–**189** (equation 102), formed by equilibration processes analogous to those of **184**. We expected that **188** would have to be the intermediate by which **187** and **189** could be equilibrated. As I have said, the NMR spectrum is qualitatively what we would expect for **187**. What about quantitatively? What can you do to decide? There was no hard evidence for the other isomers, **186**, formed by the MgBr shift. I

186a **186b**

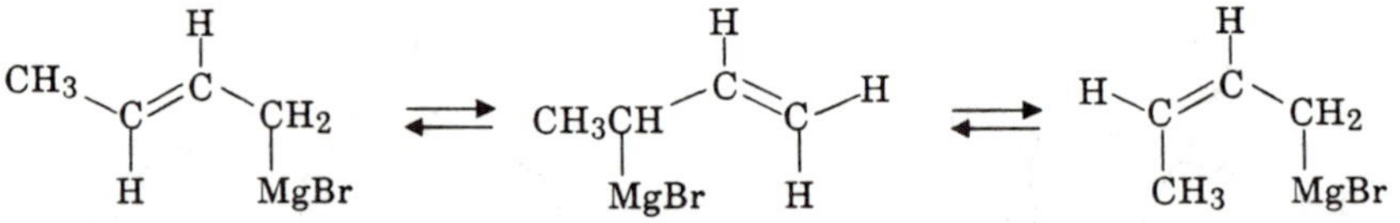

187 ⇌ **188** ⇌ **189** (102)

wondered whether what is surely an averaged spectrum, about as expected for **187**, has more than very minor contributions to that average from **188** and **189**.[179]

One substantial NMR difference expected for **187** and **189** is the $^3J_{HH}$ splitting across the double bond. The usual value for the coupling of the *trans* protons, as in **187**, is 18 Hz; the corresponding value for *cis* protons, as in **189**, is about 10 Hz. The experimental coupling is 10.1 Hz. Both **187** and **189** differ from **188** by having the CH_2 protons be $-CH_2MgBr$ rather than $=CH_2$ protons. Judging from CH_3MgBr, we might expect to have a chemical shift of –0.6 ppm relative to $(CH_3)_4Si$ (the usual standard for proton and carbon spectra). Yet, the $-CH_2MgBr$ protons of the butenyl Grignard reagent come into resonance at +0.7 ppm, and those of isopentenyl Grignard, **190**, are almost the same at +0.6 ppm.[180]

Much chemical experience leads us to expect that **190a** and **187** and/or **189** will be more stable than **190b** or **188**, respectively, because the more methyl groups you put on C-3, the more favorable it is to have the double bond between C-2 and C-3 rather than between C-1 and C-2. The fact that the shifts of the $-CH_2MgBr$ protons of **190** and the butenyl Grignard are close together suggests that the equilibrium between **190a** and **190b** (equation 103), as well as **187**, **189** and **188**, lies far on the side of **190a** and **187**, **189** (*see also* pp 29–35).

A corroborating piece of evidence is the fact that the shift of the $-CH_2MgBr$ protons of the butenyl Grignard does not change beyond

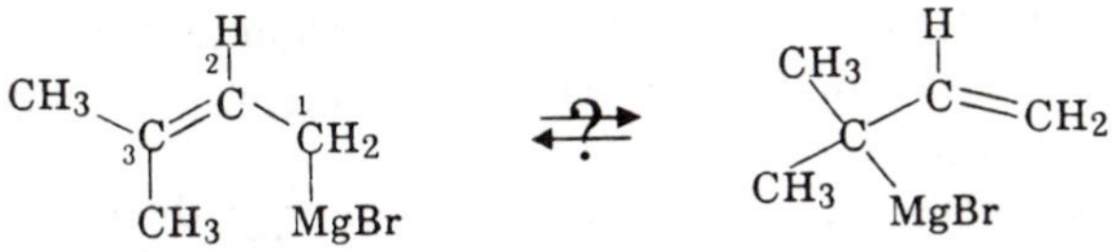

$$(CH_3)_2C{=}CH{-}CH_2MgBr \;\overset{?}{\rightleftarrows}\; (CH_3)_2C(MgBr){-}CH{=}CH_2 \qquad (103)$$

190a **190b**

experimental error over a 50 °C change in temperature. That CH_3MgBr protons come into resonance at –0.6 ppm and the $-CH_2MgBr$ of the butenyl and isopentenyl Grignards at 0.6–0.7 ppm is not unexpected because a similar, albeit somewhat smaller, difference in shift occurs between the $-CH_2Br$ protons of CH_3Br and $CH_2{=}CH{-}CH_2Br$ of about 0.5 ppm.

Comparison of the coupling constants of the allyl and butenyl Grignards gives further information, which is not as definitive. Part of the problem is the substantial variability of coupling constants with structure. Thus, the reported range of $^3J_{HH}$ for the substructure, $=CH{-}CH_2{-}$, is 5–10 Hz. If we take $^3J_{HH}$ for $-CH{=}CH{-}$(*trans*) as 18 Hz and $^3J_{HH}$ for the corresponding *cis* coupling to be 10 Hz, we can derive a value of $^3J_{HH}$ for the rapidly equilibrating allyl Grignard (equation 100), which has an average coupling of 11 Hz between the single CH and the four CH_2 protons. It turns out that we can see that the average $^3J_{HH}$ for $=CH{-}CH_2MgBr$ has to be 8 Hz to give an average of 11 Hz for the four $^3J_{HH}$ couplings. An 8-Hz contribution to the average is rather smaller than the 10-Hz value of $^3J_{HH}$ found for the $=CH{-}CH_2MgBr$ substructure of the butenyl and isopentenyl Grignards. Further, all of these couplings are much larger than the approximately 5-Hz couplings for $=CH{-}CH_2X$, where X is –Br, –OH, and –OR. If we take 10 Hz for the $=CH{-}CH_2MgBr$ coupling, then the *cis* and *trans* $-CH{=}CH-$ couplings will have to have average values of 12 Hz to give the overall average of 11 Hz. This value is rather more than the observed 10-Hz average for the butenyl Grignard reagent.

Because the *trans* $-CH{=}CH-$ coupling is surely larger than 10 Hz, this result leads to the interesting conclusion that the butenyl Grignard reagent is likely to be a rapidly equilibrating mixture of *trans* and *cis* isomers, **187** and **189**, with **188** as the intermediate, but almost surely present in low concentration. The upshot of all of this analysis is that, judging from the coupling constants alone, the proportion of the *cis* isomer, **189**, must be at least 50%, a heretofore unpublished conclusion.

George Whitesides got into this particular phase of research on allylic Grignard reagents with his studies of the proton NMR spectrum of the isopentenyl Grignard reagent as a function of temperature. As can be seen from **190a–190b**, interconversion and rotation about the sir-gle bond of **190b**, then back to **190a**, will interchange the methyl groups on the double bond. Indeed, the methyl groups give a *single* resonance at room temperature.

If we assume that the known free-energy difference of 3 kcal between $(CH_3)_2C{=}CH{-}CH_3$ and $(CH_3)_3CH{-}CH{=}CH_2$ also applies to **190a–190b**, then, even without taking into consideration the fact that –MgBr is surely more favorably bonded to a carbon carrying two hydrogens than one carrying two methyl groups (*see* p 180), we can expect that the equilibration of **190a** and **190b** will be much slower than that of the allylic Grignard reagent, because **190b** is less stable than **190a**, while **182a** and **182b** are equally stable. George showed that, below 0 °C, the methyl groups gave separate resonances, as expected for a slowed interconversion of **190a** and **190b**. He determined a $\Delta E^{\ddagger}$ of activation of 7 kcal by the Gutowsky and Holm[172] procedure of analyzing the line separation as a function of temperature. This value is most likely too small because the procedure does not take into account possible chemical-shift changes with temperature (*see* p 176).[181]

The Bottini–Nordlander–Whitesides NMR work confirmed that the butenyl Grignard reagent was essentially exclusively the primary structure, but that there is evidence of a very rapid equilibrium between the primary and secondary forms. Most likely, the primary form is very substantially present, with the *cis* configuration at the double bond (**189**). It is amazing how much NMR can contribute to solving difficult structural problems of this kind.

George Whitesides' major work as a graduate student derived from my discovery of an unexpectedly complex proton NMR spectrum of neohexyl chloride, $(CH_3)_3CCH_2CH_2Cl$ (*see* p 171). Instead of two clean triplets, I observed two sort of ragged multiplets that were the consequence of exactly the same kind of coupling situation as McConnell investigated for $CF_2{=}CH_2$.[162]

The characteristic of neohexyl derivatives that makes them not A_2X_2 (A_2X_2 represents an NMR classification with two A and two X nuclei with widely separated chemical shifts and a single J_{AX} coupling constant) but AA′XX′ (AA′XX′ is the same as A_2X_2, except the symmetry of the arrangement is such that $J_{AX} = J_{A'X'} <> J_{A'X} = J_{AX'}$) is the tendency for large groups to favor the *transoid* conformation, **191**. If this

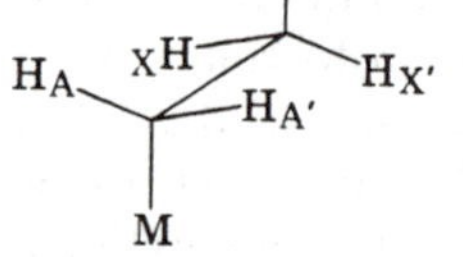

C(CH3)3
HA xH HX′
HA′
M

191

conformation predominates, then, on the average, $(J_{AX} = J_{A'X'}) \neq (J_{AX'} = J_{A'X})$, and a complex spectrum can result. The preference for conformation **191** will change with the size of the group, M, and $J_{AA'}$ will depend on the electrical character of M. Therefore, the appearance of the spectra of derivatives, $(CH_3)_3CCH_2CH_2M$, changes substantially with the nature of M.

George and I decided that making M into –MgCl offered a unique opportunity to study the *configurational stability* of simple Grignard reagents.[182] The idea was that the Grignard reagent, as **192a**, would show an AA'XX' spectrum. However, if the –MgCl were to move *rapidly* to the other side of C-1 and form **192b** (which, by bond rotation, would lead to **192c** and cause the hydrogens A and A' to exchange places), then the average spectrum would be of the A_2X_2 type (equation 104). At room temperature the spectrum was A_2X_2, but at

$C(CH_3)_3$, H_X, $H_{X'}$, H_A, $H_{A'}$, MgCl — inversion ⇌ — rotation ⇌

192a **192b** **192c**

(104)

–15 °C it changed to AA'XX'. Dineohexylmagnesium, as would be expected, lost its configurational stability, as far as NMR was concerned, much more slowly. It finally turned into A_2X_2 above about 80 °C. Michael Witanowski, a visitor from the Polish Academy's Institute of Organic Chemistry, investigated the configurational stability of a number of other neohexyl derivatives with different metallic elements as M.[183]

It was clear that the Gutowsky and Holm procedure was not going to help us determine the rates of inversion of configurations represented by the change of **192a** to **192c**. Complex changes in line shape were involved, and, up to that time, we were only equipped to deal with cases where AB changes to A_2. George, with his ineffable brilliance, was able to show that the line-shape changes for AA'XX' to A_2X_2 could be calculated from two invariant lines and coalescence of two different AB to A_2 systems. From this finding, he calculated a series of line shapes for the change of AA'XX' to A_2 with a rate of inversion that we could match to the experimental spectra taken as a function of temperature and thus allow determination of an activation energy of 11 kcal for conversion of **192a** to **192c**.[184]

Further, George's thesis[185] contains a wonderfully clear exposition of the density–matrix procedure for computation of line shapes as a function of exchange rates. I was able to understand the elements after a lot of work. Later, I gave a talk on the density–matrix approach and its application at the Reaction Mechanisms Conference at Corvallis. In general, it went over like a lead balloon. My only solace was that George came up afterwards and said, "You seem to have come to sort of understand that stuff."

Cyclooctatetraene

George was so sharp and outwardly sarcastic that he had a somewhat depressing effect on the less intellectually gifted graduate students. He was highly competitive with everyone, including me. At one time, we

John D. Roberts with George M. Whitesides in 1985. Whitesides is a professor of chemistry at Harvard University and is well-known for his entrepreneurship in developing commercial applications of chemistry and for his imaginative research achievements, such as the chemistry of organic compounds bonded to gold surfaces, organoplatinum compounds in homogeneous catalysis, formation of Grignard reagents, and enzymes as reagents for organic syntheses.

had a discussion about the inversion rate of cyclooctatetraene, **193**, a process that can go without, or with, a shift of bonds. The stable tub structure, **193a**, that cannot be significantly stabilized by resonance (such as in benzene) because it is nonplanar, could flatten to structure **193b** by opening up the C–C–C bond angles from 120° to 135° (equation 105). Now **193b** could flop over to **193c** without the bond shifts, which would lead to **193d**, or else proceed via **193d** to **193e**.

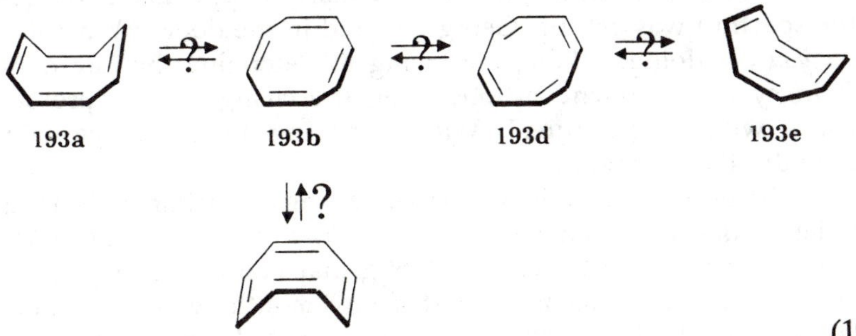

193a 193b 193d 193e

193c

(105)

I had postdoctoral fellow Don Gwynn try to make fluorocyclooctatetraene, **194**, with the idea that rapid equilibration of **194a** with **194b** (equation 106, which corresponds to **193a** ⇆ **193e**) would average $^3J_{HF}$,

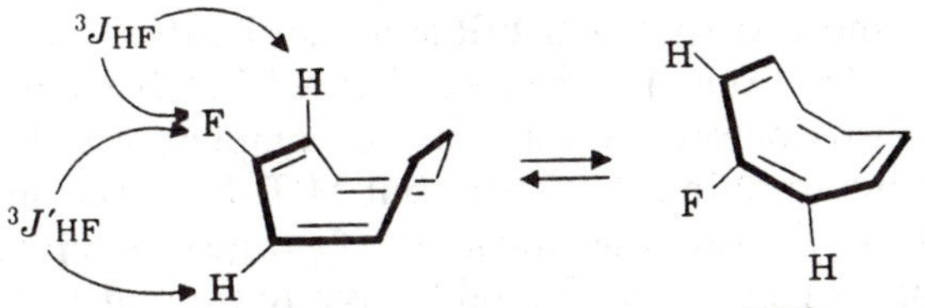

$^3J_{HF}$ $^3J'_{HF}$

194a 194b

(106)

which is across a double bond, and $^3J'_{HF}$, which is across a single bond, and give a simpler ^{19}F NMR spectrum than either **194a** or **194b** alone. The synthesis of **194** was not an easy one and was going to take some time. But then, in a flash, it came to me that the same thing could be done on cyclooctatetraene itself, with the aid of the ^{13}C present to the extent of about 1.5% in all natural terrestrial carbon compounds. Cyclooctatetraene containing ^{13}C, **195**, will have a coupling between the

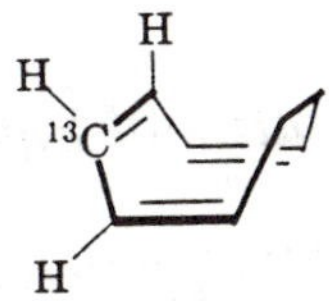

195

^{13}C atom and the directly attached hydrogen, causing the proton system to change from A_8, in the absence of ^{13}C, to approximately AXX′X″X‴X⁗X′′′′′X′′′′′′ if bond shift is slow and $AX_2X'_2$ it is fast.

This seemed to be such a good idea that I decided to run the spectra myself. I spent quite a few hours observing the very weak A_8 resonance and found that it started to become more complex below −24 °C. While I was seated at the spectrometer and feeling relatively pleased with myself, although not particularly happy about the quality of the spectra I was getting, George walked in the door. "What do you think *you* are doing?" "Oh, I'm trying to determine the rate of bond shift in cyclooctatetraene by observing the change in the proton ^{13}C satellites with temperature." With a pitying look, George said, "I already did that last week!"

Whatever we had in collective feelings of triumph were quite completely dashed when I received a *Journal of the American Chemical Society* communication to referee, stating that Frank Anet at UCLA had done the same experiment at least a few months earlier.[186] Further, Frank had devised a very clear method based on magnetic nonequivalence arising from molecular dissymmetry to determine the rate of simple inversion without bond shift (i.e., **193a** to **193c**). Amazingly, the rates **193a** → **193c** were not much different from **193a** → **193e**. With this development, we abandoned cyclooctatetraene but continued with **194a** ⇆ **194b**.[187]

George showed so much brilliance and future promise that the next time I saw Art Cope, I suggested that MIT offer George an instructorship, 18 months before he was due to complete his thesis. Art was not willing to do anything that soon, but MIT did make an offer at least 9 months before George was finished. George's acceptance removed any last vestiges of guilt that I might have felt about leaving MIT. To be sure, a few years had intervened, but MIT was going to get a man who was sure to be one of the best and most versatile chemists of the century.

George's sardonic wit can be intimidating, and some people find him hard to take. There is another, very different, side to George that many people do not encounter. A warmth and humor, quite the reverse of C. K. Ingold's journal writings (*see* pp 134–135), shows up in a special way in his letters. Another facet of George's other side is seen when he is with his children—the relationship is warm and caring.

Conformational Equilibria

One of our other major early NMR projects was concerned with conformational equilibria and equilibrations. This project took several directions, including conformations about single bonds, an area for which we used ^{1}H, ^{19}F, and later ^{13}C spectroscopy. Bruce Hawkins, a postdoctoral fellow from Ed Garbisch at Minneapolis, played a major part in this, both experimentally and in data analysis. Bruce's work on data analysis was concerned with the very practical problem of calculating even more complex line shapes than we had handled previously, with a computer program capable of making much-improved fits to the line shapes.[188] At one point, I became interested in seeing whether the rates of rotation could be calculated with a force-field procedure similar to the one used by Westheimer and Mayer[189] for the barriers to rotation in optically active biphenyls, such as **196**.

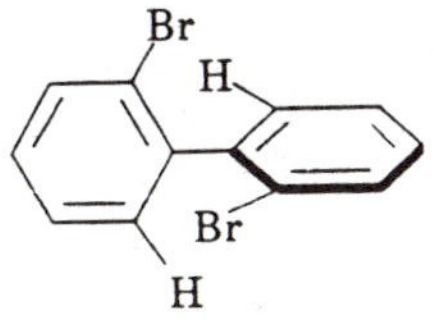

Br H Br H

196

When I was notified that I would receive the Roger Adams Medal and Prize at the Organic Symposium in Burlington, Vermont, in 1967, I wanted to talk about rotation. However, by the time I had to submit an abstract, I had not done enough work to be confident that I would have anything substantive to talk about. So I submitted an abstract based on a superb thesis by Tom Halgren on free-radical rearrangements involving interconversion of **197** and **198** (equation 107).[190–192] When we started with reactions that would generate either

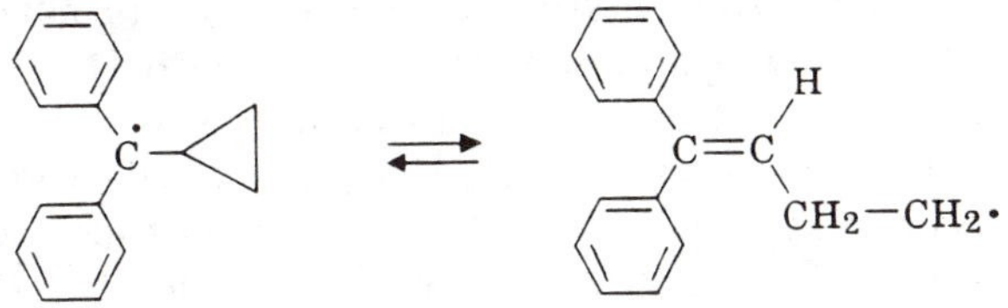

$$\text{197} \rightleftarrows \text{198: } (C_6H_5)_2C{=}CH{-}CH_2{-}CH_2\cdot \qquad (107)$$

197 198

197 or **198**, the products from the usual free-radical traps seemed to be derived almost exclusively from **198**, even though **197** would surely be expected to be the more stable free radical. Tom provided strong evidence that this expectation was indeed correct, but **198** was more reactive, even if less stable, than **197**, and only when extremely reactive free-radical traps were used did the properties of products derived from **197** increase.

I had more success with the rotational-barrier problem as the time for the symposium approached. Finally, I switched to a talk wholly different from the abstract, which was probably a mistake—it was a bit too much like my density–matrix fiasco at the Mechanisms Conference (*see* p 184). Later, I was never really comfortable enough with the rotational-barrier calculations to write them up for publication. I tried to be as rigorous as possible, using only potential functions obtained spectroscopically for stretching and bending force constants or atomic scattering experiments. I calculated a barrier for rotation about the single bond of hexachloroethane of about 12 kcal and had hopes of measuring the rates of rotation by NMR of 1,2-dibromotetrachloroethane, **199.** These hopes were based on a yet another hope that **199** would exist with detectable proportions of achiral **199a** and the chiral forms **199b** and **199c** (equation 108), and that these would have

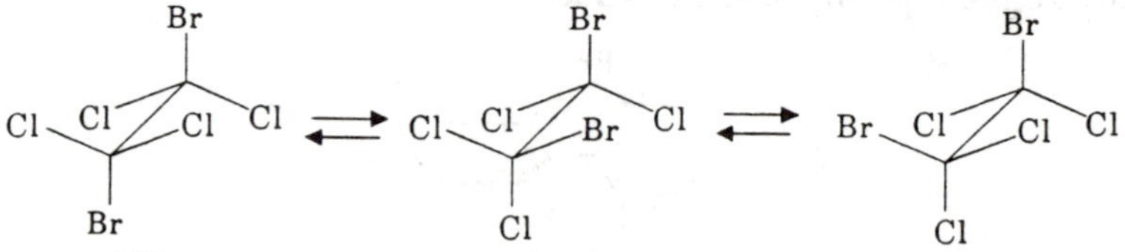

199a ⇄ 199b ⇄ 199c (108)

different ^{13}C shifts in the NMR. However, only one ^{13}C line was observed that did not change with temperature, and the project ground to an ignominious halt. Perhaps that experiment should be tried again with a 600-MHz spectrometer.

In studies of conformational analysis of simple cyclic compounds such as cyclohexane, NMR is of limited value unless some kind of substitution is made. In principle, the proton ^{13}C-satellite peaks could be analyzed as a function of temperature, but the spectra would be exceedingly complex. Frank Anet at UCLA attacked this problem in a most elegant way by making cyclohexane substituted with 11 deuteriums and one hydrogen so that, with deuterium decoupling to remove the small deuterium–proton couplings, a single sharp averaged proton resonance or two single lines will be observed if the equilibrium, **200** (equation 109), is fast or slow.[193] Data from studies like this are directly applicable

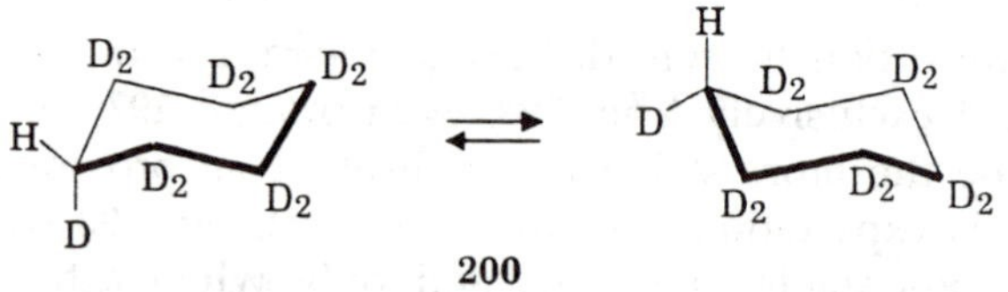

200 (109)

to cyclohexane itself, because the effect of deuterium on the rate and equilibrium of **200** should be quite small.

I decided to take quite a different approach, which utilized the fact that fluorine has much larger shift differences than hydrogen. E. J. Corey had the same idea and sent me a sample of fluorocyclohexane to investigate the cyclohexane inversion barrier. However, from my Du Pont consulting, I knew of the conversion of ketones to *gem*-difluoro compounds with sulfur tetrafluoride (equation 110).[194] I was sure 1,1-

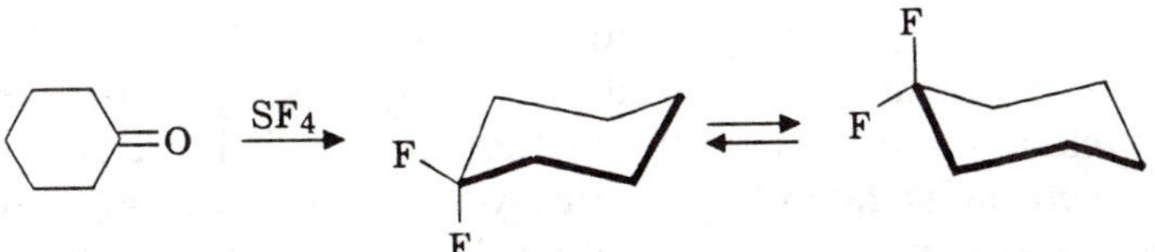

(110)

difluorocyclohexane, **201**, would be more amenable to study because there would be no problem with the position of the equilibrium, which would be the case with fluorocyclohexane. We already had experience with substituted 1,1-difluorocyclobutanes prepared from our cycloaddition products (*see* pp 100–104), and a most productive graduate student, Joseph B. Lambert (now at Northwestern), did an excellent job of investigating the large temperature effects on such fast equilibria as **202** (equation 111).[195]

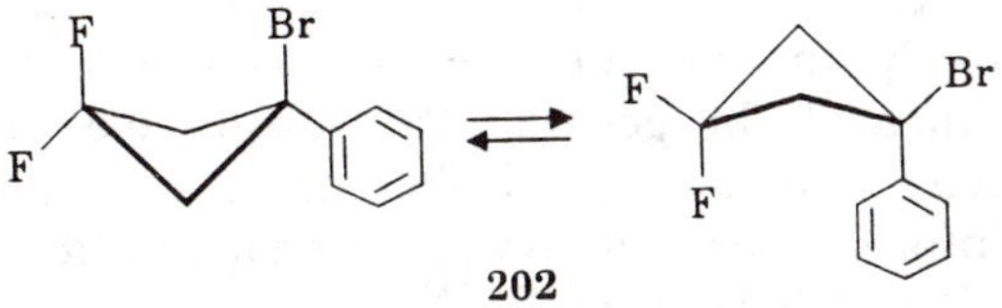

(111)

With **201**, K. Nagarajan, a very fine postdoctoral fellow from Cal Stevens at Wayne State University, was able to show that **201a** and **201b** were rapidly equilibrated at room temperature with the appearance of an A_2X_4 system in the ^{19}F spectrum, where the Xs represent the four hydrogens on the carbons directly attached to the carbon carrying the fluorine. At −100 °C, the spectrum changed dramatically to what could be excellently approximated by an ABX_2X_2' system, with a chemical-shift difference between the fluorines of 15.7 ppm and a $^2J_{FF}$ of 237 Hz.[196] We used 1,1-difluoro compounds to determine conformational equilibration rates and/or equilibrium constants for many different substituents.[197] Of special interest was the marvelous work done by postdoctoral fellow Tom Gerig (now at the University of California, Santa Barbara) on equilibration of *cis*-decalin ring systems,[198] which were especially suitable for study by using *gem*-difluoro groups as the NMR probes into conformational behavior.

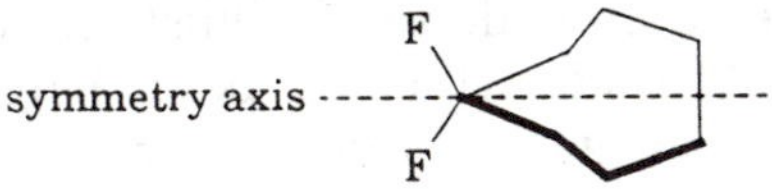

Graduate students Ed Glazer and Eric Noe and postdoctoral fellow Edgar Anderson applied the procedures to the cycloheptane, cyclooctane, and cyclodecane ring systems. The preference for *gem* substituents to occupy the unique axis position of the favored conformation of cycloheptane, **203**, meant that 1,1-difluorocycloheptane showed equivalent fluorines, even down to temperatures as low as −170 °C. We were able to change the situation by using 1,1-dimethyl-4,4-difluorocycloheptane because the methyl groups, being larger than the fluorine, were most favorably located on an axis position, leaving the fluorines to occupy a ring position beyond which they are nonequivalent, **204a–204b** (equation 112). However, pseudorotation, which

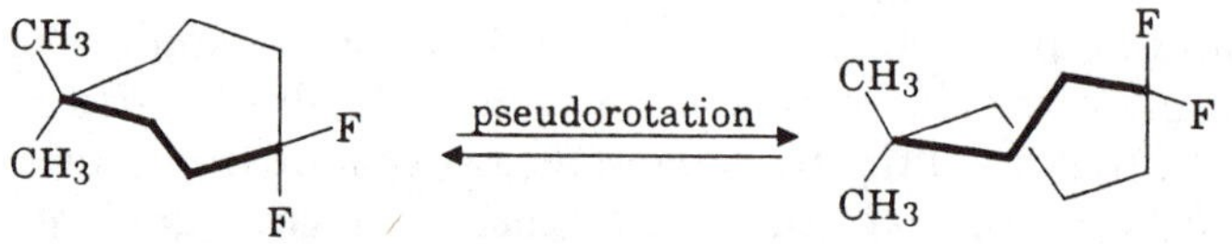

CH_3 CH_3 F F pseudorotation CH_3 CH_3 F F

204a **204b** (112)

very rapidly interchanges the positions of the carbons in cycloheptane, makes for easy interconversion of **204a** and **204b**, and, hence, averages the chemical-shift difference between the fluorines to zero. We had to go to below −150 °C to slow the interconversion down and see nonequivalent fluorines.[199] In general, the results we obtained corroborated the excellent force-field calculations of substituent effects on cycloheptane conformations and conformational equilibration by James B. Hendrickson (Brandeis University).[200]

Cyclooctane turned out to be a horse of a very different color from either cyclohexane or cycloheptane. It is a very flexible molecule and has a number of possible conformations of about the same energy. The complexity of the conformational situation is illustrated by the fact that the 1,1-difluoro derivative undergoes two transitions as the temperature is lowered. There is no $\Delta\nu_{FF}$ at room temperature, but below −100 °C the ^{19}F spectrum is of the AB type, with a small chemical shift of 3.9 ppm and $^3J_{FF}$ of 245 Hz. Then, below −150 °C, this AB pattern was found to be replaced by two overlapping AB patterns with chemical-shift differences of 14.3 ppm and 16.7 ppm, with corresponding $^3J_{FF}$ values of 240 Hz and 245 Hz, results that indicate that the fluorines get stuck in two different and almost equally populated environments on the ring. Ed Glazer and I thought we could explain the changes we observed at the lowest temperature with a twist-tub conformation that would have "inside" and "outside" *gem*-difluoro positions, **205**, and we had some encouragement on this from molecular mechanics calculations. At −120 °C, **205a** and **205b** could be in rapid equilibrium (equation 113) but remain nonequivalent as far as NMR is concerned.[201,202]

In 1965, at an organic chemistry symposium in Tempe, Arizona, I gave a talk on these equilibrations. To illustrate what was involved, I had molecular models made with 1-m bond lengths, and with balls about 6 inches in diameter representing atoms. When you make cycloheptane and cyclooctane with bonds this long, they are so large inside fluorines

outside fluorines

205a ⇌ **205b** (113)

and have so many degrees of freedom that they cannot be handled by one person alone, so Tom Gerig got up on the stage, and we worked these together. It must have been a spectacular sight, but whether it contributed to more than the amusement of the audience, I have no

John D. Roberts and postdoctoral fellow, Tom Gerig, struggle with demonstrating how pseudorotation causes magnetic equivalence of groups attached to cycloheptane at the Organic Chemistry Symposium in Tempe, Arizona, 1965.

way of knowing. Later, Frank Anet showed that our results on cyclooctane could be interpreted in a different way, in terms of equilibrating chairlike conformations,[203] and I am sure that this interpretation was an improvement.

With 1,1-difluorocyclodecane, we were able to slow equilibration below −130 °C, and the results we obtained on this system[204] fitted very well with the prior work and speculations of Vlado Prelog at the

Eidgenössiche Technische Hochschule (ETH) in Zurich,[205] with the most favored conformation being **206**. Almost all of our earlier investi-

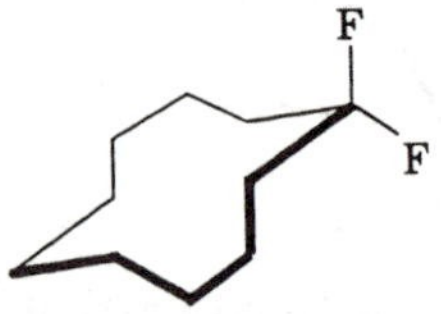

F

F

206

gations of conformational equilibration were summarized in a review article published in 1966 in *Chemistry in Britain*,[197] as a result of my being Centenary Lecturer of the Royal Chemical Society. On the whole, it seems that this work made less of an impact than I had hoped. Perhaps, it was poorly timed; but, more likely, it was because there was always a lurking suspicion that the results we obtained by labeling conformations with fluorine could not be translated directly to all-hydrogen compounds. Obviously, some suspicion is warranted, because with cycloheptane and cyclodecane, the fluorines acted as though they were occupying favorable positions. On the other hand, where comparison data are available on rates and activation barriers for conformational equilibrations, the results obtained from the fluorine derivatives and the hydrogen compounds are in very good agreement.

Natural-Abundance ^{13}C NMR Spectra

The conformational program was well along, and I was becoming interested in the potential of natural-abundance ^{13}C NMR spectra. Some interesting NMR work had been done on the ^{13}C resonances of compounds enriched with ^{13}C,[206] and we had even used its effect on the proton NMR spectrum as an indirect means of using ^{13}C as a tracer.[207] But, clearly, to be really useful, ^{13}C NMR had to be run at the natural-abundance level on ordinary organic compounds. The problem was the old bugaboo, signal-to-noise ratio.

The great NMR spectroscopist and developer of NMR imaging, P. C. Lauterbur, had shown that it was in fact straightforward to take NMR spectra using natural-abundance levels of ^{13}C, but the procedure used the rapid-passage, dispersion-mode technique[208] that was, in my hands at least, difficult to make work and sacrificed resolution to a very substantial degree. Further progress was made by my great friend and NMR consultant David M. Grant at the University of Utah, who used Lauterbur's method and also greatly simplified the spectrum by proton decoupling. About this time, computer technology and applications to data acquisition were coming along rapidly. Of special interest were computers of average transients in which one could repeat the taking of an NMR spectrum, adding the signal from one spectrum to those accumulated previously and, in the process, having random noise tend to average to zero.

The procedure corresponds to human experience. If you can't understand what someone says in a noisy room, you ask the person either to speak more slowly or else to say it again and again, until you understand. In either case, you are doing signal averaging and your brain is the computer of average transients. Unfortunately, the ratio of signal to noise does not increase linearly with the overall time of observation. Improvement depends on the square of the time involved. A factor-of-two improvement requires a longer time by a factor of four.

I was sure that some way of achieving useful ^{13}C NMR spectra could be devised, and I submitted a proposal to the National Institutes of Health for the purpose. It was a bit of a gamble, and even though it apparently got good priority scores, the funding was held up for a couple of years. Anyway, the signal-averaging technique looked promising enough to justify the purchase of a small computer of average transients. However, signal averaging can solve the problem of taking natural-abundance ^{13}C spectra only if one has a spectrometer that is sufficiently stable and free of drift so that a difficult spectrum can be taken over and over again, for hours or even days on end.

Another approach would be to increase the intensity of the signal. One way to do this would be to lower the temperature, because lower temperatures increase the equilibrium magnetization. This approach has obvious disadvantages if a large gain is necessary. Other approaches involve magnetization transfer, and one such method is to irradiate the electrons of stable free radicals at their resonance frequency and have some of that magnetic excitation transferred to ^{13}C nuclei to enhance the ^{13}C resonance signal.

This technique was demonstrated to me, at a low magnetic field, by Rex Richards (later Sir Rex) at Oxford University; it clearly had promise, but there were at least two major problems. First was the fact that, even for 15-MHz ^{13}C NMR, the electron frequency is very high, and techniques of irradiating chemical samples at those frequencies had not been developed. Second, it was clear that magnetization transfer would not be equally efficient for all types of compounds. One of Harden McConnell's excellent graduate students, Alvin Kwiram, was willing to try, but I kept hoping that some more generally useful technique would come along.

Being at the right place at the right time, at this juncture, meant perusing the ads in *Scientific American*. At just the right time, I found an advertisement by Hewlett–Packard of a frequency synthesizer, one of incredible (for then) frequency stability, with a bank of push buttons to select the desired frequency and the promise of being able to perform a frequency sweep driven by the output of an external ramp generator. These instruments were expensive, \$10,000, but they did things that I had not been aware could be done. In conjunction with a computer of average transients, which I already had, they seemed to have the potential for helping to get ^{13}C spectra at the natural-abundance level. I ordered a frequency synthesizer from Hewlett–Packard forthwith.

Then, I contacted Varian and asked if they would be willing to try to use it in a ^{13}C spectrometer. The initial response was optimistic, but this was followed by a flat, "It won't work; phase problems." Not knowing about phase problems, I said, "That's too bad. Well, I'll just have to try it myself; maybe we can get it to work." I'm not sure what transpired at Varian in the interim, but they shortly called up and said that if I would send the synthesizer up to them in Palo Alto, an effort would be made to build a spectrometer to utilize it, and would I put in an order. I was happy to do so. At Varian, Vern Berger was assigned to the building of our digital-frequency sweep (DFS) spectrometer, to operate at 60 MHz with a field-frequency lock on protons in the sample. This lock precluded its use as a proton spectrometer, but that was okay with me—^{13}C or bust!

It took quite a while to build the spectrometer, and delivery was finally made in the fall of 1966. Just a short time before, a new graduate student, Frank Weigert, had arrived from MIT, where he had done undergraduate research with George Whitesides. Frank was a small, very thin, black-haired, and hawk-faced young man with rare energy and determination. His first choice of problem was to study ^{15}N spectra, but with this new spectrometer, he was more than willing to work on ^{13}C. It was clearly the right place and the right time for him, and he was the right person at just the right time for me.

Frank took over the DFS spectrometer and lived with it day and night. His productivity was legendary. He completed his thesis work in 2 years and 8 months, with 16 papers resulting from his research. He made a broad sweep through natural-abundance ^{13}C NMR, starting with an analysis of the very complex proton-coupled ^{13}C spectrum of benzene, an $AXX_2'X_2''X'''$ spin system.[209] At the outset, we did not have broad-band proton decoupling. Frank's work was done with proton coupling, which enabled him to obtain many $J_{^{13}CH}$ coupling constants along with chemical shifts. We also measured many ^{13}C to ^{13}C couplings at natural abundance, the first time this had been done.[210] Just months before the DFS spectrometer was delivered, the prospect of taking routine ^{13}C spectra at all was iffy, and here we were, reliably measuring ^{13}C peak positions at 1/50 of the natural-abundance level.

Not only was this spectrometer as good as it was promised to be, it was even better than we had hoped! It was a massive piece of instru-

John D. Roberts inspecting an early natural-abundance ^{13}C spectrum taken by Frank J. Weigert with the Varian DFS NMR spectrometer in 1967.

mentation with built-in frequency synthesizer, digital-sweep programmer, field-frequency lock channel, and a signal-averaging computer, in addition to the usual Varian spectrometer appurtenances, including a power supply packed with the good old 304TL tubes. Varian and Vern Burger deserved immense credit for the development of the DFS spectrometer. Surprisingly, it did not become a production model, but its special features were implemented, in one way or the other, in later spectrometers.

A real breakthrough came when Frank was able to implement broad-band noise decoupling.[211] Now, peaks that were often extremely complex because of proton couplings became single peaks. The usual expectation would be that proton decoupling of a ^{13}C resonance that was split into a doublet by a single proton, as would be expected for $Cl_3{}^{13}C–H$, would give a resonance twice as high as one of the doublet peaks. In fact, because of the operation of the nuclear Overhauser effect (NOE), the resulting peak is *six* times the intensity of one of the doublet peaks! This increase happens because some of the energy used to decouple the protons leaks over into the ^{13}C magnetic states and enhances the signal intensity of the ^{13}C resonances.

After development of stable spectrometers, nothing did more to make ^{13}C spectra routine than broad-band decoupling, in which all of the proton couplings could be removed and the signals made much easier to detect because of the NOE. Broad-band proton-decoupled ^{13}C spectra are collections of sharp, strong, and more or less equally intense resonances from which all of the proton-coupling information, which might be used to identify the peaks, has been removed. That information can be regained if needed but, of course, at some cost in signal intensity.

The simplicity of decoupled spectra placed great emphasis on determination of chemical shifts for identification purposes, and there was intense pressure on deciding what compound to use as the zero for chemical-shift reference. We had used different standards, particularly carbon disulfide. Carbon disulfide gave a resonance toward the high-field end of the common ^{13}C signals, which gave a parts-per-million scale going naturally from 0 to more positive values in accord with the definition of the shielding parameter.

However, Leroy Johnson, probably the most skillful NMR spectrometer operator of all time, pushed very hard for tetramethylsilane (TMS) to be the ^{13}C standard as it had become the proton standard, and TMS won out. The result was to go with a positive scale toward higher magnetic field (even though this was in defiance of the definition of shielding). As a result, anyone using the data from our papers has to correct them to correspond to the TMS scale.

With broad-band decoupling, it became possible to obtain and interpret the ^{13}C spectra of quite large organic molecules.[212] Frank Weigert had run the coupled spectrum of cholesterol. The spectrum was very complicated because of the plethora of proton–carbon couplings, and the signal-to-noise ratio was not very favorable, even after several hours of signal accumulation. In contrast, the first decoupled spectrum showed 25 distinct resonances for the 27 cholesterol carbons. There was surprisingly little overlap of the peaks, even though cholesterol has many carbons that are very similar to one another in chemical character.

At this point Frank was nearing the end of his thesis work. When he showed me the cholesterol spectrum, he said, "I guess it will take you 10 years to identify which carbons those resonances come from." I replied that we would see about that. In fact, postdoctoral fellows Hans Reich (Ph.D. with Don Cram at UCLA and now at Wisconsin) and Manfred Jautelat, using a variety of techniques, had the assignments well in hand in less than 6 months! The paper describing their work on the ^{13}C NMR of steroids,[213] according to *Science Citations*, became the most cited of our publications. Only very few of our less certain assignments were later changed by use of more sophisticated techniques. We studied the ^{13}C NMR of many kinds of natural products, perhaps most notably through the work of Doug Dorman (postdoctoral fellow, now at Eli Lilly) on carbohydrates,[214,215] and by Jim Nourse (graduate student, now in computer science at Stanford and author of a best-selling book[216] on solving the problem of Rubik's Cube) on the very difficult problem presented by erythromycin-type antibiotics.[217]

The DFS–60 spectrometer was super state-of-the-art in 1966, when it was installed. However, its procedure for achieving good signal-to-noise ratio (S/N) was simply to add conventional frequency-sweep spectra, one on top of the other, until a satisfactory resultant was obtained. The flaw in this procedure is that in a broad sweep, where there are just a few sharp resonance lines, the spectrometer spends almost all of its time averaging the noise in the baseline between the peaks. In such situations, the pulse Fourier-transform procedure for taking spectra becomes very efficient. What is done is to excite all of the ^{13}C nuclei at once with a very short, powerful burst of radio waves at the ^{13}C resonance frequency. Then, the spectrometer's rf receiver detects the free-induction decay (FID), which is the combined relaxation wiggles of all of the excited ^{13}C nuclei. If there are differences in chemical shifts, then the FID becomes an interference pattern. The nuclei usually have different decay rates, a condition that adds to the complexity of the FID. The FID is the signal as a function of time that the Fourier transform can convert to signal as a function of frequency. In the process, the decay information, expressed by the relaxation time, T_2, shows up as line width. The faster the decay, the broader the line. Signal averaging is achieved in this mode of operation by averaging the FIDs before they are transformed.

The advent of pulse Fourier-transform NMR spectroscopy in our laboratory had its traumatic side. In the first place, I remember meeting with Varian and being told, "It won't work for ^{13}C; too long relaxation times." When I persisted, I was told there was no way it could be implemented effectively on the DFS–60. After further persistence, they agreed to supply a ^{13}C unit complete with computer for the HR–220 spectrometer that we were purchasing to set up as a southern California

NMR facility for the National Science Foundation. The HR–220 had been developed at Varian at the behest of Bill Phillips of Du Pont and was the first commercial spectrometer with its magnetic field supplied by a superconducting solenoid. The HR–220 was a frequency-sweep machine like the DFS–60, and the ^{13}C Fourier-transform unit had to be developed from scratch.

Nonetheless, Varian promised delivery in 6 months. When it became clear they were not going to make it, we had to have our purchasing department badger them for specific monthly progress reports. The unit was finally delivered, 5 months late. With the 5-mm tubes that were all we could use in the HR–220, the sensitivity was only 55% of what we could do with 10-mm tubes on the DFS–60. This caused consternation because it meant we had to run samples almost four times longer in the HR–220 than in the DFS–60. To make matters worse, the HR–220 was a shared instrument, which we could only use 1–2 days a week for ^{13}C.

I wanted desperately to move the ^{13}C Fourier-transform unit to the DFS–60, where we could use it full time. Varian said flatly that it could not be done. At that time, I had a French postdoctoral fellow, Jean-Yves Lallemand, working with me, who was a physical organic chemist, and he said that he and Bruce Hawkins wanted to do the job. When I asked him about his electronics experience, he muttered something about having worked on FM radio sets. I was concerned. It was a big job, the frequencies were off by a factor of almost four, and we needed to implement a deuterium-lock channel.

Fortunately, just then, Chris Tanzer, a former postdoctoral fellow with Jack Richards who worked for Bruker Magnetics, came on the scene, eager to get something other than Varian NMR equipment into Caltech, especially if it was Bruker equipment. Chris surveyed the situation and proclaimed Bruker's eagerness to help by providing us with modules for the lock channel as well as technical advice. In just a few months, the Lallemand–Hawkins–Tanzer team had us taking ^{13}C Fourier-transform spectra with good signal-to-noise and high reliability! So we changed the name from the Varian DFS–60 to The Brukarian Pulse–Fourier-Transform spectrometer.

One interesting direction our work took was on the ^{13}C spectra of polypropylene polymers. As the result of a ^{13}C NMR course I gave in Portugal in 1968, a Dr. Gatti suggested to Adolfo Zambelli of Milan that he should propose we collaborate on investigating the ^{13}C NMR of stereoisomeric polypropylene polymers. Zambelli could supply the polymers, and did, in January 1969. I was in the right place at the right time, but I was not as responsive as I should have been, probably because I did not think we could get good enough spectra on such high-molecular-weight materials. I did not push W. O. Crain as rapidly as I should have to give Zambelli's samples high priority. We submitted our paper on January 3, 1970, hoping to publish the first ^{13}C spectra of such materials,[218] but we got scooped by Leroy Johnson and Frank Bovey, whose paper came out in early 1970 and showed that ^{13}C NMR was to be an important way of studying polymers.[219] I did not pursue this area of work at the time; perhaps I should have.

However, in 1971 a prestigious New York law firm approached me with respect to a suit that Montecatini Edison of Milan had brought against a gaggle of major United States companies for infringement of G. Natta's patent for preparing isotactic polypropylene. The work that led to this patent was so striking that Natta had shared the Nobel prize in Chemistry with K. Ziegler in 1963. One of the key issues in this lawsuit was the degree of isotacticity of the polymer. The patent, which Natta had assigned to Montecatini, claimed the polymer to be "substantially isotactic".[220] There is always a question about just what "substantially" means in any case but, even if no more than because of scientific curiosity, one would like to know the degree to which this polymerization is stereospecific.

That it showed any high degree of specificity was surprising, because the ingredients used to make the polymerization catalyst were relatively simple substances, $(CH_3CH_2)_2AlI$ and $TiCl_3$ in toluene. A priori, these ingredients would hardly be expected to produce polymer at 70 °C with a particular stereochemical configuration, almost as if it were a biochemical peptide synthesis. Further evidence against high stereospecificity was provided by Paul J. Flory, one of the greatest polymer chemists of all time, who received a Nobel prize for his many contributions to polymer science in 1974. Flory, in company with my future colleague, John D. Baldeschwieler, reported that the degree of isotacticity was not likely to be more than 85–90%.[221]

The Montecatini lawyers wanted me to determine by ^{13}C NMR the isotacticity of the various commercial samples, as well as of materials prepared as described in the Natta patent. The project had a strong enough basic-science component to be interesting, but it would have to involve use of the institute's research facilities for consulting in a way that was against the rules. An exception was possible only because our spectrometer was uniquely able to do the job. The spectra were run with Bruce Hawkins' help. It was a time-consuming and difficult process, because there were so many samples, they were slowly soluble, and they had to be taken in deuterio-*sym*-tetrachloroethane at 140 °C.

The ^{13}C spectra showed that the polymer Natta claimed to be substantially isotactic was, in fact, highly isotactic, with probably fewer than two stereochemical errors for every 50 or so monomer units on the average. The law suit was a very large one, involving as it did Dart Industries, Avisun, Phillips, Esso, Amoco, and Eastman Kodak. The lawyers were very slow about coming to trial. It seemed that, except for the substantial legal expenses, it was, in fact, in the defendants' interest to come to a late decision, even if they could not get the complaint dismissed.

Whether that was true or not, the preparations dragged on. In one of the law offices there were said to be 250,000 documents relating to the case, which was to be tried before a single federal judge in Wilmington, Delaware. I was essentially ready for the trial in the spring of 1973, but the defendants wanted to do more discovery and, later, to take more depositions at Montecatini, in Italy. The judge allowed them to take a lot more time. By the end of the summer of 1974, the report was that only about two-thirds of the necessary depositions were in hand.

Apparently, the judge finally lost patience. The trial was to begin early in 1975, with my testimony to come in the rebuttal phase. I was all prepared for the issue of the degree of isotacticity but, after the plaintiff had laid out the essence of the case against the defendants, the suit was settled out of court. The terms were not revealed to me, but apparently they were quite favorable to Montecatini. It was not clear how effective our ^{13}C NMR data were in influencing the decision to settle. They might not have been as important as the analyses of the prior defendant depositions I worked with the attorneys to develop in preparation for the trial, which were very heavy on proton NMR of polypropylene. In fact, I was told that the defendants were shocked when they heard, in a pretrial conference with the judge, that NMR evidence given in one of their own depositions on the proton NMR of polypropylene was going to be used as proof of infringement. They then wanted to claim that NMR was not good for structure determinations at all, but I guess they finally decided that that argument would not fly. The case was legally, scientifically, and financially interesting. I was glad to have had the experience.

^{15}N NMR Spectroscopy

Another NMR area that was ripe for plucking was natural-abundance ^{15}N spectroscopy. Long before we got into ^{13}C NMR, we had started a program using samples enriched in the ^{15}N isotope to determine the general features of this form of spectroscopy. There was little hope then of natural-abundance ^{15}N giving an observable signal. The magnetic moment of the nucleus is about one-tenth that of protons, and the proportion of ^{15}N in naturally occurring nitrogen is 0.31%, while the ^{1}H abundance is 100%. These comparisons make natural-abundance ^{15}N about a millionth of the signal strength of protons, something like the difference in intensity of starlight and sunlight. But with samples enriched to 90% or so in ^{15}N, it was possible to get signals with our original 60-MHz spectrometer operating at 6 MHz.

The basic difficulty was that compounds enriched in ^{15}N are expensive. At the outset, we made a series of compounds in which A would be made and its spectrum taken; then A would be converted to B and the spectrum of B taken, and so on. Thus, one such series started with nitric acid as A and nitrobenzene as B. From there, well-known procedures allowed for preparation of nitrosobenzene, phenylhydroxylamine, azobenzene, azoxybenzene, hydrazobenzene, and aniline as conversion products. These substances displayed an amazing range of chemical shifts that spanned about 900 ppm, whereas the ordinary shift range of protons is only about one one-hundredth as much.[222]

Thanks to the efforts of W. E. Moffit (a brilliant British theoretical chemist and a student of C. A. Coulson, who became a Harvard professor but died at a very young age of a heart attack), I came to have a rudimentary knowledge of the *second-order paramagnetic effect* on chemical shifts. Thus, we had a useful rationale for the observed shifts.[223] Performance of the synthetic work with expensive enriched ^{15}N took nerves of steel, but Joe Lambert and Gerhard Binsch were more than equal to the task.

Accumulating evidence indicated a useful parallelism between $^{1}J_{^{13}CH}$ coupling constants and the hybridization of the carbon. The important factor is the degree to which the carbon 2s orbital involves the C–H bond. With no involvement of the 2s orbital, $^{1}J_{^{13}CH}$ is expected to be zero. By Pauling's criteria,[224] the series CH_3–CH_3, CH_2=CH_2, HC≡CH should have 25, 33, and 50%, respectively, of 2s character to the C–H bond. The respective $^{1}J_{^{13}CH}$ values are 125, 156, and 249 Hz. We thought it would be interesting to see if there is a similar relationship for $^{1}J_{^{13}C^{15}N}$ values, but, at the time, this determination could be done only with compounds isotopically enriched in *both* ^{13}C and ^{15}N. Gerhard Binsch undertook the synthesis of these breathtakingly expensive compounds, where the relay technique was not only useful but essential. In all, he made 10 compounds labeled with both ^{13}C and ^{15}N and measured their $^{1}J_{^{13}C^{15}N}$ values. The correlation with the postulated 2s orbital contributions to the bonding was a bit ragged, but the trends were clearly there.[225]

Frank Weigert used the DFS–60 in 1969 to take the first ^{15}N spectrum at the natural-abundance level. His sample was neat hydrazine, which is 93% nitrogen and about the most favorable case one could imagine. The S/N was actually quite good, and the experience got us off to a start with natural-abundance ^{15}N spectra. Robert L. Lichter, a very fine postdoctoral fellow from Wisconsin, played a critical role in getting this program going. Among many other things, he defined the conditions for taking the spectra and worked on the development of the nuclear Overhauser effect (NOE) for ^{15}N resulting from proton decoupling.[226] With ^{13}C proton decoupling, the associated NOE makes the carbon signals stronger (*see* p 196). However, because the ^{15}N nuclear moment has the opposite sign to that of ^{13}C, the NOE actually reverses the direction of the signals, making them go downward instead of upward, as with ^{13}C. At maximum NOE efficiency, the absolute value of the ratio of the intensity of the negative signal to the same signal without NOE is about 30% greater than for ^{13}C. A low-efficiency ^{15}N NOE can be of no help at all. Indeed, it can even cancel out the normal upward signal.[227]

Spielvogel and Purser[228] had reported that there were some useful parallelisms between ^{13}C and ^{11}B chemical shifts in analogously constructed compounds. Bob Lichter investigated the possibility of a corresponding correlation between amines and hydrocarbons. The idea was to plot the ^{13}C shift of hydrocarbons like $CH_3CH_2CH_2{}^{13}CH_3$ and $CH_3CH_2{}^{13}CH_2CH_3$ against the corresponding ^{15}N shifts of $CH_3CH_2CH_2{}^{15}NH_2$ and $CH_3CH_2{}^{15}NHCH_3$, respectively. Bob found marvelous correlations—the primary amines fell along one line and the secondary amines fell along another line with a slightly different slope.

The only compound studied that was far off the line was aniline, 207, which would be expected to be exceptional because its unshared pair is delocalized over the benzene ring (equation 114), something not possible for toluene, **208**.[229] The sensitivity of the spectrometer was not

$$C_6H_5{-}^{15}NH_2 \longleftrightarrow {}^{\ominus}{:}C_6H_5{=}^{15}NH_2^{\oplus} \qquad (114)$$

207

$$C_6H_5{-}^{13}CH_3$$

208

great enough to detect the ^{15}N resonances from tertiary amines, with their long relaxation times. It was some years before we were able to show that these gave the same kind of correlation, but not with quite the same accuracy.[230]

Sensitivity accounts for a lot in ^{15}N NMR, but availability of time to run the spectra is almost as important. For this reason, I was already planning, in 1970, for a wholly dedicated ^{15}N spectrometer to operate at the highest possible field. I knew that using a high field would increase sensitivity, but also increase the cost. Varian and Bruker were the only possible vendors for what I had in mind. In 1970, Varian quoted a 300-MHz proton machine with ^{15}N capability in 10-mm tubes at $366,000. Clearly, they were eager to develop such an instrument for its potential proton and ^{13}C capabilities, but I don't think that they cared much about natural-abundance ^{15}N. Delivery was promised for 18 months after receipt of the order. The price was high, but at least it gave me something to work on in terms of an NIH proposal.

At the end of 1971 we began to have a close relationship with Bruker, through Chris Tanzer's help with making the DFS–60 into a pulse Fourier-transform spectrometer. When they were asked to quote on ^{15}N capability, they came through with the same kind of proposal as Varian for a 270-MHz proton machine with a ^{15}N probe for $300,000. The NIH would only make a grant of $150,000 for the spectrometer, so something had to give.

Over the years I had become progressively better acquainted with David M. Grant at Utah, a student of Herb Gutowsky and a pioneer in natural-abundance ^{13}C NMR, especially of hydrocarbons. David and I talked frequently about how best to solve the sensitivity problem. He convinced me that the high-field approach would probably not give as much improvement as using larger samples. "Use a milk bottle!"

At first blush, this seemed pretty far out in terms of resolution. Remember, we were running protons in 5-mm tubes, ^{13}C and ^{15}N in 10-mm tubes. The thought of using 25–30-mm tubes at first conjured up visions of grave difficulties in obtaining the necessary homogeneity. However, Dave pointed out that the shift range of ^{15}N was large and the requirements for 1-Hz resolution for ^{15}N would be one-tenth of that for protons because of the difference in resonance frequencies. A 30-mm tube could have 10–20 times more sample in the active coil volume than a 10-mm tube, and it would be beyond the technology of the time to achieve a 10–20-fold increase in sensitivity by going to higher fields.

So, in the summer of 1972, we asked Varian and Bruker to submit new proposals for a large-sample ^{15}N–^{13}C spectrometer with a superconducting solenoid for $150,000 total or less. Varian proposed a 20-MHz frequency for ^{15}N, 30-mm sample tubes, 12–14 months delivery, for $146,000. Bruker suggested using a 15-MHz observation frequency for ^{15}N and 25-mm tubes to start, with 30-mm to come later; all for $131,500.

We put in our order with Bruker, where it seemed to sink at once to the bottom of their priority list. In October 1973 the Caltech purchasing group began to write nasty letters, but not much seemed to happen until mid-1974 when Craig Bradley, a chemistry Ph.D. who had worked some years earlier with Frank Anet at UCLA, where he had considerable experience with spectrometer construction, was assigned by Bruker to work on the project. Craig decided that the easiest way to go was to take a Bruker WH–90 spectrometer console and double the frequency to 180 MHz for the proton-decoupling channel, run ^{15}N at 18 MHz and ^{13}C at 45 MHz, and use quadrature detection and a superconducting solenoid with a 3-inch bore. Craig was extraordinarily skillful, even if almost unbearably abrasive. In January 1974 he finally delivered

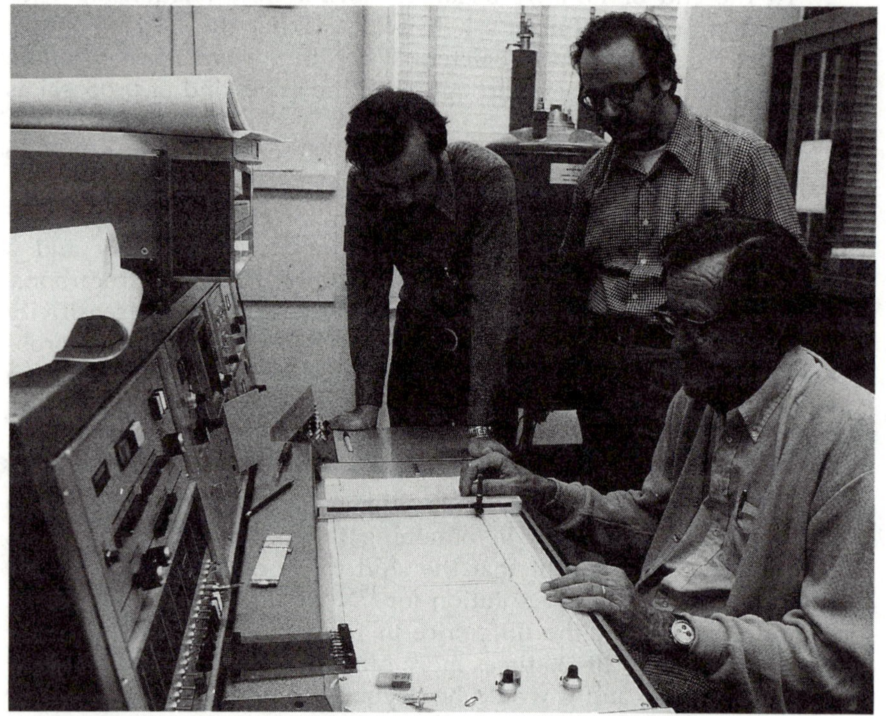

The Bruker WH–180 NMR spectrometer wholly dedicated to use for ^{15}N, mostly at the natural-abundance level. (Left to right) postdoctoral fellows Devens Gust and Donald Giannini, and John D. Roberts.

our WH–180. Final assembly was done in Switzerland. I was a bit put out by the fact that Bruker allowed Ed Randall of Queen Mary College, London, to be the first to do research on the spectrometer we were paying for, without so much as a by-your-leave. We got this straightened out and published a joint paper on the shift of liquid dinitrogen in early 1975.[231]

The WH–180 was a fantastic spectrometer. As with the DFS–60, it worked not only better than we had specified, but better than we had even hoped. To be sure, we had to have two changes of magnet because of drift in the magnetic field. Further, Craig had done very little documentation of the circuit changes he had made in the console. We had to write our own instruction manuals, and there were other major and minor problems, but after the somewhat rocky start, our third magnet held its field for 8 years. There were a minimum of electronic problems with the console.

The 25-mm tube size turned out to be optimal. The specified 30-mm sample probe was finally supplied, after a long wait. Despite having 50% more sample, the homogeneity was not good enough to increase S/N, so we went with the 25-mm size henceforth.

The luxury of having a dedicated ^{15}N spectrometer has to be experienced to be appreciated. We could, and did, run spectra over as long as 72 hours with resolution at least as good as 1 Hz. Of course, there was a problem with sample size. The total volume had to be at least 18 mL and the concentration of many nitrogen compounds had to be on the order of 1 M to get a spectrum in 2–3 hours. One problem was that the T_1 relaxation time of ^{15}N was often very long, especially when there are no hydrogens bonded to the nitrogen being observed. Thus, *N*-methylpyrrolidine, **209**, has something like a 1000-s or even

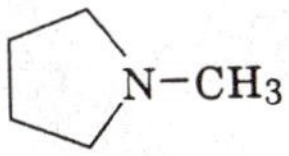

N−CH_3

209

longer T_1. Such a T_1 means that you can only pulse the ^{15}N nuclei on the order of once per hour—a process hardly conducive to efficient use of the spectrometer. Fortunately, T_1 diminishes with increasing molecular weight, which means that much lower concentrations can be used, if the time between pulses can be shortened.

Two highly skilled and productive postdoctoral fellows, Richard Moon and Devens Gust, made a superb survey of possible applications of ^{15}N NMR to medium-molecular-weight, biochemically important compounds. They showed that useful spectra could be obtained of molecules such as vitamin B_{12}; enzymes, including lysozyme and trypsin; protamines, such as salmonine; and yeast *t*-RNA.[232] The grams of samples used surely caused true biochemists to wince, but the demonstration of utility was there.

An important finding was the extreme variability of the NOE, even within the same molecule. Thus, with lysozyme, the peptide nitrogens in the backbone gave upright peaks, although those from the arginine and lysine nitrogens were sharply downward. The difference is in the degree and speed of molecular motions associated with the backbone nitrogens relative to the side-chain nitrogens. The NOE turns out to be quenched when the molecular motions are slow. The ^{15}N spectra of solutions of yeast *t*-RNA show the same kind of phenomena.

Cartoon by Ingeborg Schuster showing the Roberts research group in 1978, pondering the latest results in ^{15}N NMR with Schuster at the board used for displaying spectra.

Some peaks are upright, some downward. When the temperature is increased, the NOE tends to increase and make some of the upward peaks go just so far downward as to almost disappear in the baseline noise. This effect has obvious utility in studying molecular motions of large molecules. With the help of Alex Wei, an excellent undergraduate researcher now doing graduate work in chemistry at Harvard, we have recently made considerable progress in quantifying the ^{15}NOE for small molecules as a function of structure and solvent viscosity.

The Moon–Gust paper was a landmark that gave me great satisfaction because it showed what we wanted to show—that ^{15}N spectra could have an important place in the study of biochemically interesting molecules of substantial molecular weight.

As with ^{13}C NMR, we did a lot of work (more than 100 papers) on various applications of ^{15}N NMR, the largest part of which was primarily directed to broadening the base of understanding of shifts, couplings, relaxation times, and exchange rates, all of which are vital to the practical use of this form of NMR. As proud as I have been of our work on ^{13}C NMR, the ^{15}N research has, in one way, been more rewarding because we were able to solve some important chemical problems through ^{15}N NMR.

For almost all of these problems, we have taken advantage of the fact that nitrogens, which are doubly or triply bonded and also carry unshared electron pairs, show sizable shift changes when the unshared pairs are hydrogen-bonded or otherwise complexed, and a large shift change when protonated (equation 115). These shift changes are man-

$$\text{C}_5\text{H}_5\text{N:} \xrightarrow{CH_3OH} \text{C}_5\text{H}_5\text{N:}\cdots\text{HOCH}_3 \quad +18\text{ ppm}$$

$$\text{C}_5\text{H}_5\text{N:} \xrightarrow{HCl} \text{C}_5\text{H}_5\overset{\oplus}{\text{N}}\text{:H} + \overset{\ominus}{\text{Cl}} \quad +85\text{ ppm} \qquad (115)$$

ifestations of reducing the second-order paramagnetic effect[223] as the result of making it less favorable for mixing certain excited-state wave functions into the ground-state wave function of the molecule. These excited-state wave functions more or less correspond to excitation of one of the electrons of the unshared pair into the p–π orbitals of the unsaturated nitrogen, known spectroscopically as $n \rightarrow \pi^*$ transitions. Any contribution of such excitation makes for some increase in overall stability by having an unpaired electron circulate around the ^{15}N nucleus, thereby causing a marked paramagnetic effect on the chemical shift.

Because the degree of stabilization depends linearly on the strength of the applied magnetic field, the second-order paramagnetic effect, like other chemical-shift influences, is proportional to the size of the magnetic field. Interestingly, the second-order paramagnetic effect operates oppositely to conventional wisdom in these situations. Usually, when an atom becomes positively charged, its resonance is expected to shift *downfield* (or up frequency). Indeed, the shift of the nitrogen of ammonia does just that; it moves downfield by 8 ppm. However, quenching of the second-order paramagnetic effect results in a much larger *upfield* effect, even though the nitrogen involved becomes positively charged. Because the second-order paramagnetic effect involves mixing excited-state wave functions into the ground-state wave functions, it is not surprising that ^{15}N chemical shifts often correlate with the color of the compound they represent. Thus, the ^{15}N resonance of nitrobenzene (light yellow) is several hundred parts per million separated from that of nitrosobenzene (green).

A typical example of use of the second-order paramagnetic effect is provided by our determination of the protonation site of vitamin B_1 (equation 116). Just one of the pyrimidine nitrogens undergoes a large upfield chemical-shift change on protonation. Anne Cain and Glenn

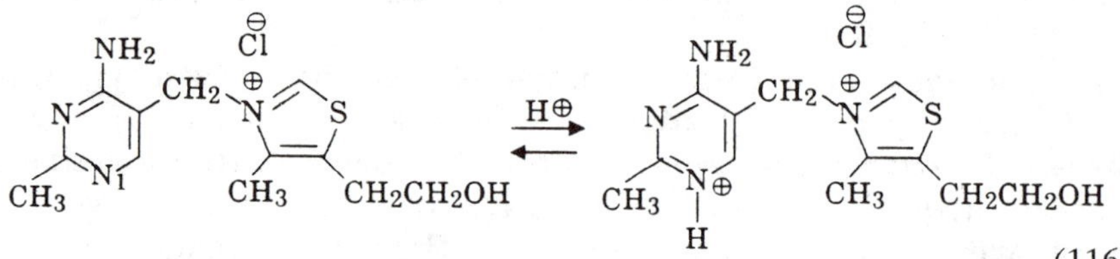

NH2 Cl⊖ CH2 N⊕ S N N1 CH3 CH3 CH2CH2OH $\xrightarrow{H^{\oplus}}$ NH2 Cl⊖ CH2 N⊕ S N N⊕ H CH3 CH3 CH2CH2OH

(116)

Sullivan were able to show that the pyrimidine N-1 accepts the proton.[233] We made several similar studies of the protonation of nucleotide bases, nucleotides, and nucleosides,[234] as well as of *cis*-platinum complexes of amines and nucleosides.[235]

Probably the most significant part of our ^{15}N research as applied to specific problems was concerned with enzymes and nitrogen metabolism of molds and bacteria. The enzymatic studies got started in an attempt to confirm the ^{13}C results obtained by Hunkapillar and Richards.[236] These results supported the "charge-relay" mechanism postulated by Blow[237] for the hydrolysis of peptide bonds by the enzymes known as *serine proteases,* of which chymotrypsins, studied by Carl Niemann (*see* pp 146–147), is an example.

These enzymes have a serine residue in their peptide chains at the active site, which acts in the *slow* step to cleave a peptide bond and accept the acidic fragment of the peptide to form a peptidyl ester. Hydrolysis of the ester then regenerates the enzyme (equation 117)

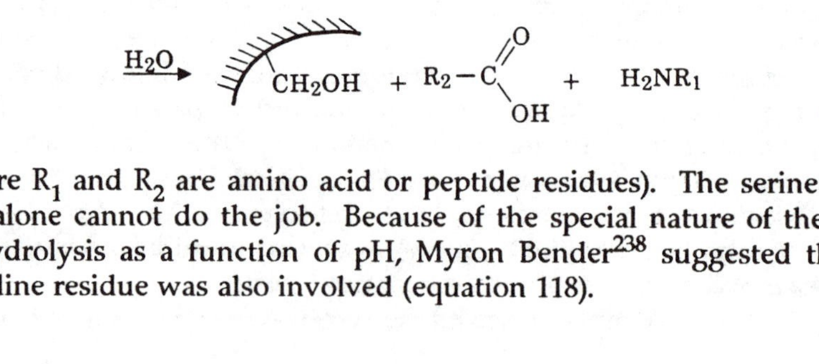

enzyme CH2OH serine residue H N R1 C R2 O peptide bond → CH2—O—C(=O)—R2 + H2NR1

$\xrightarrow{H_2O}$ CH2OH + R2—C(=O)OH + H2NR1

(117)

(where R_1 and R_2 are amino acid or peptide residues). The serine residue alone cannot do the job. Because of the special nature of the rate of hydrolysis as a function of pH, Myron Bender[238] suggested that a histidine residue was also involved (equation 118).

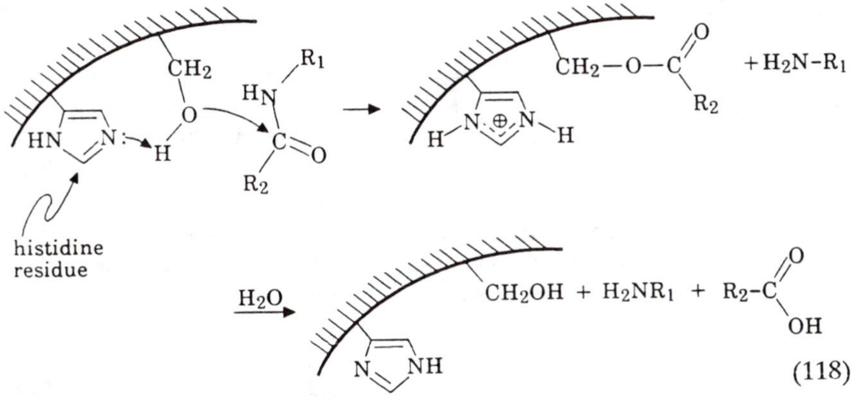

(118)

The histidine nitrogens, being moderately basic, were postulated to facilitate the removal of the serine OH hydrogen, and, thence, the attack of the serine oxygen on the peptide bond. Blow's crystal-structure determination of trypsin showed that an anionic aspartate residue is in close proximity to the histidine. He suggested that the aspartic anion may further facilitate the initial step by a process whereby a proton is transferred from serine to histidine and, at the same time, the other histidine proton is "relayed" to the aspartate anion (equation 119). The rest of the cleavage mechanism is as before.

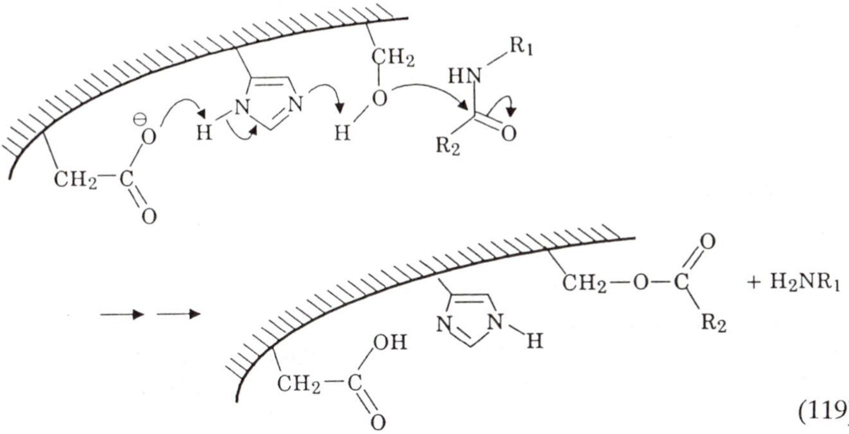

(119)

The mechanism is called "charge relay" because the charge that would be generated on the histidine is transferred to the aspartate anion. The combination of serine, histidine, and aspartate (or some similar carboxylate anion) is featured in all serine proteases and is called the catalytic triad of amino acids. The triad amino-acid residues are not contiguous to one another in the enzyme peptide chain, but are actually substantially separated and brought into proximity by folding of the chain.

Blow's speculation seemed so reasonable that it enjoyed an extensive vogue, even though it suffered from the fact that the aspartate anion is a much weaker base than histidine. The question, then, is what good can it do to transfer the proton from the stronger-base histidine to the weaker-base aspartate anion. Hunkapillar and Richards reasoned that the aspartate and histidine residues were so located in hydrophobic pockets of the folds of the enzymes as to cause the basic and acidic properties to be quite different from those that would be measured in ordinary aqueous solution.

To test this hypothesis, they made a histidine labeled with ^{13}C at the 2-position of the ring between the nitrogens and incorporated it into a particular simple serine protease called α-lytic protease, which contains only one histidine and has a molecular weight of about 15,000. The ^{13}C–H coupling constant of histidine is known to change when histidine adds a proton (equation 120). From measurements of $^{1}J_{^{13}CH}$ of the

H, N, ^{13}C–H, R, N: $\xrightleftharpoons{H^{\oplus}}$ H, N, $\oplus$ ^{13}C–H, R, N, H

$$^{1}J_{^{13}C\text{-}H} = 208 \text{ Hz} \qquad\qquad J_{^{13}C\text{-}H} = 218 \text{ Hz} \qquad (120)$$

labeled histidine in α-lytic protease as a function of pH, Hunkapillar and Richards concluded that the histidine residue was, in fact, an abnormally weak base.

There is a change in ^{15}N shift of about 80 ppm upfield of the nitrogen of histidine in water, which is not carrying the hydrogen in the protonation of equation 120. It seemed reasonable to check the ^{13}C experiments by labeling the nitrogens of histidine in the enzyme and following the shift as a function of pH to obtain the base constant of the histidine nitrogens. This project was taken on by Bill Bachovchin, who had just finished his Ph.D. with Jack Richards on a different problem and wanted some ^{15}N NMR experience.

It was not an easy thing to do. To be able to tell the resonance of one histidine nitrogen from the other in the NMR, it was desirable to have the histidine labeled at just one specific ring nitrogen, and also either at the other or at both of the nitrogens. The required ^{15}N-labeled histidines were available at several thousand dollars per gram. Bill wanted to be *sure* that they were used efficiently, without being diluted by ordinary histidine synthesized from other nonlabeled nitrogen sources by the *Myxobacter* strain that made the enzyme. To accomplish this goal, he set about to make a mutant for which histidine would be an essential amino acid.

He was successful after 250 tries. Then he used meticulous care to check that the mutant did not become contaminated with wild-type bacteria or back-mutated to something that could synthesize histidine from other nitrogen sources. Furthermore, instead of throwing all of our precious ^{15}N-labeled histidine in with the mutant, he carried out the incorporation reaction in 10 different batches. He made careful checks on the progress of each batch, combining them only when he was certain they contained the desired labeled enzyme. The measurements of the changes in the shifts of each histidine nitrogen of the enzyme were fabulously informative[239] when taken in conjunction with studies using the ^{15}N resonances of imidazole, **210**, and *N*-methylimidazole, **211**, as a model.[240]

N N H N N CH_3

210 **211**

Without going into detail about the logic involved, we drew a number of important conclusions. First, contrary to the Hunkapillar and Richards conclusion, the histidine of α-lytic protease had a wholly normal base strength. Second, when the histidine was protonated, it became hydrogen-bonded to the aspartate anion, **212**. Third, when the pH was increased to the point where the proton was lost from the histidine, the lost proton was the one bonded to N-1, **213** (equation 121).

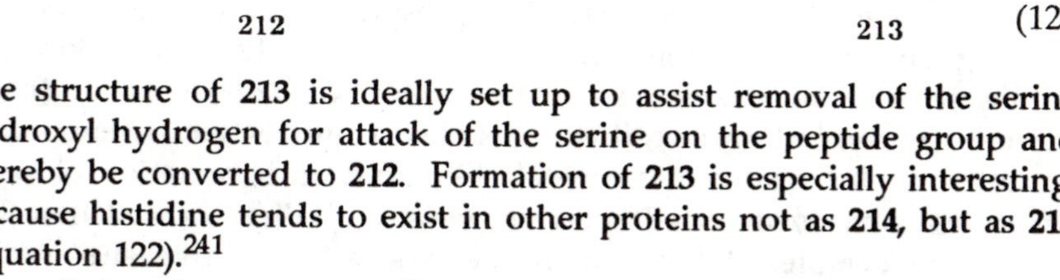

$$\mathbf{212} \underset{}{\overset{-H^{\oplus}}{\rightleftharpoons}} \mathbf{213} \qquad (121)$$

The structure of **213** is ideally set up to assist removal of the serine hydroxyl hydrogen for attack of the serine on the peptide group and thereby be converted to **212**. Formation of **213** is especially interesting, because histidine tends to exist in other proteins not as **214**, but as **215** (equation 122).[241]

Later, it was rewarding to be able to demonstrate that *cis*-

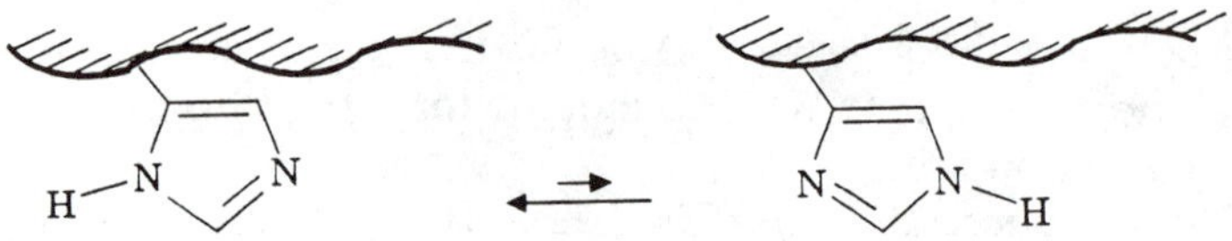

214 ⇌ 215 (122)

urocanic acid, **216**, but not its *trans* isomer, shows changes of its ^{15}N shifts with changes of pH almost identical to those of the histidine of α-lytic protease.[242] Bill Bachovchin has been extending these studies at

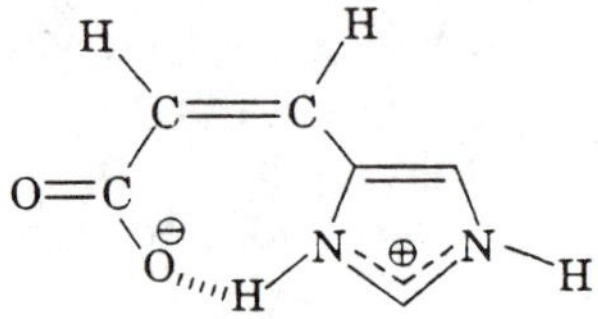

216

Tufts University in recent years and has had absolutely wonderful success in further delineating the species present in the enzyme and in its complexes with peptidylboronic acid inhibitors.[243] Bill also repeated Hunkapillar's work and showed that, in fact, the $^{1}J_{^{13}CH}$ couplings in the histidine do change with pH in the same way as the ^{15}N shifts of the histidine nitrogens.[244]

Our second major enzyme project was carried out by Bill Bachovchin while he was a postdoctoral fellow with Bert Vallee at Harvard Medical School. Bert was engaged in a rather acrimonious debate with W. N. Lipscomb of the Harvard Chemistry Department with respect to the spatial location of a particular tyrosine residue (Tyr 248) of carboxypeptidase. Lipscomb had used X-ray crystallography to show that this tyrosine was substantially removed from the active zinc atom in the native enzyme, but it swung in close to the zinc when the enzyme was complexed with an inhibitor.[245] The change in conformation was quite spectacular and may have been helpful in getting a favorable decision from the Nobel prize committee for chemistry. To be sure, Lipscomb was not going to back away easily from the significance of the change.

Vallee modified this particular tyrosine of carboxypeptidase by coupling on an azoarsanilic group, **217** (equation 123). This kind of modification of an enzyme would seem to be the kiss of death to enzyme activity, but, in fact, it remains almost fully active. Compounds like **217** complex readily with metals like zinc, and formation of such complexes is easily detected spectroscopically. Vallee showed that **217** was in the complexed form, with the zinc at the active site in the absence of inhibitor.[246]

Tyr 248 — CH_2 — OH → CH_2 — OH — β α $N{=}N$ — AsO_3H

217

(123)

We got involved because it was straightforward for Bill Bachovchin to label the β-azo nitrogen of **217** with ^{15}N. We could then run the ^{15}N spectrum to check for complexing as a function of pH and inhibitor concentrations. This part of the work was carried out with brilliance and painstaking attention to detail by Keiko Kanamori, another of those graduate students you hope to have at least once in

(Left to right) Hiroo and Keiko Kanamori with John D. Roberts in March 1989, at a 70th birthday symposium and party at Caltech arranged by colleagues and friends.

your career. Keiko, a native-born Japanese, speaks and writes impeccable English. She is outwardly self-effacing and infinitely considerate, yet she knows how to get what she wants. The wife of the world-famous geophysicist and Caltech professor Hiroo Kanamori, Keiko started graduate work later than most, after having had two sons. When she completed her thesis, she stayed on with me as a postdoctoral fellow and was extraordinarily helpful during the time I was provost. Later, she worked with Richard Weiss at UCLA in biochemistry and finally was able to set up a reasonably independent program at UCLA. She has excellent taste in choosing problems, is meticulous in execution, and is a wonderful teacher. I am pleased that she is currently associated with the Huntington Medical Research Institutes in Pasadena.

To summarize the results briefly, it appeared that the disagreement between Vallee and Lipscomb could be just a matter of pH. The Lipscomb crystal structure with remote Tyr 248 was done with the enzyme on the acidic side. In contrast, Vallee's experiments were done close to the pH where the peptidase activity was at its highest. Keiko's experiment[247] showed a strong pH dependence of complex formation, with the maximum near the pH where Vallee showed strong complexing. It was ironic that it has been shown in the last few years that Tyr 248 is not essential to the catalytic activity of carboxypeptidase.[248] It looks as though all of us were pursuing a will-o'-the-wisp.

The work with carboxypeptidase involved an enzyme with a molecular weight of 36,000, and the ^{15}N resonances of the azo derivative, **217**, were often quite broad. The experiments worked, nonetheless, because the chemical-shift changes involved were quite large. Later, Nancy Becker and Keiko suggested looking at the pyrazole inhibition of liver alcohol dehydrogenase (LAD). I was skeptical when I found that LAD has a molecular weight of almost 80,000. I was sure that any ^{15}N resonance lines would be too broad to be useful but, as often happens, I was wrong about that.

The problem was to corroborate by ^{15}N NMR the structure of the complex between pyrazole, **218**, *N*-nicotinamide–adenine dinucleotide (NAD^+), **219**, and LAD, with special attention to the degree of bonding between the pyrazole and the zinc atom at the active site. The structure proposed for the complex has pyrazole bonded to both the NAD^+ and the LAD zinc, **220**. Nancy was able to prepare the LAD complex labeled

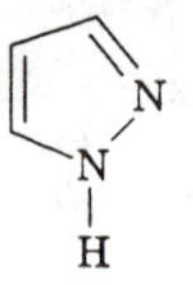

218 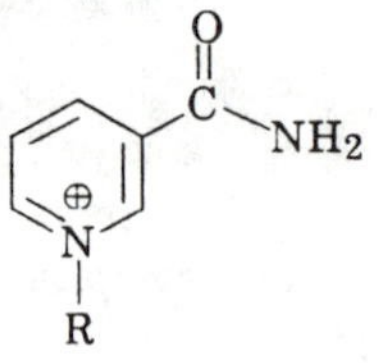 **219** (R = adenine dinucleotide)

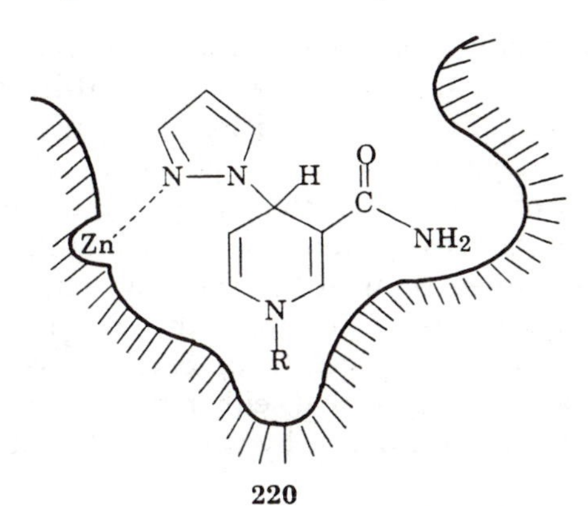

220

with ^{15}N at both pyrazole nitrogens and the ring nitrogen of the NAD^+ as well. The ^{15}N spectra obtained from the complex was consistent in every detail with structure **221**, by comparison with an extensive set of model compounds.[249]

Beyond the simple elegance of Nancy's structure proof, the important general finding of this research was that the ^{15}N NMR resonance lines were extraordinarily sharp. Indeed, with both nitrogens of the pyrazole labeled with ^{15}N and having different chemical shifts in the complex, it was possible to observe the $^{1}J_{^{15}N^{15}N}$ coupling of 12 Hz. Furthermore, the resonance of the ring nitrogen of the NAD^+ in the complex was comparably sharp.

Of course, one possibility would be that the material, given the spectrum, was not actually complexed with the enzyme. This possibility could be ruled out on two counts. First, the chemical shift of one of the ring nitrogens was such that it had to be complexed to zinc. Second, dialysis of the mixture removed the excess pyrazole present in the solution, but not that in the material, and gave the resonances ascribed to **221**. The conclusion seems unescapable—one should approach application of NMR to biochemical problems in the spirit of optimism, abandoning prejudices as to what will or won't work until proven otherwise. Another example of the validity of this maxim follows.

In 1980, Richard Weiss of UCLA asked about the possibility of following the course of nitrogen metabolism by the mold *Neurospora*, using ^{15}N as a tracer and ^{15}N NMR to follow the progress of formation of the amino acids involved. Again, or perhaps as usual, I was a bit skeptical of the feasibility of the study. Largely, I think, I hesitated because of the work Bob Lichter had done with ^{15}N-labeled glycine,[226] which showed extreme sensitivity to metal-ion contamination at near-physiological pH values.[250] There was also the matter of sensitivity. Would there be enough free amino acid within the cells? Or would we get signals only from that which was outside?

Whatever my misgivings and, indeed, regardless of some work in the literature[251] in which ^{15}N spectra were taken of cells in such a way as to obtain peaks only from those peaks with short T_1 and T_2 values, Keiko waded in. She found this to be a most fertile field for application of ^{15}N NMR to the study of biochemistry. Not only were the resonance lines extremely sharp, they came from amino acids inside the cells. With a high level of isotopic enrichment in ^{15}N, useful spectra could be obtained in minutes. Furthermore, the growth and metabolism of *Neurospora* could be studied right in the spectrometer. Much of the work Keiko has done, and continues to do, in this area involves how various strains of *Neurospora* and nitrogen-fixing bacteria assimilate ammonia or nitrate and produce glutamic acid, either by the glutamine synthetase pathway, equation 124, or directly by way of ketoglutaric acid, equation 125.[252]

$$HO_2C\underset{\underset{NH_2}{|}}{C}HCH_2CH_2CO_2H \xrightarrow{^{15}NH_3} HO_2C\underset{\underset{NH_2}{|}}{C}HCH_2CH_2C(=O)^{15}NH_2$$

$$\xrightarrow{HO_2C\underset{\underset{O}{\|}}{C}CH_2CH_2CO_2H} HO_2C\underset{\underset{^{15}NH_2}{|}}{C}HCH_2CH_2CO_2H + HO_2C\underset{\underset{NH_2}{|}}{C}HCH_2CH_2CO_2H \quad (124)$$

$$HO_2C\underset{\underset{O}{\|}}{C}CH_2CH_2CO_2H \xrightarrow{^{15}NH_3} HO_2C\underset{\underset{^{15}NH_2}{|}}{C}HCH_2CH_2CO_2H \quad (125)$$

It turns out that the ratio of these pathways depends markedly on the strain of the organism used, on whether nitrate or ammonia is the substrate, and on the preconditioning of the diet of the organism. With regard to the last factor, large differences were found when the organisms were starved or given preliminary diets of unlabeled ammonia, nitrate, or amino acids before being given a dose of ^{15}N-labeled ammonia or nitrate.[253,254]

Keiko also carried out a remarkable study of intracellular pH and NMR relaxation times. Her data indicated sequestering of particular amino acids in vacuoles within the cell at different pH values and that the mobilities of these amino acids were different from those of other amino acids in the cytoplasm.[255]

One other chemical application of ^{15}N NMR may be of interest. It was certainly interesting to me, and it was truly a fun research project in every sense of the word. It came about like this. During our broad sweep through the factors influencing ^{15}N chemical shifts, we investigated a number of diazo compounds to see if there was, indeed, a correlation between color and ^{15}N chemical shifts (*see* p 207). Back at MIT, I had been impressed with the permanganatelike color of diphenyldiazomethane, and I knew that **221** was also deeply colored. However, I also knew from my consulting at Du Pont that **222** was colorless.

We measured the ^{15}N shifts of many compounds of these types. Of particular interest, as part of the series, was **223**, a yellow compound.

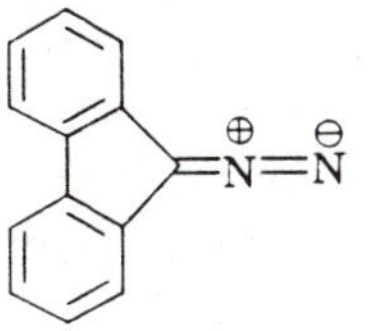

221 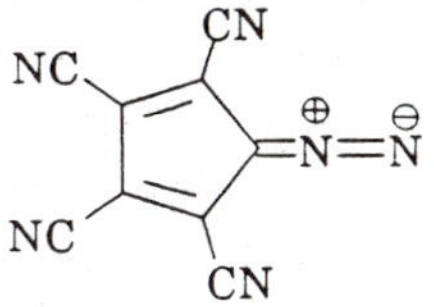 **222** 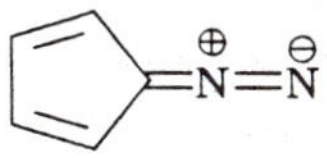 **223**

We did pretty much find the general relation between color and shift that we expected for these compounds. However, with **223**, as well as the other substances in this series, we had a special problem in that it was hard to tell which nitrogen resonance came from which nitrogen of the $=N^{+}=N^{-}$ group. So, we decided to make sure our inferences were correct by labeling one of the nitrogens of **223** with ^{15}N; (the spectra up to then were taken at the natural-abundance level). We started by labeling **224** (equation 126)

$$HO^{15}NO + NH_2NH\text{-}Ts \longrightarrow \underset{\mathbf{224}}{^{15}N{=}N{=}N\text{-}Ts} + H_2O \qquad (126)$$

where Ts is *p*-toluenesulfonate. For labeling of **223**, the sequence of equation 127 was used. By its accepted mechanism, this sequence should result only in labeling of the inside nitrogen of **223**.

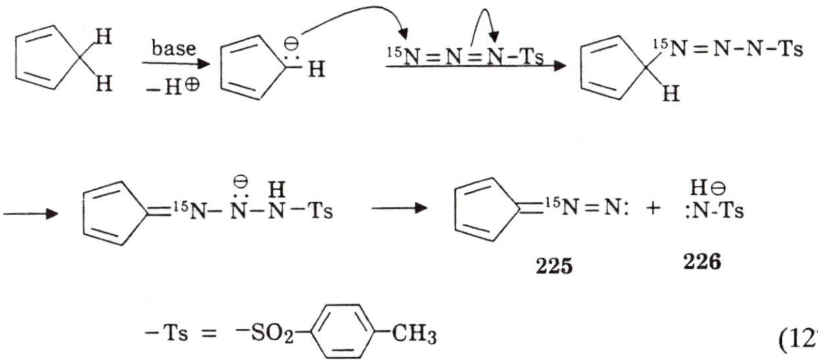

base, $-H^{\oplus}$; $^{15}N{=}N{=}N\text{-}Ts$; $^{15}N{=}N\text{-}N\text{-}Ts$; $^{15}N\text{-}\ddot{N}^{\ominus}\text{-}N(H)\text{-}Ts$; $^{15}N{=}N{:}$ + $H^{\ominus}{:}N\text{-}Ts$

225 **226**

$$-Ts = -SO_2\text{-}C_6H_4\text{-}CH_3 \qquad (127)$$

All of this worked as it should, giving **225**, labeled as shown. There was a fly in the ointment, in that about 5% of **227** was also formed.[256] Of course, if **224** contained 5% of **228**, this would account for the **227** formed. This possibility seemed quite unlikely by equation

$$\text{(cyclopentadienylidene)}=N={}^{15}N \qquad N={}^{15}N=N\text{-Ts}$$

227 **228**

127, and, furthermore, the ^{15}N NMR of the tosyl azide showed only the presence of **224**. I suggested to Carla Casewit, a graduate student, that she look into this problem. It was one of those situations where the answer could be trivial, but also might be quite rewarding. At least at the outset, we had no inkling or ideas of why or how **227** was formed.

After some checking on possible rearrangements of the starting materials, without getting any idea of what was happening, we decided to see whether some rearrangement of **224** might be induced by the reaction product, **226**. Carla put **224** in, together with (unlabeled) **226**. In a figurative sense, all hell broke loose. The ^{15}N spectrum changed from the one line of **224** to at least 10 different lines, with an enormous spread of chemical shifts. Almost all of the lines were sharp and changed slowly in intensity with time as the reactions progressed.

This plethora of resonances arises *only* from ^{15}N-labeled substances. The concentrations were all too low to allow any natural-abundance ^{15}N lines to be observed. At the outset, we knew the identity of only one resonance line, that from the starting material, **224**. It took months to identify each of the other resonance lines and, to our surprise, one of them was from dinitrogen, $^{15}N{\equiv}N$. A scheme of reactions that accounts for most of the observed products is shown in equations 128–133. Equations 128–133 explain formation of seven different

$$\underset{\mathbf{226}}{\text{TsN}\overset{H\ominus}{:}} + \underset{\mathbf{224}}{{}^{15}N=N=N\text{-Ts}} \longrightarrow \text{TsNTs} + \underset{\mathbf{229}}{{}^{15}N=N=\overset{\ominus}{N}} \tag{128}$$

$$^{15}N=N=\overset{\ominus}{N} + {}^{15}N=N=N\text{-Ts} \rightleftarrows {}^{15}N=N=\overset{\ominus}{N} + \underset{\mathbf{230}}{N=N={}^{15}N\text{-Ts}} \tag{129}$$

$$\text{Ts-}\overset{H\ominus}{N:} + N=N={}^{15}N\text{-Ts} \rightleftarrows \text{Ts-}\overset{H}{N}\text{-}N=N\text{-}{}^{15}\overset{\ominus}{\ddot{N}}\text{-Ts}$$

$$\rightleftarrows \text{Ts-}\overset{\ominus}{\ddot{N}}\text{-}N=N\text{-}{}^{15}\overset{H}{N}\text{-Ts} \rightleftarrows \text{Ts-}N=N=N + \underset{\mathbf{231}}{:{}^{15}\overset{H\ominus}{N}\text{-Ts}} \tag{130}$$

$$\text{Ts-}{}^{15}\overset{H\ominus}{N:} + {}^{15}N=N=N\text{-Ts} \longrightarrow \underset{\mathbf{232}}{\text{Ts}{}^{15}\overset{H}{N}\text{-Ts}} + {}^{15}N=N=\overset{\ominus}{N} \tag{131}$$

$$\text{Ts-}\overset{H\ominus}{N:} + {}^{15}N=N=N\text{-Ts} \rightleftarrows \text{Ts-}\overset{H}{N}\text{-}{}^{15}N=N\text{-}\overset{\ominus}{\ddot{N}}\text{-Ts}$$

$$\rightleftarrows \text{Ts-}\overset{\ominus}{\ddot{N}}\text{-}{}^{15}N=N\text{-}\overset{H}{N}\text{-Ts} \rightleftarrows \underset{\mathbf{233}}{\text{Ts-}N={}^{15}N=N} + :\overset{H\ominus}{N}\text{-Ts} \tag{132}$$

$$\mathrm{Ts\text{-}\overset{H\ominus}{N}:} + \mathrm{N{=}^{15}N{=}N\text{-}Ts} \longrightarrow \mathrm{Ts\text{-}\overset{H}{N}\text{-}Ts} + \mathrm{N{=}^{15}\overset{\ominus}{N}{=}N} \quad \underset{\mathbf{234}}{} \qquad (133)$$

resonances that correspond to **224** and **229–234**. Two lines remain unaccounted for besides that of N_2. As to N_2, I recalled a paper published many years earlier on the kinetics of the reaction of bromine with azide ion, N_3^+, which results in formation of N_2. No mechanism was given, but it seemed that something like equations 134–136 would be wholly

$$\mathrm{N{=}\overset{\ominus}{N}{=}N} + \mathrm{Br_2} \longrightarrow \mathrm{N{=}N{=}N\text{-}Br} + \overset{\ominus}{\mathrm{Br}} \qquad (134)$$

$$\mathrm{N{=}N{=}N\text{-}Br} + \mathrm{N{=}\overset{\ominus}{N}{=}N} \longrightarrow \underset{\mathbf{235}}{\mathrm{N{=}N{=}N\text{-}N{=}N{=}N}} + \overset{\ominus}{\mathrm{Br}} \qquad (135)$$

$$\mathrm{N{=}N{=}N\text{-}N{=}N{=}N} \longrightarrow 3\mathrm{N_2}$$

(via cyclic N_6 ring, **236**) (136)

reasonable. My graduate student, Ken Servis, now at the University of Southern California, looked into this for a brief period in the early 1960s, without definitive results. However, because of the similarity of Ts (*p*-toluenesulfonate) in some of its reactions to Br, there seemed a possibility of tosyl azide reacting with azide ion to form ^{15}N-labeled N_6, **235**, equation 137, and liberate Ts:⁻ (toluenesulfinate ion). There are, of

$$\mathrm{N{=}\overset{\ominus}{N}{=}N} + \mathrm{^{15}N{=}N{=}N\text{-}Ts} \rightleftarrows \mathrm{N{=}N{=}N\text{-}^{15}N{=}N\text{-}\overset{\ominus}{\ddot{N}}\text{-}Ts}$$

$$\longrightarrow \mathrm{N{=}N{=}N\text{-}^{15}N{=}N{=}N} + :\mathrm{Ts}^{\ominus} \longrightarrow 2\mathrm{N_2} + \mathrm{^{15}N{\equiv}N} \qquad (137)$$

course, other labeled materials that could be involved, because the azide ion can be labeled in two ways, **229** and **234**, and the tosyl azide can be labeled in three different ways, **224**, **230** and **233**. Equation 137 can account for the N_2 resonance, and equations 138 and 139 the two remaining peaks that arise from two differently labeled addition products of Ts^- to labeled tosyl azides.[257] Whether or not cyclic N_6 is involved is an open question. We have not been able to devise a labeling experiment that could distinguish between **235** and **236**.

One labeling experiment Carla carried out was significant in telling something about possible intermediates.[258] This experiment involved unlabeled tosyl azide and labeled azide ion, **229**, in methylene chloride, a solvent that does not lead to formation of dinitrogen and **237**

$$\mathrm{Ts\!:^{\ominus}} + {}^{15}\mathrm{N{=}N{=}N{-}Ts}\ (\mathrm{or\ N{=}N{=}{}^{15}N{-}Ts}) \rightleftarrows [\mathrm{Ts{-}{}^{15}N{=}N{-}\ddot{N}^{\ominus}{-}Ts} \leftrightarrow \mathrm{Ts{-}{}^{15}\ddot{N}^{\ominus}{-}N{=}N{-}Ts}] \quad \mathbf{237} \qquad (138)$$

$$\mathrm{Ts\!:^{\ominus}} + \mathrm{N{=}{}^{15}N{=}N{-}Ts} \rightleftarrows [\mathrm{Ts{-}N{=}{}^{15}N{-}\ddot{N}^{\ominus}{-}Ts} \leftrightarrow \mathrm{Ts{-}\ddot{N}^{\ominus}{-}{}^{15}N{=}N{-}Ts}] \quad \mathbf{238} \qquad (139)$$

or **238**. The azide ion quickly equilibrates its label with tosyl azide, equation 140. These reactions are also part of the scheme of equations

$$\underset{\mathbf{229}}{{}^{15}\mathrm{N{=}\overset{\ominus}{N}{=}N}} + \mathrm{N{=}N{=}N{-}Ts} \rightleftarrows \underset{\mathbf{224}}{\mathrm{Ts{-}{}^{15}N{=}N{=}N}} + \underset{\mathbf{230}}{\mathrm{Ts{-}N{=}N{=}{}^{15}N}} + \mathrm{N{=}\overset{\ominus}{N}{=}N} \qquad (140)$$

128–133. What is different is that, in the *absence* of TsNH:⁻, there is slow formation of **233** and **234**. The only means we have of explaining this result is *reversible* formation of **239** or **240**, equations 141–143. The formation of **239** has a close parallel in the cyclization of $C_6H_5N{=}N{-}N{=}N{=}N$ observed by Huisgen and co-workers,[259] equation 144.

$$\mathrm{TsN{=}N{=}N} + {}^{15}\mathrm{N{=}\overset{\ominus}{N}{=}N} \rightleftarrows \mathrm{Ts\ddot{N}^{\ominus}{-}N{=}N{-}{}^{15}N{=}N{=}N} \rightleftarrows \mathrm{Ts\ddot{N}^{\ominus}{-}N}\ (\text{ring: } \mathrm{N{=}{}^{15}N{-}N{=}N}) \quad \mathbf{239} \qquad (141)$$

$$\left(\text{or } \mathrm{TsN}\ (\text{ring: } \mathrm{N{=}{}^{15}N{-}\ddot{N}^{\ominus}{-}N{=}N})\ \mathbf{240}\right) \rightleftarrows \mathrm{TsN{-}\ddot{N}^{\ominus}{=}N{-}N{=}{}^{15}N{=}N} \rightleftarrows \mathrm{TsN{=}N{=}N} + \underset{\mathbf{234}}{\mathrm{N{=}{}^{15}\overset{\ominus}{N}{=}N}} \qquad (142)$$

$$\mathrm{N{=}{}^{15}\overset{\ominus}{N}{=}N} + \mathrm{N{=}N{=}NTs} \rightleftarrows \mathrm{N{=}{}^{15}N{=}N{-}Ts} + \mathrm{N{=}\overset{\ominus}{N}{=}N} \qquad (143)$$

$$\mathrm{C_6H_5\overset{\oplus}{N}{\equiv}{}^{15}N} + \mathrm{N{=}\overset{\ominus}{N}{=}N} \rightarrow \mathrm{C_6H_5N{=}{}^{15}N{-}N{=}N{=}N} \rightarrow \mathrm{C_6H_5{-}N}\ (\text{ring: } {}^{15}\mathrm{N{=}N{-}N{=}N}) \rightarrow \mathrm{C_6H_5N{=}{}^{15}N{=}N} + \mathrm{N{\equiv}N} + \mathrm{C_6H_5N{=}N{=}N} + {}^{15}\mathrm{N{\equiv}N} \qquad (144)$$

Applications in Medicine

When I was about to retire from the vicissitudes of being provost at Caltech, I decided that the applications of NMR to medicine through

The first four provosts of the California Institute of Technology, 1984. (Left to right) Rochus E. Vogt, a cosmic-ray physicist; Robert F. Bacher, a physicist whose early research on atomic spectroscopy had important chemical applications (later in World War II, Bacher was second in command at the MIT Radiation Laboratory [radar development] and then head of the Physics Division at Los Alamos, where he helped in the final assembly of the first atomic bomb with his own hands); John D. Roberts; and Robert F. Christy, a distinguished theoretical nuclear physicist.

imaging and spectroscopy looked very promising and interesting. However, I was sure that I would not really work very hard at it unless I gave up my ongoing NMR programs. So I dropped my NSF and NIH support and accepted an offer by the Huntington Medical Research Institutes (HMRI) in Pasadena to be an unpaid consultant and research participant, as well as a member of the board, treasurer, and chairman of their NMR research committee (now called MR research committee in deference to those who do not like to associate NMR with the nucleus).

The imaging facilities at HMRI are superb. The Institutes acquired the first commercial MR imaging unit in the world, and a marvelous program of clinical and scientific work has been brought to fruition under the direction of Dr. William G. Bradley (Caltech undergraduate, Princeton Ph.D. in chemical engineering, and M.D. radiologist from the University of California at San Francisco). There is unlikely to be another imager anywhere in the world that can match the more than 10,000-patient throughput of the Diasonics unit at HMRI.

Whole-body MR spectroscopy holds much promise for the future. HMRI has acquired a 4.7-T (200-MHz) proton spectrometer with a 30-cm bore for imaging and spectroscopy of small animals and neonates. An active program of research is underway that uses this instrument to take localized phosphorus spectra. At the same time, a vigorous effort is being made to obtain a high-field instrument large enough to accommodate adult humans for whole-body spectroscopy.

The technical difficulties are substantial, the biggest problem being the old bugaboo of sensitivity. Proton imaging and proton spectroscopy are relatively easy because living organisms contain large amounts of water, although the spectroscopy of water itself is not very interesting. The relatively limited range of proton shifts is also not helpful in the use of proton spectroscopy. Phosphorus NMR is potentially much more useful, with substantial chemical shifts and great biochemical interest. However, when you take into account the magnetic moments and concentrations in the body, phosphorus signals are about 1/10,000 as strong as proton signals. In proton imaging, it is possible to get a useful signal in a reasonable time in the human brain from about 3×10^{-3} mL. For the same S/N, phosphorus would require 30 mL of sample. To be really useful, no more than 2–3 mL should be required. That requirement is being approached, but is not yet routine.

On "promotion" to Institute Professor of Chemistry, Emeritus, Caltech withdraws the privilege of having a research group of graduate students and postdoctoral fellows or of obtaining research support at all, except for research that one does with one's own hands. As provost, I helped to write those rules, so it is a little awkward to complain. Fortunately, nothing was promulgated with respect to undergraduate research, and remembering how good that was for me as a student at UCLA as well as a young faculty member at MIT, I decided to take advantage of this loophole. And it has been rewarding—indeed revitalizing—because I had almost given up the idea of being able to do laboratory research at all at Caltech. To be sure, undergraduates require a lot of help to get going, especially when you don't have any infrastructure of postdoctoral fellows and graduate students to gain teaching experience by helping out. However, in the long run, I suspect undergraduates will be more appreciative than graduate students and postdoctoral students, because what help you can give them comes at a more formative stage of their careers. Through this experience, I have come to appreciate much more the problems faced by professors in liberal arts colleges, who often have heavy teaching loads by Caltech standards and still carry on a lot of undergraduate research. It is one thing to work with undergraduates when you are just starting, as I did at MIT, and quite another when you have the extra responsibilities and demands on your time that are associated with a full professorship.

Some of our current research has been directed to collaboration on imaging research at HMRI, largely because the contrast of MR images is a complex function of the *chemical* properties of the tissues involved through proton content and composition, along with the corresponding T_1 and T_2 relaxation times. An example is a very fine study by an undergraduate student, Rachael Clark, now in the

Rachael Clark, a Caltech undergraduate research student, measuring the NMR relaxation times of blood as a function of the extent of oxygenation, hematocrit, clotting time, and field strength (1989).

M.D./Ph.D. program at Harvard Medical School, on the factors that are expected to cause blood to appear hypointense in T_2-weighted images of cerebral hemorrhage.

The future of NMR is still very bright. Over the more than 30 years during which I have worked with NMR, some entirely new advance occurs every few years—usually at just about the time I decide perhaps I should be doing something different.

Mechanistic Studies of Iodonium Hydrolysis

In 1954, when we were deep in our studies of amine–nitrous acid reactions, I began to speculate as to whether or not such reactions could actually be reversible, whether they might somehow be involved in bacterial nitrogen-fixation processes. Aryldiazonium cations, such as $C_6H_5N^+{\equiv}N$, are reasonably stable. They do not lose nitrogen almost instantly, as do the simple alkyldiazonium cations such as $CH_3CH_2N^+{\equiv}N$. The reason lies in the difficulty of generating an aryl cation like $C_6H_5^+$. In 1954 there were no reactions in solution for which one could say with confidence that a phenyl carbocation is the intermediate, except very possibly the hydrolysis that occurs on heating of solutions of the aryldiazonium cations themselves (equation 145).

$$C_6H_5\text{-}\overset{\oplus}{N}{\equiv}N \xrightarrow[?]{-N_2} C_6H_5^{\oplus} \xrightarrow[-H^{\oplus}]{H_2O} C_6H_5\text{-}OH \qquad (145)$$

If phenyl carbocations are unstable and phenyldiazonium ion is stable, then the possibility exists that, if one could generate a phenyl carbocation, it might react with dinitrogen to give the phenyldiazonium ion. Of course, reaction with dinitrogen would have to be made competitive with other likely reactions of the carbocation. One way to make this reaction competitive would be to see if equation 145 is reversible by carrying it out in the presence of labeled dinitrogen. At this juncture, with no mass spectrometer and no ^{15}N NMR, this step looked harder than finding a suitable reaction to generate an aryl carbocation, seeing if it formed a cation, and testing to determine whether that cation would react with dinitrogen to give a diazonium compound. In retrospect, the idea does not seem very good *at all*, but, at the time, I was excited about it and was able to raise $10,000 from the Guggenheim Foundation to test it.

The first problem was to find a reaction to generate phenyl carbocations. An ideal choice would be to lose xenon from $C_6H_5{-}Xe^+$, which was a nonpractical enterprise at the time, but the type of approach seemed reasonable. Another much simpler possibility would be decomposition of a diphenylhalonium cation, equation 146. For

$$C_6H_5\overset{\oplus}{X}C_6H_5 \xrightarrow[?]{-X-C_6H_5} C_6H_5^{\oplus} \xrightarrow[-H^{\oplus}]{H_2O} C_6H_5OH \tag{146}$$

X = F, Cl, Br, I

starters, it seemed desirable to determine whether the reaction shown by equation 146 would go with any of the possible X atoms. The easily available one, known for many years, was with X = I, the diphenyliodonium compounds. Nucleophilic displacements of such substances had been studied, especially by F. M. Beringer,[260] but there was apparently no search for S_N1-like processes.

Don Glusker, a postdoctoral fellow, took on the project of seeing whether S_N1 was possible. He found very quickly that it was necessary to get away from the counterions such as iodide ion, because these gave diphenyliodonium salts that are not very soluble and also resulted in S_N2-like nucleophilic displacements (equation 147). Much better results

$$C_6H_5\overset{\oplus}{I}C_6H_5 + I^{\ominus} \xrightarrow{H_2O} C_6H_5I \tag{147}$$

were obtained with anions of low nucleophilicity such as toluenesulfonate, trifluoroacetate, or especially tetrafluoroborate. With these anions, what looked like an S_N1 reaction occurred in equation 146 with X = I. Nonetheless, the rates were far from reproducible, although the products were phenol and iodobenzene in accord with equation 146, but Don had to leave before the matter was resolved.

Some months before, I had received a rather unusual letter of inquiry about a postdoctoral fellowship from a Marjorie Beckett, who had obtained a Ph.D. with Ernst Berliner at Bryn Mawr. Such letters are seldom specific or very informative, or revealing of any thought as to why the writer wants to work in your research group. The Beckett letter had a lot of information and showed a lot of motivation. With an excellent letter from Berliner, it was not difficult to decide to offer her a position. She was English, with an extensive education in both Britain and the United States. Not only did she have talent in writing, but also talent in the laboratory and in dealing with people.

When she was confronted with the dilemma of why the rate measurements were so erratic in the hydrolysis of diphenyliodonium salts, Marjorie soon found out that the reaction was enormously subject to cuprous ion catalysis. The more she purified the reagents, the slower the hydrolysis reaction went. It was never clear whether even the slowest rate we observed was still not cuprous ion catalyzed. There may well have been enough trace copper in the Pyrex tubes we were using as reaction vessels. In any event, as she reduced the cuprous ion concentration, the reaction slowed and slowed. More detailed study of

Marjorie and Fred Caserio about 1958. Fred worked in industry for many years before finally retiring as vice-president of Magna Products. In 1989, Marjorie was a professor of chemistry and chairman of the Chemistry Department of the University of California at Irvine.

the hydrolysis showed some unusual substituent effects for something I hoped would form a phenyl carbocation. Thus, starting with **241**, the cleavage of the two possible C–I bonds occurred almost equally (equation 148). This and other evidence was convincing that the hydrolysis was not a carbocationic reaction like equation 146, but involved a free-radical mechanism resulting in the overall effect of an S_N1-like reaction (equation 149).[261–262] So that nitrogen-fixation project got shoved onto the back burner.

$$\underset{\mathbf{241}}{CH_3O{-}C_6H_4{-}\overset{\oplus}{I}{-}C_6H_4{-}NO_2} \xrightarrow[-H^{\oplus}]{H_2O} CH_3O{-}C_6H_4{-}OH + I{-}C_6H_4{-}NO_2 + CH_3O{-}C_6H_4{-}I + HO{-}C_6H_4{-}NO_2 \qquad (148)$$

$$C_6H_5{-}\overset{\oplus}{I}{-}C_6H_5 \xrightarrow{Cu^{\oplus}} C_6H_5{-}\dot{I}{-}C_6H_5 + Cu^{2\oplus} \xrightarrow[-H^{\oplus}]{H_2O}$$

$$\longrightarrow C_6H_5\cdot + I{-}C_6H_5 + Cu^{2\oplus} \xrightarrow[-H^{\oplus}]{H_2O} C_6H_5{-}OH + I{-}C_6H_5 + Cu^{\oplus} \qquad (149)$$

Marjorie married Fred Caserio, also a very able postdoctoral fellow who came from UCLA, where he worked with Bill Young. Fred went to work in local industry because, as a fanatical surfboarder, he was not interested in employment very far away from the waves. Marjorie stayed on at Caltech for several years before going to a very successful career at the University of California at Irvine.

The Textbook of Organic Chemistry

One of W. A. Benjamin's goals when he started his publishing company was to get me to do an organic chemistry textbook for the first-level course. This was something I wanted to do and had the missionary spirit to do. However, even though, at Edwin Buchman's behest, I had prepared some written material to hand out in the organic chemistry course at Caltech, I knew it was going to be more of an undertaking than I wanted to do alone. Don Cram and George Hammond had approached me earlier to join with them to do a book. Still, although I knew that we could do a good job, I was sure it would be a struggle of epic proportions to bring the project to fruition, because we had such disparate personalities and interests.

The prospect of having Marjorie as a coauthor was much brighter, because we had already worked together. She had excellent ideas and could express them clearly, and she would be close at hand. Also, as more or less the senior author, I had more opportunity for control in case serious disagreements should arise. So we signed a publishing contract with Benjamin's company. The decision to sign with Benjamin was made easier because I had been involved in helping to start his company as director and also as what the company's prospectus for the Securities and Exchange Commission called a promoter. The company had by now graduated from joint tenancy in Bill and Orly's apartment to quarters above a bowling alley on Broadway, above Columbia University. It was a nice office, but noisy from the crash of the bowling pins as strikes were rolled below. The contract was a very favorable one, for a textbook, and Bill complained bitterly about the terms I extracted from him.

It took about three years to write and publish the book. We did a syllabus to test the text on students at Caltech and in a summer course at Purdue. Marjorie and I were able to work very smoothly. Each of us would originate a chapter. We would then exchange them, rewrite, exchange, and rewrite until it was probably hard to tell who originated what.

The first edition was, I think, a landmark book with many features that are standard today.[263] It was also a shocker to many, espe-

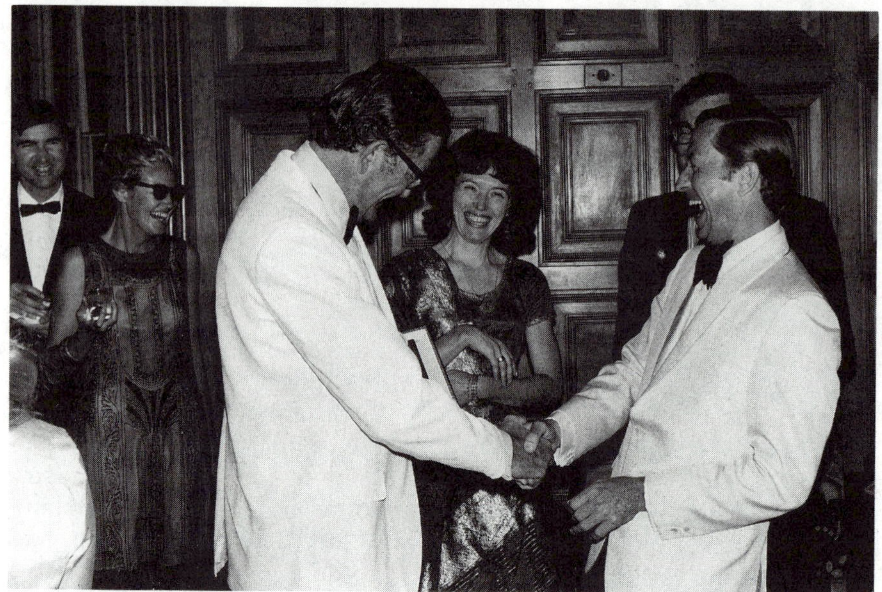

John D. Roberts being greeted by William A. Benjamin at a surprise party in 1970, arranged by Benjamin and Edith Roberts at Caltech's Athenaeum to celebrate 10 years of the W. A. Benjamin publishing company and the production of a commemorative volume, Thirty Years of Research, *which contained a number of humorous commentaries by friends and colleagues of John D. Roberts, as well as reprinting all of his published papers up to 1970. It is a tribute to Edith Roberts's powers of persuasion and guile that she was able to inveigle her husband to come in formal dress to what was claimed would be a small party in the summertime. In the background are Caltech's premier electrochemist, Fred Anson, his wife, Roxana, Shirley Gray, and the famously influential inorganic chemist, Harry Gray.*

cially through the very early introduction of spectroscopy; the many follow-up spectroscopic problems; the inclusion of problems, as appropriate, right within the text; and, perhaps, more than anything, the unusual length for an elementary book. When I asked my organic chemistry colleagues, who complained about the length, for a list of things they would cut out, I never got much help on that score. The usual response was that perhaps we might "somewhat enlarge" the sections that covered the suggestor's own field of interest. Book writing is a lot of work, but we got into it at the right time with the right publisher, while the bloom was still on the publishing business and there was not so much competition that you couldn't make enough off royalties to send your children to a private college, and ours went to Stanford.

We finished a second edition in 1977.[264] Unlike some second editions, it was a complete rewrite. We hoped that we could be like a cafeteria or buffet, providing a wide variety of topics from which the users could choose. Perhaps unfortunately, we tried to introduce systematic nomenclature for the first time in an elementary text. To bridge the gap between new and old, we indicated preference for system, but included trivial names in parentheses. Thus, there was ethanoic acid (acetic acid), propanone (acetone), and so on. This was said to be a great deterrent to the use of the book, but more likely it provided a convenient excuse to choose a less rigorous tome. Anyway, for whatever reason, the book was hardly a best seller, even though it still sells quite a few copies per year, mostly overseas, almost 10 years after publication. I thought, and still think, it was a fine text. Perhaps, in one sense it was, because the most common comment I had from teachers was, "a very useful book to prepare lectures from."

George S. Hammond and Sylvia Winstein being greeted by Edith Roberts and John D. Roberts at the surprise party in 1970.

Mechanistic Studies of the Diels–Alder Reaction

One other mechanistic study we made, which had its amusing aspects, was related to the Diels–Alder reaction, wherein some kind of an alkene adds 1,4 to some kind of a diene. The prototype example is 1,4-addition of ethylene to butadiene, equation 150, although this particular

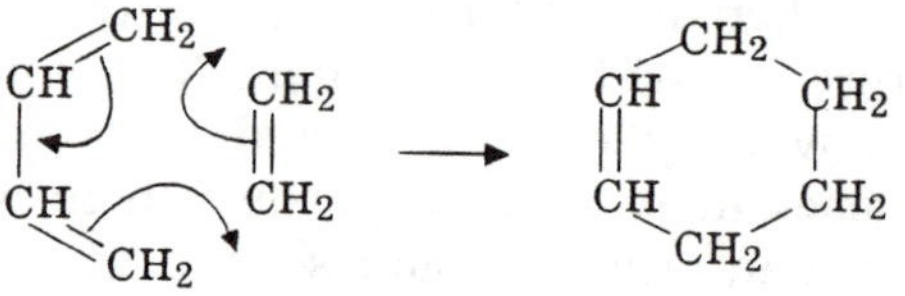

$$CH_2{=}CH{-}CH{=}CH_2 + CH_2{=}CH_2 \longrightarrow \text{cyclohexene} \qquad (150)$$

reaction does not proceed very readily. There have been questions for years as to the degree to which this reaction is concerted. Are all the bonds made *at once*, as in equation 150? Or is there an intermediate diradical, **242**, formed, which then closes very rapidly to give the final product, as in equation 151?

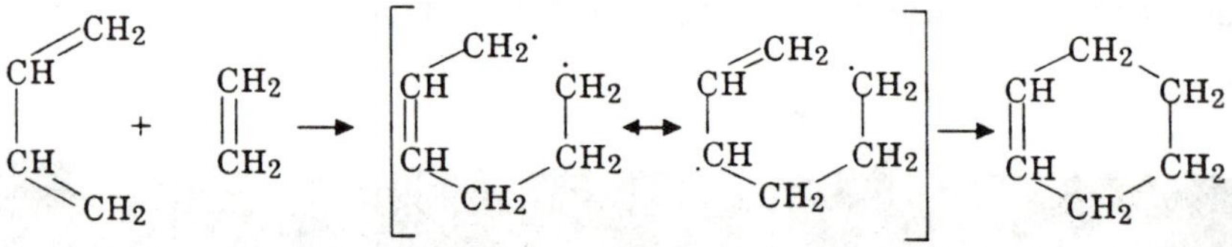

$$CH_2{=}CH{-}CH{=}CH_2 + CH_2{=}CH_2 \longrightarrow [\mathbf{242}] \longrightarrow \text{cyclohexene} \qquad (151)$$

The possibility of an ionic intermediate seemed ruled out by the fact that the reaction will go in the gas phase, where ions are very difficult to form. I have always regarded concerted mechanisms as rather sterile and essentially uninteresting chemistry, and I was not about to accept the Diels–Alder addition as concerted without a hard look. One way to test for formation of radicals, such as **242**, is to see whether or not there is a loss of stereochemistry of a substituted alkene by rotation about the single bond next to the radical center, **243** (equation 152), faster than the ring closes to form **244**. Rotation gives a new radical, **245**, which can close to give **246**, a somewhat more stable product than **244**. If the initial formation of **243** is reversible, it should be so of **245** as well, and the starting material would isomerize from the *cis* to the *trans* configuration of the chlorines.

I was able to induce Joe Lambert to investigate the reactions of equation 152 with some care because we had some evidence earlier at MIT when working with the addition of *cis*-dichloroethylene to cyclopentadiene that some **246** was formed along with **244**. However, this reaction may have been catalyzed by iron salts, because we carried out these additions in an iron bomb. Be that as it may, Joe could detect no trace of isomerization of the *cis*-dichloroethylene or formation of **246**.[265]

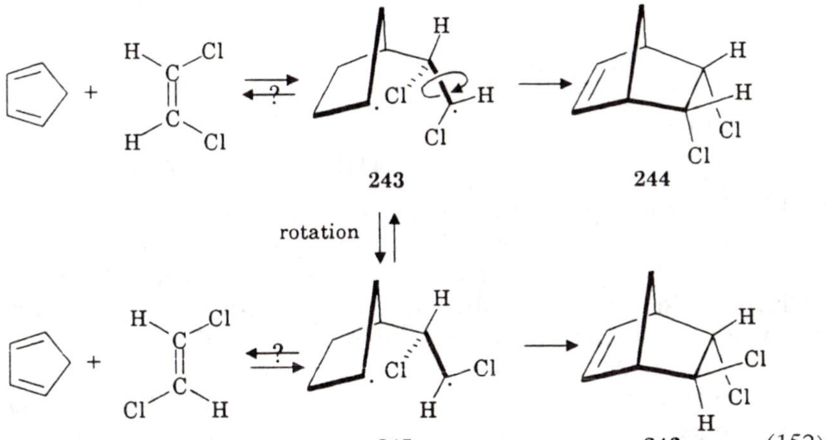

243 244

rotation

245 246 (152)

Graduate student R. E. Lutz, now at Portland State University, did some beautifully clean work on another approach to the question of concertedness of the Diels–Alder reaction that was inspired, in part, by the work of Woodard and Katz.[266] The idea was to investigate a possible rearrangement of the dimer, **247a**, of methacrolein, **247b**, which, when formed by a Diels–Alder-type addition and strongly heated, dissociates, possibly by way of a diradical intermediate, **247c** (equation 153).

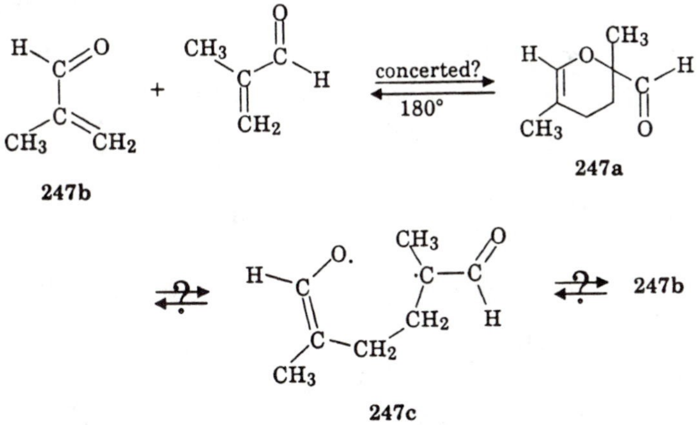

concerted? 180°

247b **247a**

247c (153)

This mechanism is relatively straightforward, but assessment of the intermediacy of **247c** requires a closer look at the possibilities. This may be best illustrated by looking at the actual experiment Ray Lutz performed. Consider the compound **247d**, which can be synthesized from **247a** (equation 154). The characteristics of **247d** are, first, that it has a deuterium in place of hydrogen on the carbonyl carbon and, second, it is chiral in that there is a specific configuration of the CH_3–

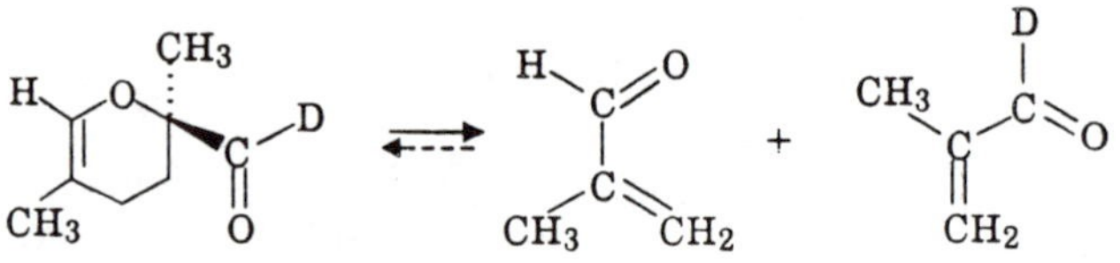

247d ⇄ 247a + 247e (154)

and –CD=O groups connected to one of the carbons bonded to the ring oxygen. On heating, **247d** can simply dissociate to give one molecule of **247a** and the corresponding deuterium-labeled **247e**. Alternatively, **247d** could form the diradical **247f**, for which two extreme resonance forms are shown. Cyclization, as expected for the upper one, would return to **247d**; dissociation would give **247a** and **247e**.

Rotation about the CH_2–C• carbon–carbon bond (the reorientation is somewhat more complex, but rotation is close to the right idea) would lead to a conformational isomer, **247c**, with a different stereochemical configuration. Cyclization, as expected for the lower resonance form of **247f**, would lead to **247h**, which results with the deuterium on a ring carbon (which was originally the carbon of a –CD=O group) and with the *same* relative configuration of CH_3– and –CH=O groups (equation 155). Correspondingly, **247g** can cyclize to **247i** and

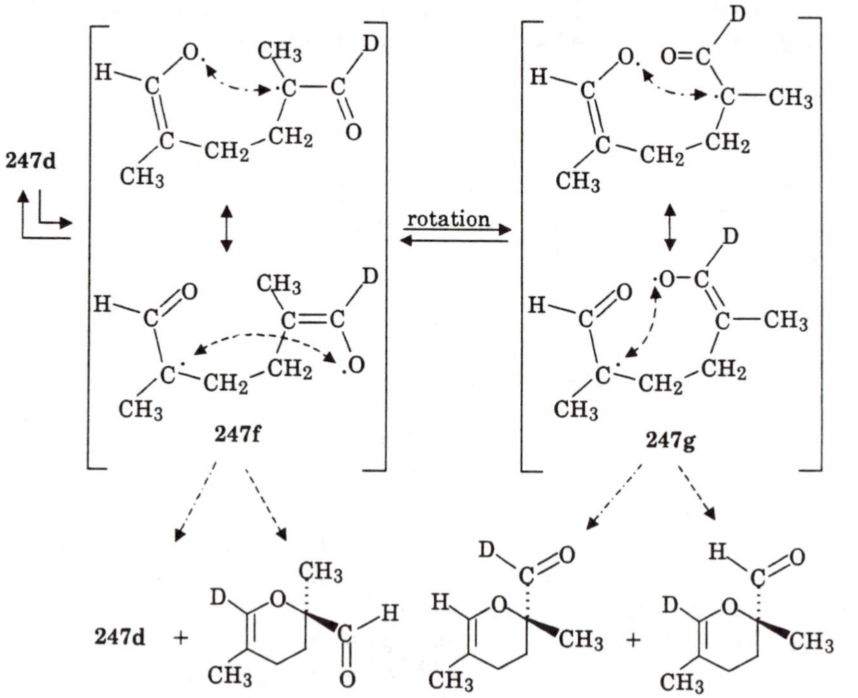

247d → 247f ⇄ (rotation) 247g; 247f → 247d + 247h; 247g → 247i + 247j (155)

247j, one of which has a different position for the deuterium; however, both have opposite configurations of the carbon with methyl and aldehyde groups attached from the configuration of **247d**.

The object of this experiment was to see if rearrangement of the deuterium occurred and whether or not there was a configurational change associated with it. Obviously, one wants to avoid the complications that would be introduced by having the products form by dissociation–recombination, equation 154, rather than by way of equation 155 or its equivalents. This was easy to do because at 180 °C the methacrolein, bp 68 °C, distills out of the mixture before it can recombine to give the dimer. The undesirable results of the recombination reaction can easily be seen when the reaction is carried out in a sealed tube and the methacrolein cannot escape.

What Ray found was that the rearrangement occurred at a rate comparable to dissociation, but with *no* change in the methyl and aldehyde configurations. Thus, along with dissociation, **247d** went cleanly to **247h**.[267] The key question is whether the results obtained in this reaction have any relevance to the mechanism of the Diels–Alder reaction, because the reaction can be written as an oxy Cope rearrangement, equation 156. In that regard, it suffers from the same problem as

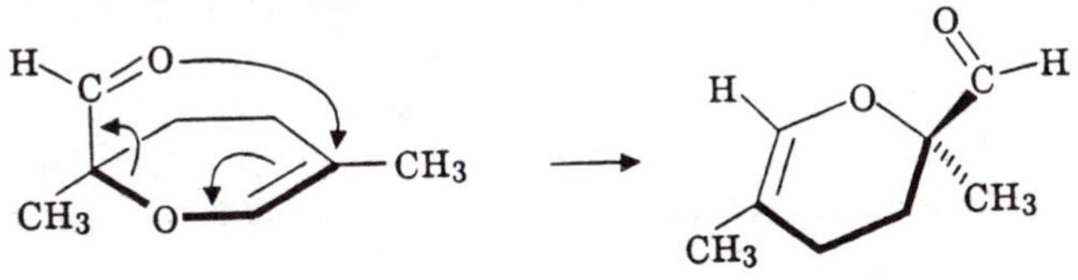

(156)

the Woodward and Katz experiments, except that it is clearly somewhat closer to the Diels–Alder reaction, because rearrangement and dissociation have very nearly the same free energy of activation. It would have been nice to find some other criterion that would show whether the oxy Cope rearrangement that Ray Lutz studied proceeds on the same, or widely divergent, potential-energy surfaces.

Another question about the Diels–Alder reaction was how *endo*- and *exo*- addition products to cyclopentadiene are interconverted. A classic example is the change of **248a** to **248b** at 180 °C (equation 157).

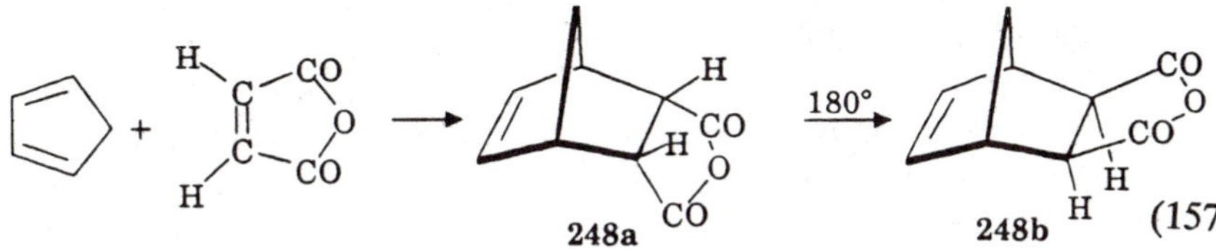

248a **248b** (157)

This reaction had been investigated by Jerry Berson and co-workers while Jerry was still at USC, by the device of running the reaction in the presence of labeled maleic anhydride and finding that the label was not incorporated into **248b** as fast as **248a** was converted to **248b**. This finding suggests that the isomerization does not involve reversion of **248a** to maleic anhydride and cyclopentadiene followed by readdition to give **248b**.[268]

Isomerization without reversion can imply three different mechanisms. The first possibility is a breakdown of **248a**, but instead of having the maleic anhydride and cyclopentadiene coming apart, they could stay briefly in a solvent "cage", then recombine to give **248b** (equation 158). Second, the maleic anhydride and cyclopentadiene

248a → [cyclopentadiene + maleic anhydride (H–C=C–H, CO–O–CO) in solvent cage] → 248b

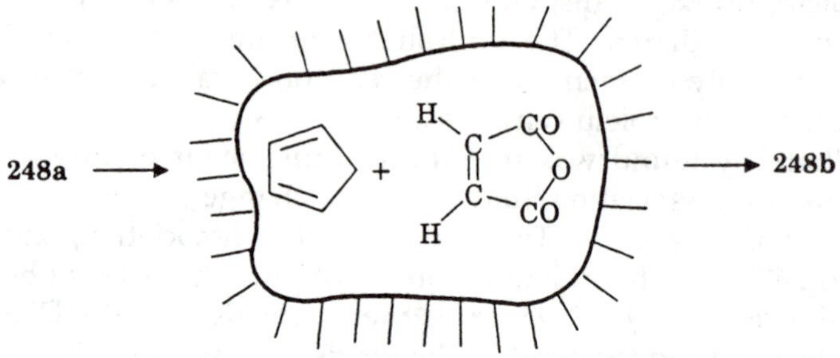

(158)

could be held together in parallel planes by "π-complex" formation, first in the configuration that would revert to **248a** and then, by a simple rotation of the planes, come into the configuration that would lead to **248b** (equation 159). Third, the bond connecting the two CH–CO groups could break to give a diradical, **249a**, which could flip over to a different diradical, **249b**, which then could go to **248b** (equation 160).

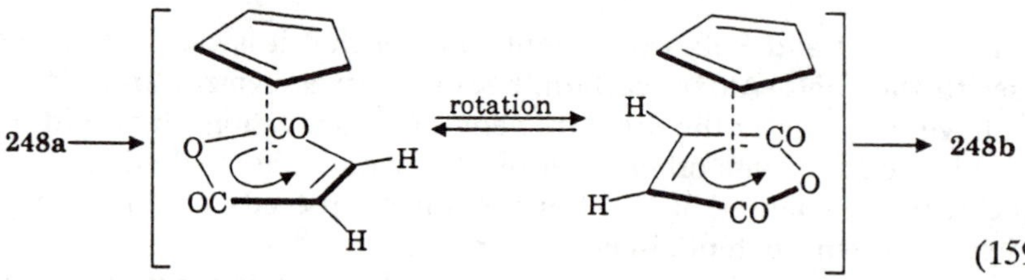

248a → [π-complex ⇌ (rotation) π-complex] → 248b

(159)

Bob Woodward seemed intensely interested in this problem, and I was inspired to show experimentally what was happening. The idea was to make **248a** labeled with ^{14}C in one particular CO group (marked with *), **250a**, but not make its mirror image, **250b**. Then, depending on the reaction mechanism and starting with only **250a**, we could get either **251a** or **251b** or a mixture of them. The route of equation 158, with maleic anhydride and cyclopentadiene presumably bouncing around in a solvent cage, would be expected to give a mixture of **251a** and **251b**. In contrast, exclusive operation of the mechanism of equation 159 would

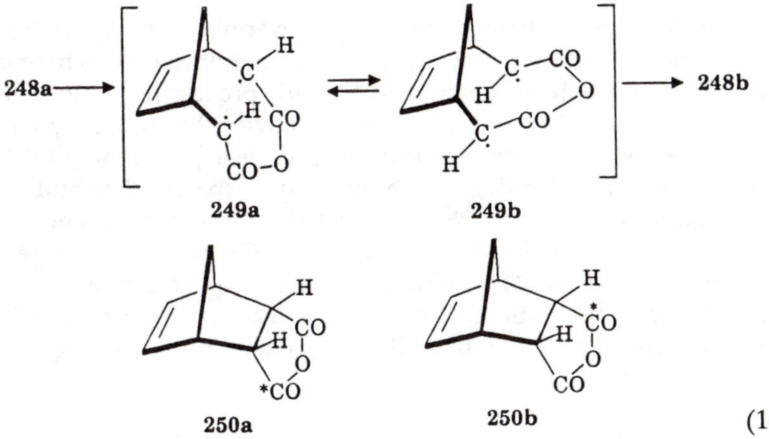

248a → [249a ⇄ 249b] → 248b

250a 250b (160)

give only **251b**, while exclusive operation of equation 160 would give only **251a**.

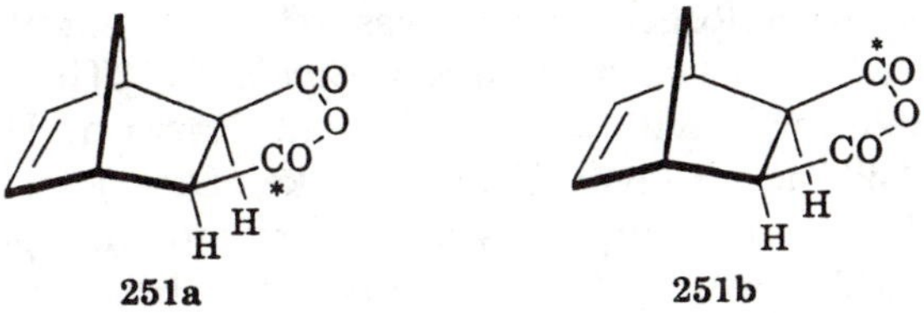

251a 251b

The problem with all of this is that preparation of **250a** is far from trivial, and the same is true of finding out whether the product is **251a**, **251b**, or a mixture of them. Ulrich Scheidegger, an immensely skillful Swiss postdoctoral fellow, was able not only to synthesize **250a**, but to prove that it contained no **250b**, by a route that could also be used to distinguish between **251a** and **251b**. This was a year's work and required both a many-step synthesis and a many-step degradation procedure.[269a]

Then Camille Ganter, another talented Swiss postdoctoral fellow, came aboard to carry out the actual isomerization experiment and analyze the product for its content of **251a** and **251b**. Being an extremely careful worker, Camille first wanted to check the isomerization conditions. Berson and co-workers had used decalin as the solvent at 180 °C. Camille repeated the conditions of their experiment and noticed that not all of the specified excess of maleic anhydride was in solution! Little globules of liquid maleic anhydride were drifting around.

This result was potentially devastating, because the conclusion that **248a** goes to **248b**, at least partly, without reversion to maleic anhydride, and cyclopentadiene depended on everything being in solution. Indeed, when Camille ran the reaction in *t*-pentylbenzene, where everything was in solution, reversion was the *only* process he could detect by which **248a** goes to **248b**.[269b] Berson was obviously not happy with the result, but we weren't either, because it was clearly not going to do any good to isomerize Scheidegger's beautifully prepared **250a** and have it give a mixture of **251a** and **251b** by reversion. We had expended enormous time and effort on a will-o'-the-wisp. On balance, I suspect that even if **248a** did go to **248b** without reversion and we could determine the mechanism, the ratio of effort expended to significance achieved would not be very favorable. But the manner of investigating the mechanism seemed to be *such a nice idea*!

Organic Syntheses

My association with Roger Adams was relatively slight until I was elected to the editorial board of *Organic Syntheses.* "The Chief", as his colleagues called Roger, with James B. Conant (Harvard), Hans T. Clarke (Eastman Kodak) and Oliver Kamm (Parke, Davis), started *Organic Syntheses* in 1920. The purpose was to facilitate the preparation of research chemicals, which, for the most part, had previously been imported from Germany. The idea was to publish a series of books, each containing 20 or so tested recipes for preparing particularly useful compounds, as well as illustrating good laboratory practices and providing model reactions for preparing related substances. The enterprise was fabulously successful and became Organic Syntheses, Inc., a nonprofit corporation that is run by a board of directors with Adams, Richard T. Arnold, Nelson J. Leonard, and Blaine C. McKusick as successive presidents. The volumes describing the preparations are assembled by an editorial board of eight members, a new member being elected each year, and each serving as editor-in-chief during the sixth year of service on the editorial board. On completing service, one graduates to the editorial advisory board. The affairs of the editorial board are supervised by the secretary, a 10-year position with some remuneration. At the completion of this 10-year term, the secretary starts editing a collective volume that includes almost all of the preparations in the annual volumes of the past 10 years.

Each preparation published in *Organic Syntheses* is checked meticulously, and the directions are made as clear as possible to anyone reasonably skilled in organic preparative work. Particularly during recent years, precautionary notes are appended with respect to potential hazards. Many very hazardous preparations are included, but the known perils of each are carefully detailed. Of course, many submissions are rejected as not being sufficiently general, or not of important compounds, or else as uncheckable. When difficulties are involved, substantial efforts are made to reconcile suggested changes with the submitters, and, many times, negotiations drag on for years. Each editor-in-chief, who needs 20 checked and edited preparations ready to go at press time, can usually count on a legacy of five to eight preparations from the previous editor. If prospects for having enough look poor, the editor-in-chief may take on, for checking by his or her own research group, many more preparations than required, preferably sure-fire ones, in hopes of filling the volume, with some excess. As might be expected, not all of them work out, and this is why the editor-in-chief is given a few extra years to clean up any backlog.

Most of the cream of U.S. synthetic chemists had at one time been involved with *Organic Syntheses,* the members of which, up to 1940, included many names already mentioned in other connections: Adams, Conant, Fieser, Gilman, Johnson, Marvel, and Whitmore. As a physical organic chemist, I was greatly surprised to be invited to join the editorial board, even though, at Art Cope's behest, I had submitted a preparation of tetraethyl orthocarbonate, $C(OC_2H_5)_4$, several years before. The invitation was delivered by Nelson Leonard, a man for whom I had increasing respect and liking. I was not too happy to take on further responsibilities, but it seemed wrong not to take advantage of this opportunity.

The modus operandi of *Organic Syntheses* is relatively simple. Much of the work is carried on by mail, and the editors meet twice a year for most of a day to assign preparations, work out difficulties, plan for future volumes, choose new editors, and the like, with the current editor-in-chief presiding. Following that meeting, the editors join with the editorial board for cocktails and dinner at one of the best (and most expensive) restaurants in town. In the old days, the dinner would be followed by announcements, general discussions of new proposals, and usually by an Adams report on whatever was on his mind at the time. He often ribbed his colleagues, without malice, but nonetheless, severely. After Adams finished, the meetings often degenerated into the semblance of a stag smoker, with rounds of off-color jokes, the most tasteless of which usually came from the representatives of the publisher, John Wiley & Sons. Rather prodigious quantities of spirits, wines, and brandies were consumed, and some of organic chemistry's most renowned practitioners had to be helped off the scene. My first encounter with these bacchanalian festivities was in Atlantic City, in the spring of 1956. I made it back to my hotel under my own steam, but I spent the next day in bed, a lesson I did not forget. Adams complained endlessly, but not really seriously, about the expense of the dinners, so at one of the very best dinners, he was served a hamburger, to great merriment. The fact is that the editors received no compensation for their work, just good fellowship and the semiannual dinners. In more recent times, the format of the dinner has been changed to include spouses and, every other year, the current Roger Adams Medalist is honored.

The submission of preparations for checking and publication was on a catch-as-catch-can basis. By the time I came on board, synthetic chemistry had recovered from the wartime emphasis on antimalarial drugs and specialty materials and was coming into full flower, but the quality of submissions to *Organic Syntheses* was not generally keeping up with this development. Furthermore, the sales of volumes had begun to fall off, and the level of interest among graduate students had diminished. It was a good time to reassess the operation and see what could be done to liven things up again. Not that *Organic Syntheses* was in any financial difficulties. The relatively meager royalties from the sale of books had been shrewdly invested in Eastman Kodak, IBM, and other companies, and the resulting corpus was out of proportion to book earnings. The treasurer and the members of the board, who supervised the investments collectively, had made an almost unending stream of superb decisions. So far as I know, not one of them did so well for any personal account.

We needed to make some changes, and it seemed strange to the newcomers that there was no program of soliciting preparations that would be at the leading edge of current synthetic needs and techniques. When asked about this, the old hands said that solicitation had been tried but had not been successful. I was to find out more about that later. But there was additional skepticism about the editors being committed to testing and accepting solicited preparations that might not be really satisfactory. Another problem was a stereotyped publishing format; we made some improvement by having each submitter supply a brief section as to the value and generality of the preparation, updating the *Organic Syntheses* style guide, and having each editor-in-chief write an introduction to the specific volume to summarize its special features.

Being, by this time, influenced by W. A. Benjamin's incentive programs for generating book manuscripts, sales, and so on, I suggested that the volumes of *Organic Syntheses* be sold to graduate students at half price. This proposal caused a lot of controversy. Our publisher, John Wiley, was very negative, and the company wanted to distance itself completely from any aspect of administering such a program. To my surprise, although Adams and Marvel were strongly enthusiastic, other members of the advisory and corporate boards were greatly concerned about the possible effect on the fiscal health of the organization. That concern, plus revulsion against the notion of a graduate student "handout", caused W. W. Hartman of Eastman Kodak, then treasurer of

Organic Syntheses, Inc., to resign, making clear in the process that he was opposed to squandering the corporation's assets, which he had a large part in developing. Despite these fears and the lack of Wiley enthusiasm, when the program was initiated, it and the other changes that were made brought about greatly increased interest in *Organic Syntheses.* Instead of being decimated, the corporate assets increased steadily.

The growth of assets came about during the period in which the American Chemical Society was mounting a vigorous campaign for donations toward a new headquarters building in Washington. Roger Adams and Speed Marvel were very important people to the ACS and, in fact, Marvel was a leader in the campaign drive. The relative affluence of the corporation and its increase without any programs tending to limit growth (except the half-price book sales that turned out to have the opposite effect) led to suggestions that a very sizable grant be made to the ACS for a library to be named after Roger. This was not to the liking of the younger members of the group—many of whom had little or no connection with the ACS, except through its publication of research journals. Furthermore, the ACS had no connection with *Organic Syntheses,* and there was little incentive to work so long and hard for a facility that few, if any, would ever be in a position to use. Nelson Leonard, one of Marvel's colleagues, came up with the alternative of creating a Roger Adams Medal and Award for contributions to organic chemistry, with a monetary value that would be spectacular relative to the nearly uniform $1000 prizes then associated with the various awards made by the ACS. To broaden support, Nelson enlisted me and also Art Cope and David Curtin, two stalwarts of Organic Reactions, a somewhat similar corporation set up by Adams in 1939 to publish reviews covering important synthetic reactions with a uniformity of style and quality of coverage not achieved by *Chemical Reviews,* which was a pretty catch-as-catch-can operation. Of course, Cope was also a veteran of *Organic Syntheses,* but by 1958 he was heavily involved in both Organic Reactions and the ACS. The original plan was hatched over breakfast in San Francisco after an *Organic Syntheses* dinner, at which Marvel had pushed hard for the donation to the ACS. Setting up an award such as the Roger Adams Award is a lot of work, establishing who administers it, how often, how much, chosen for what, where awarded, when to start, how international, and so on.

Nelson carried on the negotiations with skill, but there was a stumbling block in that an obvious first awardee (other than Adams himself, who, though clearly pleased with the idea, distanced himself altogether from the process of setting it up) was the redoubtable Robert Burns Woodward. Some of the older members of *Organic Syntheses,* particularly some of those at the University of Illinois (notably C. S. Marvel and R. C. Fuson, but not Roger Adams), were far from Woodward fans. This prejudice appeared to stem from their interactions with him during his summer teaching period at the university in 1941. These worthies were strong in their opposition to the choice of Woodward for the initial award. The organizing committee decided to forestall ill feelings by choosing Roger Adams as an honorary recipient, along with D. H. R. Barton of Imperial College, London, as the first regular awardee to receive a suitable gold medal and a cash prize of $5000. Not only was Derek Barton of the requisite chemical stature, but there was a very strong desire to establish the award firmly as a truly international one, right from the beginning. The award ceremony, held at the National Organic Chemistry Symposium at Seattle in 1959, was a smash success.

In 1959, $5000 was a very considerable sum, but did *Organic Syntheses* go broke? Not at all. Every plan to spend more money on good causes has wound up benefiting the fiscal balance sheet, which, even after the collapse of the stock market in October 1987, showed assets of about $2 million. After the first Adams Award to Barton, the administration of the award was turned over to the American Chemical Society, with the predictable result that Bob Woodward was the second recipient. Thus, firmly established, the award has done very well indeed. I was fortunate in that the prize was increased to $10,000 the year before I received it. I got a lot of kidding about that, and all I could say was that I was sorry not to have voted for an increase to $25,000. My $10,000 mostly went into buying half of Don and Jean Cram's sailboat, but that is another story, for another time.

The Adams Award was the premier award of organic chemistry until after Art Cope died and his estate was finally settled. Art had done well in the stock market and left a very substantial sum to the American Chemical Society for the establishment of the Arthur C. Cope Award for organic chemistry. For many years, Art was clearly Adams' heir as the de facto leader of U.S. organic chemistry; he followed, even surpassed, Adam's example in service to the American Chemical Society. He also followed Adams as editor of *Organic Reactions*, and his research was enormously successful, but, although highly respected, he never really achieved the same level of personal popularity as Adams. In later years, there seemed to be some difficulties between the two men, and it may have been that Art intended the Cope Award to eclipse the Adams Award. Certainly there was money enough to do that, and, in the early years, the ACS was thinking in terms of a $50,000 award. A number of us who were involved and concerned with awards in organic chemistry put on a lot of pressure to keep the Cope award more or less on a par with the other ACS awards. The ACS responded well by setting up the Cope Award with a personal prize comparable to the Adams, but going beyond by also giving a sizable research grant as well as establishing a number of Cope Scholar Awards designed primarily for promising young organic chemists. Cope's money is helping and encouraging many chemists.

With respect to upgrading *Organic Syntheses* by especially desirable preparations, the senior members of the board maintained that it had been tried but failed, without giving much detail. The editors, however, thought it was a good idea, well worth trying, and I suggested that we solicit the classic Bartlett–Knox[270] preparation of apocamphanecarboxylic acid from camphor (equation 161). Everyone was

CH_3 CH_3 CH_3 O camphor $\xrightarrow{H_2SO_4}$ CH_3 CH_3 CH_2 SO_3H O **252** $\xrightarrow{PCl_5}$ CH_3 CH_3 CH_2 SO_2Cl O **253** $\xrightarrow{KMnO_4}$ CH_3 CH_3 CO_2H O **254**

$\xrightarrow[HCl]{Zn(Hg)}$ CH_3 CH_3 CO_2H **255** (161)

enthusiastic about doing so, but I had some worries about approaching Paul, because I remembered that whenever I had mentioned *Organic Syntheses* to him, he would snort (and Paul was a good snorter) something about "that low-grade outfit". I found out why. He had been solicited by the senior members before the war, and indeed he had been solicited to submit the Bartlett–Knox preparation of equation 161. It is not a trivial task to get a preparation in shape for submission to *Organic Syntheses*, but it was duly submitted in 1939. Later, Paul got a letter from Ralph Shriner, then the secretary of the editorial board, saying that the board had decided that the synthetic sequence would be more useful if it used D-camphor rather than D,L-camphor and would Bartlett and Knox start over again with D-camphor? This practically blew Paul's mind; he was unwilling to make the effort, even though the board had a point in that **252**, **253**, and **254** have potential use as agents to resolve racemic mixtures into the separate enantiomers, even though the final product, **255**, is achiral. Paul relished telling me this story. Later, when I urged him to dig up the directions that he had submitted before, I assured him that not only would we take them starting with D,L-camphor, but I would check them personally if necessary (that youthful enthusiasm that usually leads to later problems). Paul's reply was in the vein of his characteristic humanness and humor.

> As an enthusiastic collector of Robertsiana, I am delighted with the masterpiece of tact which shows that you have become as adept with the silk glove as you used to be with the bludgeon. How can I fail to admire the vision and enlightenment of the current *Organic Syntheses* board in extracting these jewels from its musty files and giving them a wipe with its polishing cloth? As you request, I am enclosing the yellowed sheets, bearing the original directions, considerably more promptly than we submitted them after the board's previous request some 20 years ago. The only thing you can do to heighten my pleasure is to put in a footnote giving the dates of submittal and of checking of these preparations.

As I have intimated, there were problems and difficulties. The outcome of all of this is perhaps best appreciated by way of my long plaintive letter to Henry Baumgarten, then secretary of the editorial board. Most of the letter, which I also sent to Paul, follows:

> Herewith edited and retyped copies of the three Bartlett preparations which checked:
>
> D,L-10-Camphorsulfonic Acid
> D,L-10-Camphorsulfonyl Chloride
> D,L-Ketopinic Acid
>
> I regret to say I struck out badly on apocamphane-1-carboxylic acid and I am convinced that the procedure is deficient. A certain amount of internal evidence makes me feel that the preparation was not actually run by the submitters as submitted. For example, it is recommended to remove the desired acid from a combined toluene–ether extract with 25% sodium hydroxide solution. When you add 25% sodium hydroxide solution to the extract, a voluminous white precipitate clogs the separatory funnel and the operation ends right there; 10% sodium hydroxide works much better although not well. Furthermore, despite other directions in the original paper, the product is supposed to be crystallized four times from 50% ethanol.

Unfortunately, 20 g of product dissolves in 25 mL of warm (not hot) 50% ethanol and it takes a better man than me to perform this operation four times and have any significant amount of product left.

Personally, I feel that the use of the Clemmenson–Martin procedure for the reduction of ketopinic acid was a mistake because ketopinic acid is pretty soluble in hot water and is not decarboxylated on heating. The trouble with the reaction as described is basically that it just doesn't go far enough. I tried it two ways—four half-scale runs in all. In the first runs, I used a round-bottomed flask and got a fair amount less of crude (but thoroughly dried) product than reported. After *one* recrystallization, I fell pretty far behind the advertised yield for *four* recrystallizations, but the really unnerving part was the NMR analysis (the *gem*-methyls are really dead giveaways), which showed the recrystallized product from one flask to be 75% reduced while the product from, as far as I know, a wholly identical flask was only 50% reduced.

At this point, I went to *Organic Reactions* and boned up a little on Mr. Clemmenson and decided that perhaps I should use an Erlenmeyer flask to bring the zinc and toluene layers into juxtaposition. Two identical runs made this way gave crude products corresponding to 45% and 53% extents of reduction, and I am sure that after four, or some like number, of recrystallizations, these products could be freed of ketopinic acid, which is rather more soluble in polar solvents than apocamphanecarboxylic acid. Nonetheless, even if the yields could be duplicated (and I don't think I can do it), this does not seem to be the kind of preparation *Organic Syntheses* wants—the reduction should be at least 85% complete, not just half way.

Even though I recognize the risk of removing Paul forever from any possibility of being buddies with *Organic Syntheses,* I would still like to twit him a bit about the Knox preparations because they illustrate so beautifully the gap between preps which work when you just follow the directions (usually because real thought has been given to where the checker could experience trouble) and preps which work only when you read between the lines and guess at what the submitters neglected to tell you.

I had a bright, but inexperienced, Indian graduate student try to check the preps on the first time around and all he had was trouble. Everything worked to a point, which was usually about half as good as the preps should have worked. After I started repeating the work myself, I realized that the directions were really not nearly as good as they read and trouble is everywhere for the unwary.

In the sulfonation of camphor, the product is supposed to be washed with ether and air dried. From this I deduce that Knox must have run the reaction in a really cold winter with the normal atmospheric moisture completely frozen out on the windows, because we found the product to be rather reasonably deliquescent in southern California summer air. I assume you can imagine the messy cake one gets when the solid is washed on a Büchner funnel with ether.

The sulfonyl chloride prep, if run as submitted, outside of a hood (not mentioned), could really ruin a laboratory (and a chemist) with HCl and $POCl_3$ fumes. My Indian student, after finding that the vermiculite packing around a PCl_5 bottle will just not convert a sulfonic acid to the corresponding chloride, took the submitters quite literally and stored his crude *moist* sulfonyl chloride over calcium chloride in an ordinary desiccator at summer room temperatures—he never could figure out why his yields of ketopinic acid in the next step were low!

The ketopinic acid preparation worked fine, but it was a little disconcerting to find that, when the directions call for purifying the product by recrystallization from "a little hot water", only about 1 g of the acid dissolves in 15 mL of hot water, which makes quite a bit of water for 38–45 g of crude product.

I regret to have failed so ignominiously at pushing through successfully as the last full measure of my devotion to *Organic Syntheses* and my revered friend—Harvard's brightest light.

Interestingly, the Indian graduate student who attempted to check some of the preparations was extremely intelligent, even if then inexperienced in the laboratory, and has subsequently had a distinguished career as a biochemist.

Paul's reply to my remarks about checking the preparations is typical.

One of the most reassuring manifestations in this year of tropical storms and reactionary crackpots has been the copy of your exuberant outpouring of youthful vigor on

> the subject of apocamphanecarboxylic acid. Nothing can be too wrong in a world where the broad Roberts shoulders momentarily lay aside the responsibilities of state to bend over innocuous reactions of vermiculite and nocuous reactions of phosphorus pentachloride.
>
> How different history might have been had the men who staffed *Organic Syntheses* 25 years ago been made of such stuff. I also appreciate the feeling of outraged reverence which I read between the lines of your letter to Henry Baumgarten. It is obviously as if you had discovered a lot of misspelled words in *Paradise Lost* and suddenly realized that Milton was still alive and you should watch your language in calling attention to these.

Organic Syntheses continues to flourish right up to date under the current presidency of Blaine McKusick, vastly aided by the investment choices of the more fiscally astute members of the board. Most of my input into the board deliberations is to encourage further spending, but the corpus continues to grow despite a program of lectureships given at the institutions of the members of the active editorial board and the furnishing of free paperback copies of current volumes to members of the Organic Chemistry Division of the American Chemical Society as well as to corresponding organizations in several foreign countries.

More on the Structure of the $C_4H_7^+$ Carbocation: A Chemical Chimera

When I moved to Caltech, I gave high priority to continuing to investigate the structure of the $C_4H_7^+$ carbocation, on which Bob Mazur had done such excellent work (*see* pp 63–74, 79–84). Bob's results had aroused a great deal of interest, and even Linus Pauling decided to turn his hand to assigning a structure. Thus, in the third edition of *The Nature of the Chemical Bond*, written before I came to Caltech, Linus suggests that structure **256** is likely to be favored on the basis of resonance

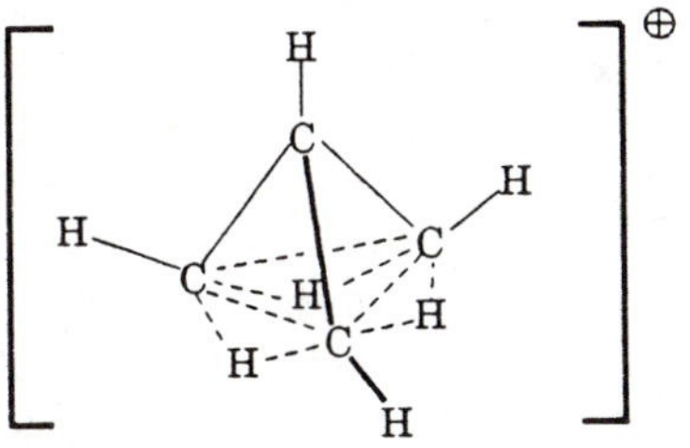

256

theory and his version of the structures of boron hydrides.[271] In fact, his structure fits a lot, but not all, of the experimental data.

We had deduced our possible structures for $C_4H_7^+$ on the basis of unusually high reactivities, the nature of the products formed, and the extensive isotope-position rearrangements that took place on treatment of cyclopropylmethylamine-α-^{14}C with nitrous acid. It now seemed appropriate to go in two directions, first to check out the ^{14}C results in several different ways and then to see how well the results translated into related systems.

Our review of Mazur's ^{14}C work was quite thorough and did not change our conclusions as to the degree of isotope-position rearrangement. In addition, we found that the formation of 3-butenyl chloride from cyclopropylmethanol-α-^{14}C with zinc chloride and hydrochloric acid gave an even distribution of the ^{14}C label over each of the three $-CH_2-$ groups of the chloride, with less than 0.1% of the ^{14}C at the $=CH$ carbon.[272] Furthermore, in some later work,[273] when we went in the reverse direction and treated 3-butenylamine-1-^{14}C with nitrous acid, besides the extraneous hydrogen-shift products (*see* pp 69–70), we got cyclopropylmethanol and cyclobutanol products, which again were not labeled quite as expected for complete equilibration. Thus, you get near, but do not achieve, complete ^{14}C equilibration, even though in this reaction you would certainly not expect to have S_N2-like direct displacements that would reduce the extent of isotope-position rearrangement.

We had a further problem with Mazur's results through the claim by Brown and Borkowski[274] that the kinetics of the solvolysis of cyclopropylmethyl chloride showed no "internal-return" isomerization (*see* p 71). Herb Brown ruled out internal return, because his co-worker had found good first-order kinetics for the solvolysis of this chloride, which had been prepared by liquid-phase chlorination of methylcyclopropane (*see* p 74). Herb believed that Mazur's chloride was impure, and that was why Mazur's solvolysis did not go to completion. We had several hot arguments on this score, the last in the presence of Frank Westheimer. Frank was amused and finally insisted on hearing just how we each had determined our rates. I explained and Herb explained. Then Frank resolved our differences. What Herb's man had done was the customary thing, that is, he measured the rate of solvolysis by determining the quantity of acid liberated as a function of time until the reaction was effectively complete, at which point an "infinity titer" was recorded. The percent of reaction at any point was then the amount of acid liberated to that point, divided by the "infinity titer". The drawback with this approach is that it cannot detect internal return. If one uses 1 mol of the chloride and 50% undergoes internal return to an unreactive chloride, the infinity titer would be 0.5 mol, and a reasonable facsimile of good first-order kinetics would be obtained for the solvolysis of the material that did actually solvolyze. Similar mistakes in detecting internal return in solvolysis have been reported by others.[274]

In the course of reinvestigating the purity of the cyclopropylmethyl chloride used in the solvolysis and the rearrangements, Marjorie Caserio performed some very nice experiments using deuterium labeling. Her work showed with great clarity that isotope-position rearrangement was taking place, not only in the overall solvolysis, but even in the internal-return isomerization. Furthermore, she and another postdoctoral fellow, W. H. Graham, were able to demonstrate very extensive isotope-position rearrangements, both in the reaction of cyclopropylmethanol and cyclobutanol with thionyl chloride and in the acid-catalyzed rearrangement of cyclopropylmethanol to cyclobutanol.[55] Clearly, near equilibration of the $-CH_2-$ groups is not just a property of the amine–nitrous acid reaction.

During this period, I was rapidly backing away from the pyramidal carbocation as the principal intermediate. It is not a logical intermediate for the observed formation of a large proportion of cyclobutyl products, and it predicts a greater degree of rearrangement than is actually found experimentally. Bicyclobutonium carbocations somehow equilibrating in a way that leads to extensive, but not complete, isotope-position rearrangement, perhaps through the pyramidal structure **63**, seemed much more reasonable (equation 162).

$$\left[\begin{matrix} & {}^{14}CH_2 & \\ CH_2 & & CH \\ & CH_2 & \end{matrix}\right]^{\oplus} \rightleftarrows \left[\begin{matrix} & CH_2 & \\ {}^{14}CH_2 & & CH \\ & CH_2 & \end{matrix}\right]^{\oplus} \rightleftarrows \left[\begin{matrix} & CH_2 & \\ CH_2 & & CH \\ & {}^{14}CH_2 & \end{matrix}\right]^{\oplus} \qquad (162)$$

But, even as an intermediate, the pyramidal cation did not look so good when graduate student Merlin Howden took a hard look at simple molecular orbital formulations. He showed that the electronic energies were not enough to overcome the substantial strain of bringing the orbitals into close enough proximity to give favorable overlap. However, Merlin did show that there was a fairly good geometry for the bicyclobutonium carbocation.[275] I did not see how to prove this hypothesis experimentally, but I hoped that study of some closely related systems might help to clarify matters.

The way we approached this problem was to substitute one or more groups, particularly methyl groups, on each of the possible different carbons and see what happened in various kinds of reactions. Some beautiful chemistry came out of this, not only of carbocations, but also of radicals and anions.[190,191,276–278] It would take many pages to even summarize, much less describe, the results. The one really important message that came through was that substitution of groups drastically alters the way the positive charge is distributed over the carbons of the cations and, in so doing, substantially alters their structures.

The concept can be illustrated by some specific examples. First, consider the substitution of a single methyl group on the CH carbon of the unsymmetrical bicyclobutonium carbocation, **257**, which would for-

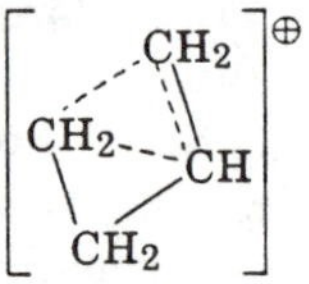

257

mally lead to **258**. In **258**, the positive charge will be located preferentially on the carbon carrying the methyl group. This structure could make **258** less stable than the symmetrical bicyclobutonium carbocation, **259**, if that entity, as expected, were to have more charge on the CH carbon (equation 163). Proponents of Bishop Occam's Razor might very

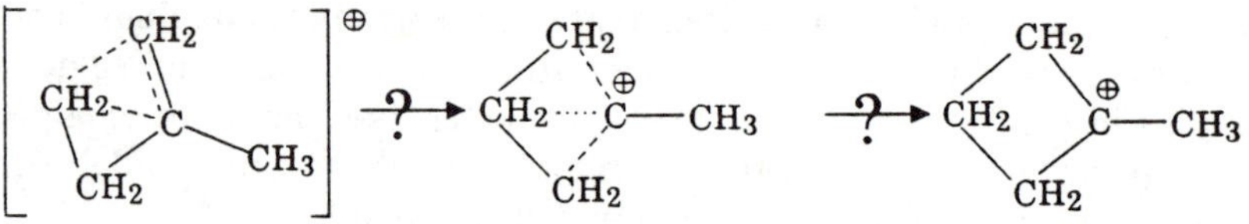

258 **259** **260** (163)

well contend that the classical methylcyclobutyl–carbocation formulation, **260**, is simpler and thence preferable.

In fact, **260** generally does account for the reaction products, because if you start with either **261** or **262**, then you do get the 1-methylcyclobutanol expected from the simple methylcyclobutyl cation, **260**, if you have a typical carbocationic mechanism operating, as in equation 164. However, there is a serious problem with **260**, which has

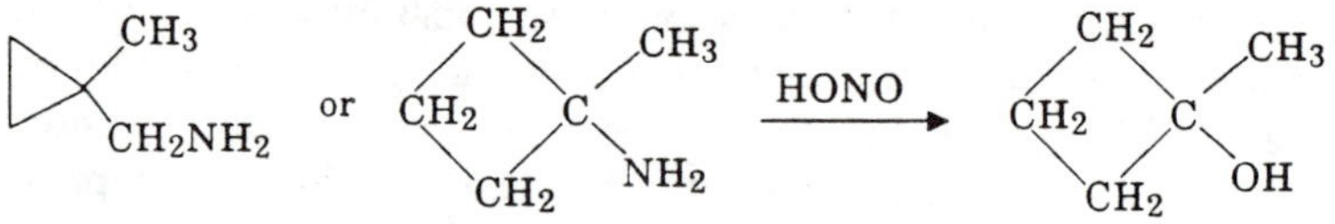

261 or **262** HONO (164)

to do with its ease of formation as measured by solvolysis rate studies. If we apply the idea that there should be substantial angle strain in classical cyclobutyl carbocations (*see* pp 70–72), then it is surprising that 1-methylcyclobutyl chloride is only about one-fifth as reactive as 1-methylcyclohexyl chloride, albeit about one five-hundredth as reactive as 1-methylcyclopentyl chloride.[279] There are other problems as well.[280a,b]

Another, somewhat more complex, example is substitution of a methyl group on a CH_2 group of **257**, for which there are three possibilities, shown by **263**, **264**, and **265**. Here, we expect that **263** and **264** will be less stable than **265**, where the methyl group will be located on a carbon carrying more positive charge, and rearrangement of **263** or **264** to something resembling **265** should occur (equation 165). Again, we

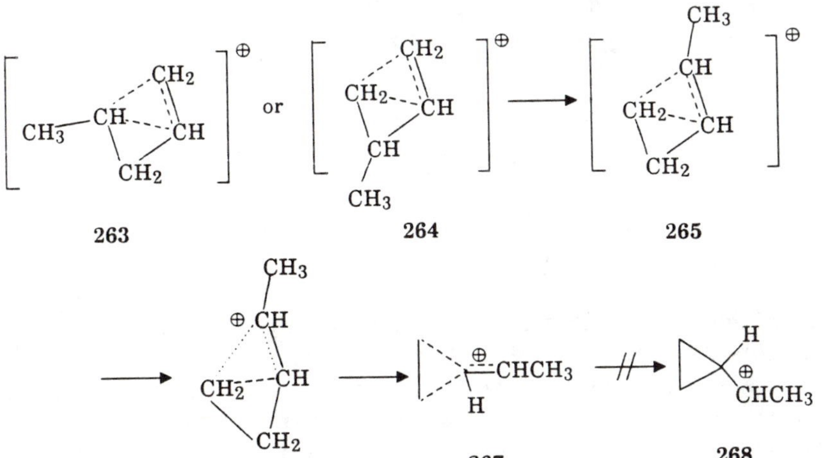

263 or **264** → **265**

→ **266** → **267** ⇸ **268**

(165)

would anticipate that the methyl group would tend to make the carbocation more like a classical cation. One possibility would be an unsymmetrical cyclopropylmethyl-like bridged carbocation, **266**, or a methyl-substituted bisected cyclopropylmethyl cation, **267**. There is no reason to suspect that the classical cation, **268**, would be the favored intermediate, because it would be a simple secondary carbocation.[276]

With these ideas, we could predict how the bicyclobutonium cation would respond to substitution at its various carbons, and we could generally predict quite well what would happen in carbocationic reactions, under conditions of both kinetic and thermodynamic control.[276]

One very nice result, which I do not think many of us expected, was the lack of isotope-position rearrangement that graduate student Bruce Kover found in the ring closure of the cation from treatment of the deuterium-labeled amine, **269**, with nitrous acid. The possibilities were for the resulting cation, **270**, to undergo a homoallylic ring closure to give the cyclopropylmethyl-type cation, **271**, thence, rearrange to the cations, **272a**–**272b**, and, finally, go on to the isotope-position isomers of 1-methylcyclobutanol, **273a**–**273b**, in the ratio of 2:1 (equation 166). The alternative is direct closure of **270** to **272a** and, thence, to give only **273a** and no **273b** (equation 167).

$$CH_2{=}C(CH_3){-}CH_2{-}CD_2{-}NH_2 \ (\mathbf{269}) \xrightarrow[-N_2,H_2O]{HONO} CH_2{=}C(CH_3){-}CH_2{-}CD_2^{\oplus} \ (\mathbf{270}) \longrightarrow \mathbf{271}$$

$$\longrightarrow \mathbf{272a} + \mathbf{272b}$$

$$\longrightarrow \mathbf{273a} + \mathbf{273b} \qquad (166)$$

$$\mathbf{270} \longrightarrow \mathbf{272a} \longrightarrow \mathbf{273a} \qquad (167)$$

Direct closure was, in fact, the way in which the reaction proceeded with *direct* formation of the most stable cation.[281]

Much of the work that we did on the reactions of the bicyclobutonium carbocation was done at the start of the public debate over nonclassical carbocations, especially between Herb Brown and Saul Winstein. Why did Herb Brown take on this particular crusade? His version is well documented[282a] and is a model of altruism. Herb was a zealot for steric effects and, indeed, he used all of his *FBI* strains (*see* p 74) to account for phenomena that others held to be better attributed to nonclassical ion formation. Also, Herb complained often and bitterly about not being part of the physical–organic establishment, which was very largely under the influence of Paul Bartlett. Not that Bartlett sought a leadership role; but, because of his superb work, his position at Harvard, and his widespread connections through students, postdoctoral fellows, and visitors, he was clearly the leading figure in the field. Louis Hammett was enormously respected, but, because he was not an organic chemist, his influence was much less. Winstein was quite close to Bartlett, and it was not long before Saul became the guru of the solvolysis field in the United States. Of course, Ingold had that position abroad. Herb Brown's "exclusion" from whatever it was that could be called the inner circle seemed almost self-imposed. One reason may have been that Herb did not seem very interested in other people's work and he was dogmatic about his own.

Another factor that appeared to motivate Brown was his relationship with Saul Winstein. There was clearly some early competitiveness between them. Herb relished telling how, when there was an opening for an instructor at the Illinois Institute of Technology for which both he and Saul applied, Saul got the position, but the department chairman told Herb that he was "overqualified" by virtue of the success of his earlier work at Chicago. As the "nonclassical" battle wore on, relations appeared to get rather more acrimonious than expected from just a healthy spirit of competition.

The early published salvos in the battle were fired by Herb. At first, he was rather indiscriminate as to what and whom he went after and he also oversimplified the experimental situation as well. His 1962 Transition State Symposium lecture[282b] was almost foolhardy in its approach to the problems posed by, and the work already done on, nonclassical cations. The thrust of this paper was to spread a measure of doubt on everybody's work in the field, without going into detail. Quite a bit earlier, there had been a Reaction Mechanisms Conference at Brookhaven National Laboratory. I remember, when driving with Paul Bartlett back to Cambridge, how Brown's contributions to the discussions made Paul uncharacteristically angry and disappointed. Later, Brown narrowed his attention on the norbornyl and cyclopropylmethyl-related systems. However, with $C_4H_7^+$, he never really seemed to grasp the subtleties of the system, which made it hard for him to explain what was going on in a consistent way. The norbornyl system

Herbert C. Brown and Saul Winstein at an American Chemical Society meeting in Los Angeles in 1953. The photograph was taken by V. Prelog and lent by Sylvia Winstein.

has its subtleties too, with internal return and hydrogen migration. But Herb pretty much ignored these also in his attacks on Winstein's arguments for the intermediacy of **65** in the solvolysis of **64** (pp 85–86). As this fracas wore on, it was not a pleasant experience; it reached its zenith at the National Organic Chemistry Symposium in 1969. Both men were on the program, although Herb talked about boron hydrides. Saul took the occasion to reply with substantial vigor to Herb's repeated public assertions that the concept of nonclassical cations, like the legendary emperor, was bereft of clothes. The ensuing discussion could at the least be described as candid.

By this time, I had become tired of the contentiousness between the principal protagonists and decided that further effort in the area was not going to be worthwhile unless I had a really good idea. There seemed to be plenty of people to take up the cudgels. Besides, I had already had a long, hard session with Herb Brown and his colleagues at Purdue with respect to our work on the norbornyl system. Furthermore, there was later a rather formal debate with Herb, arranged by the Columbia University Chemistry Department when Herb and I were simultaneously Falk–Plaut lecturers at Columbia in 1957, a debate that I am sure did not set any records for polite propriety.

In the course of writing about my experiences in this controversy, I have been chided for not being more magnanimous with respect to Herb Brown. Frankly, it is not easy, even 20–30 years later. To be sure, the whole episode had its humorous side, as has been wonderfully documented by Derek Davenport.[134] But there was much more behind the scenes that rankled deep, particularly so for Saul Winstein, who was an unusually motivated seeker after the truth. He was genuinely troubled by the fact that he could not get Herb to understand the subtleties and complexities of behavior of systems that Saul felt there was no way to account for by what was known of classical carbocations. After Saul had tried to explain these problems to Herb, Herb would just blithely keep on going public with his criticisms, giving little, if any, indication that he had read what Saul had published or had heard what Saul had tried to explain to him.

Part of my paucity of magnanimity may be understood by the following incident that occurred about 1960. Herb had come to the Los Angeles area and was to give a seminar at Caltech following a visit to UCLA. He came over in the afternoon, and I talked with him in my office about recent research on nonclassical cations. He told me about some experiments relating to the norbornyl carbocation that I had not heard before, and which I had to admit to him seemed strong evidence against the nonclassical structure.

A few minutes before Herb's seminar, Saul arrived at my office. I told him that Herb had told me about these results that I agreed seemed pretty damaging, and what did he think? Saul said, "Yes, Herb also told me about those results yesterday, but I explained to him that we had looked into the situation and what he described was experimentally incorrect. The situation is not that way at all." I said to Herb, "You mean that you spent all that time telling me about those results and you had already heard from Saul that they were invalid! Why didn't you tell me what Saul had said?" I was furious to have been gulled by Herb in that way, and it was not with enthusiasm that I introduced Herb at the seminar a few minutes later. I am willing to give Herb his due for his pathbreaking research in the field of boron hydrides and hydroboration, and indeed I have.[283] But his behavior in the nonclassical carbocation field was often not collegial. It was not calculated to make friends, and it did not.

In the mid-1960s, a real bombshell was thrown into the carbocation field by the wonderful Nobel-class work of George Olah, now at the University of Southern California. George discovered how to make concentrated solutions of carbocations in what he calls "magic acid" (and magic it is!). If the solutions are kept suitably cold, one can take both proton and ^{13}C NMR spectra of carbocations.[133] Thus, past mistakes in speculating about the structures of carbocations can be brutally exposed through direct experimental test. The proton spectrum of $C_4H_7^+$ reported in 1970[284] is a dilly! It shows a *low-field* CH proton, which indicates that this proton is bonded to a rather positive carbon. There are two equally populated *higher-field* proton groups with different chemical shifts, which means that there must be *nonequivalent* methylene hydrogens.

This nonequivalence demands that we throw out the particular pyramidal cationic structure of $C_4H_7^+$ that is more favored by molecular orbital theory, even as the transition state for shuffling the carbons around, because it would have *equivalent* hydrogens on each methylene group. It is also significant that the methylene protons, as appropriate for their nonequivalence, have different couplings to the CH proton. They are, apparently, not coupled to one another, but this is not surprising because many kinds of methylene hydrogens have small $^2J_{HH}$ methylene couplings.

How does all of this jibe with what we found by chemical means? Not too badly. A set of bicyclobutonium cations that undergoes rapid equilibration, as in equation 162, fits the NMR rather well. Furthermore, Olah showed that the ^{13}C NMR chemical shifts predicted for some possible rapidly equilibrating $C_4H_7^+$ structures give best agreement for the unsymmetrical bicyclobutonium cation, **257**.[284] A necessary condition to fit with the observed spectra is that the bicyclobutonium cation be nonplanar and undergo the configurational change, **274a** to **274b** (equation 168), at a rate that is slow compared to the NMR time scale.

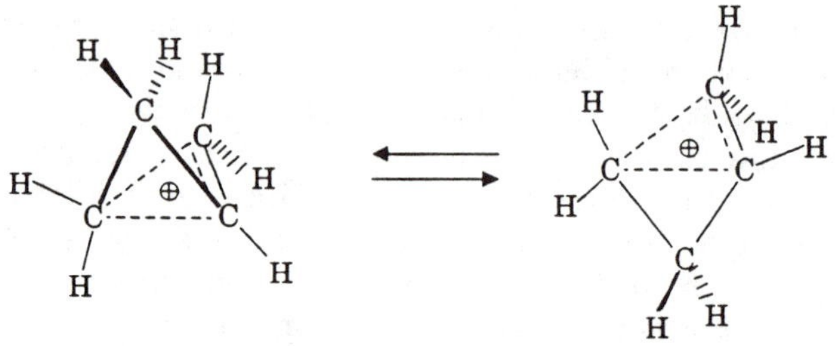

274a **274b** (168)

George Olah also found that if you put two methyl groups on the α-carbon of cyclopropylmethanol and make the corresponding $C_6H_{11}^+$ cation, then the low-temperature NMR shows the CH_3 groups to be nonequivalent, with a barrier of 13.7 kcal/mol to their changing places with one another by rotation.[285] All of the evidence here is strong for a dimethyl-substituted bisected cyclopropylmethyl carbocation **275** (or possibly rapidly equilibrating unsymmetrical forms of that carbocation). However, some of the workers in the field saw the

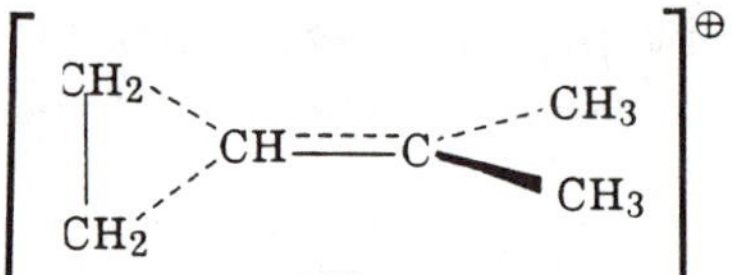

275

success of the bisected structure in accounting for the NMR of the $C_6H_{11}^+$ cation and decided that it must also be the structure for $C_4H_7^+$. In making this decision, they ignored the chemistry of $C_4H_7^+$, because it reacts with nucleophiles to give substantial amounts of cyclobutyl products. But *no* cyclobutyl products are formed at all from $C_6H_{11}^+$.

In 1978 the plot began to thicken more, when my postdoctoral fellow, John Staral, who had been trained by George Olah, began to investigate the ^{13}C spectra of $C_4H_7^+$, which had one of its methylene groups labeled with ^{13}C. One thing John found was that, if you raised the temperature as high as –70 °C, then you began to see a relatively slow migration of ^{13}C into the CH (methine) position.[286] We had not observed this reaction previously in any other $C_4H_7^+$ chemistry. It must have a free energy of activation of substantially more than 10 kcal/mol and may well be analogous to the hydrogen migration that occurs with the nonclassical norbornyl cation with a free energy of activation of about 6 kcal/mol (*see* pp 85–86).[287]

A much more significant finding was that the ^{13}C chemical shifts of the $C_4H_7^+$ carbocation changed about 8 ppm with the temperature change from –132 to –60 °C, although those of the dimethyl-substituted cation, $C_6H_{11}^+$, changed by only about 0.5 ppm.[288] This result provides strong evidence for two or more $C_4H_7^+$ species present in rapid equilibrium. From the temperature dependence of the shifts, we concluded that the free-energy difference between the forms was about 1 kcal/mol. Furthermore, from the ^{13}C shifts of related carbocations, it appeared that the more stable form is likely to be the unsymmetrical bicyclobutonium carbocation. This means that the observed reactions of $C_4H_7^+$ are those arising from rapidly equilibrating bicyclobutonium and bisected cyclopropylmethyl cations, which have nearly the same energy. Each is converted, respectively, to cyclobutyl and cyclopropylmethyl products. The small amount of 3-butenyl products formed could come either from the bisected cyclopropylmethyl cation by a ring-opening homoallylic reaction or from the bicyclobutonium cation by reverse of the ring closure exemplified by equation 166.

A wonderful test for determining whether one has equilibrating classical ions or a single nonclassical cation has been devised by Martin Saunders at Yale.[289] The idea is that the equilibrium between classical ions can be perturbed by isotopic substitution (a zero-point energy effect) and that the change in equilibrium constant can be determined by ^{13}C NMR. The chemical-shift effects on the NMR spectra can be very large indeed. A clear example is provided by putting two deuteriums on a carbon next to one cationic center, but not the other, for the equilibrating 1,2-dimethylcyclopentyl cation, **276a**–**276b** (equation 169). Here,

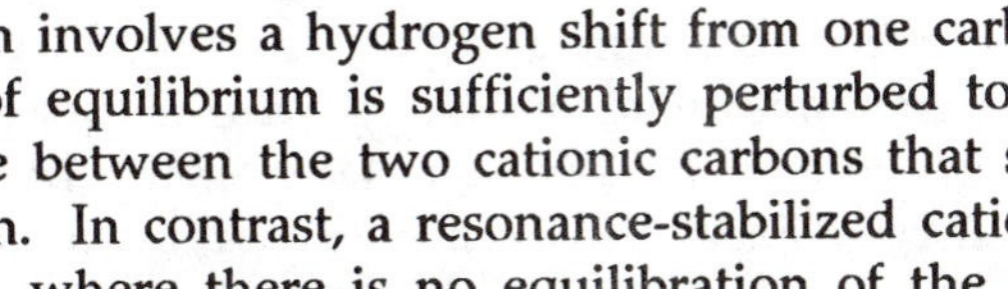

CH_3 D_2 CH_3 H H:$^\ominus$ shift CH_3 D_2 CH_3 H

276a ⇌ **276b** (169)

the equilibrium involves a hydrogen shift from one carbon to another. The position of equilibrium is sufficiently perturbed to give a *91-ppm* shift difference between the two cationic carbons that share the vagabond hydrogen. In contrast, a resonance-stabilized cation, such as **277** (equation 170), where there is no equilibration of the same kind, has only a 0.3-ppm shift difference between the corresponding carbons.

Application of the Saunders technique to a monodeuterated methylene carbon of $C_4H_7^+$ produces a proton-decoupled ^{13}C spectrum showing a rather small deuterium perturbation, which causes a shift

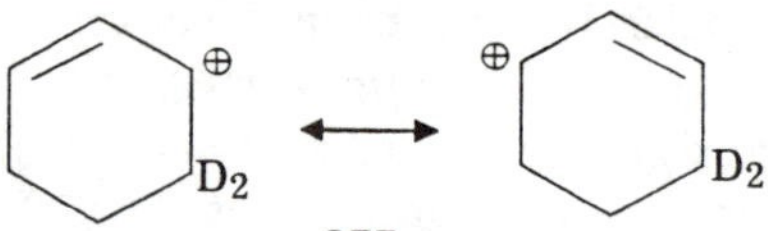

$\oplus$ D2 ⟷ $\oplus$ D2

277 (170)

difference between the monodeuterated and undeuterated methylene resonances.[290] This shift difference depends on the configuration of the deuterium on the monodeuterated methylene carbon. At least some perturbation of an equilibrium is expected here, because there is an equilibrium between the two different cations in the solution. However, the perturbation is much smaller than for an equilibrating mixture of classical cations. The zero-point energy perturbation is now intermolecular rather than intramolecular. If two deuteriums are substituted on one of the methylene carbons, then the undeuterated methylene carbons are shifted to higher magnetic field by 2 ppm, although the doubly deuterated methylene carbon is hardly perturbed at all.[291]

The epitome of complexity in this sort of experiment is shown by the ^{13}C spectrum of $C_4H_4D_3^+$ obtained at Caltech by Bill Brittain and Mike Squillicote, for which a fairly specifically trideuterated cyclopropylmethanol, **278**, was used as the starting material (equation 171).[292]

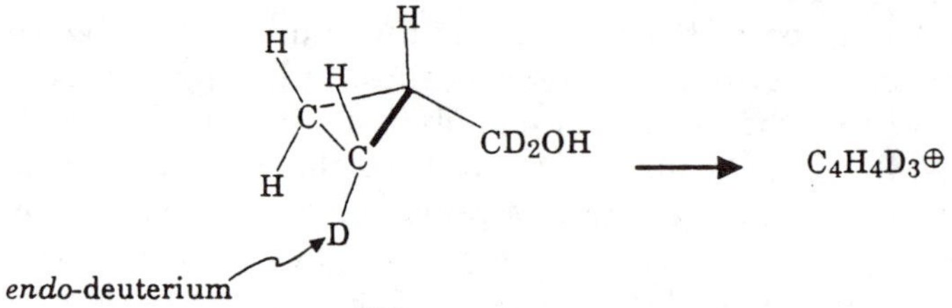

CD_2OH ⟶ $C_4H_4D_3^{\oplus}$

endo-deuterium

278 (171)

The orientation of the extra ring proton on the methylene carbon of **278** was predominately in the *endo* position. We could not explain a number of quite small methylene shift differences. The isotope perturbation acted to move some peaks upfield and others downfield, in no very consistent way.

However, the really important message from this experiment was that, if we start with deuterium in a particular orientation, it will leak over into the other orientation, but the process is more than just slow on the NMR time scale. At –95 °C it takes many hours and, consequently, must have at least a 14-kcal free energy of activation. This means that if anything like a classical, but puckered, cyclobutyl cation, as suggested by H. C. Brown,[293] is present in the equilibrium mixture, it must have a wholly nonclassical resistance to ring inversion and loss of stereochemistry. The conclusion that we can reach from the existing NMR studies is that there is an equilibrium mixture of cations present that differ only slightly in energy and, as far as the chemical shifts go, the bicyclobutonium cation is likely to be the prevalent species present in solution.

The number of atoms and electrons in $C_4H_7^+$ is by no means outrageously large, and you can expect that it has received a good bit of theoretical attention from the quantum theorists. Although I respect the power of theoretical chemistry, I feel compelled to report that, in this area, theoretical results seem to have an uncanny ability to track, not lead, experiments. Theory has yet to predict something new about $C_4H_7^+$. And the theorists vacillate. The situation reminds me of a story about the famous theoretical chemist, Henry Eyring. During a discussion at a chemical meeting, Eyring explained how theory could account for the way that hydrogen bromide adds to propylene by a free-radical mechanism. The problem was that Henry had the way it adds backwards. When this was pointed out to him, he said, "Oh, that is even better!" and went on to explain why theory would predict it would add in just the opposite way.

With that grain of salt, some illustrative results of earlier theoretical calculations are shown in Figure 4.[294,295] The optimized energy minima predicted for different $C_4H_7^+$ configurations, which correspond to the bisected cyclopropylmethyl and unsymmetrical and symmetrical bicyclobutonium cations, are not comparable between methods of calculation. Therefore, only the most favorable form is cited in each series. There is not very good agreement on what form is the most stable or even what the bond lengths should be for the same form. Figure 5 shows the results of recent calculations of the ^{13}C shifts for the $C_4H_7^+$ cation.[296] These are in moderate agreement for the symmetrical bicyclobutonium structure.

Among the recent ab initio theoretical calculations appear to be those of McKee.[297] At what is known in the quantum-mechanical trade as the "MP4SDQ/6-31G*//MP2/6-31G level", the bicyclobutonium cation is predicted to be 0.7 kcal more stable than the bisected cyclopropylmethyl cation. The author concludes by saying, "It is somewhat ironic that, contrary to previous work which suggested that the bicyclobutonium ion might exist in a shallow minimum, this work forces one to contemplate whether the cyclopropylmethyl cation is an energy minimum." It looks as if putting salt on the tail of a chemical chimera is not easy, even at the theoretical level!

What will the final answer be? We will need to see what Philip Myhre of Harvey Mudd and C. S. Yannoni of IBM at San Jose get for the solid-state ^{13}C spectrum of the $C_4H_7^+$ cation at 5 K. Their work was quite definitive for the norbornyl cation.[298] I think it is time for the theoreticians to hustle along and get their own *definitive* answers before the experiments are complete! If their latest efforts[297] are to be regarded

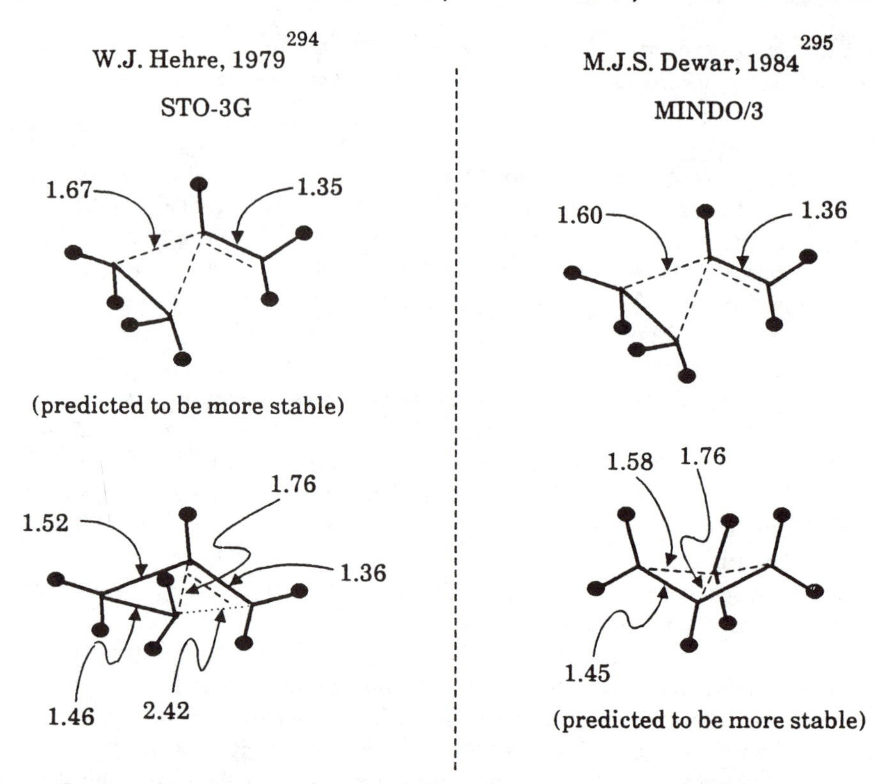

Figure 4. Results of some ab initio calculations of relative energies and predicted bond lengths (in angstroms) of $C_4H_7^+$ configurations.[294,295]

as definitive, they should not be able to abandon them later without good cause.

Epilogue

Being at the right place at the right time, usually with my hands at least half open to latch on to the goodies that sail by on fortune's wind, has led to a fun-filled, variety-filled career in education and research. There have been other great satisfactions as well. Although I would never be a candidate for even a leather medal in fathering, Edith easily merits a Nobel prize for mothering. She is proof positive that love, patience, understanding, and caring pay great dividends. All four of our children are beginning successful careers in married life, as well as in radiology, surgery, and computer engineering. It is a special pleasure that the children and I have somewhat complementary professional interests, and it

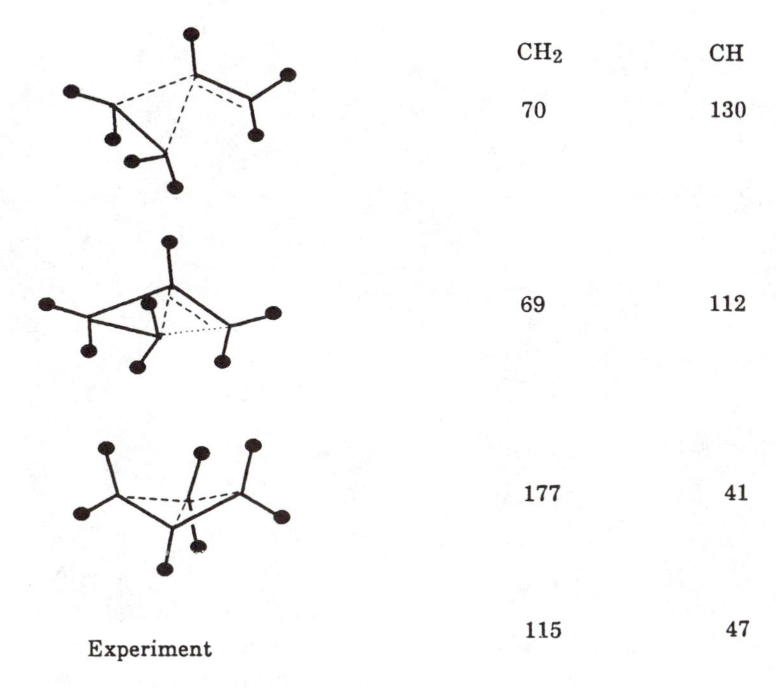

Figure 5. Calculated average ^{13}C shifts (in parts per million from TMS) of $C_4H_7^+$ configurations; rapid equilibration of equivalent forms is assumed.[296]

(Left to right) Ronald C. D. Breslow, Mrs. Paul Bartlett, John D. Roberts, Sylvia Winstein, and our Japanese host at dinner in Kyoto, Japan, as part of a National Science Foundation-sponsored summit conference on physical organic chemistry in 1965.

(Left to right) George S. Hammond, T. Nozoe, S. Winstein, and Ronald C. D. Breslow at a U.S.–Japan Conference in Kyoto on physical organic chemistry, 1965. The other U.S. participants were Paul D. Bartlett and John D. Roberts.

is very nice to have published papers with all but one.[171,299] After many years of being in more or less center stage, it is an odd feeling to run into "Roberts? Who? Oh, I thought you were much older." (Perhaps dead?) But that is the inevitable course of history. In this respect, scientists are no different from football or baseball players, and genuine Hall of Famers are very rare indeed. The memories even of many who seemed indelibly engraved on the edifice of organic chemistry in my early days, such as von Baeyer, Wieland, Willstatter, Ruzicka, Karrer, Kohler, and Meerwein, are not well-known, if at all, to today's organic chemistry students.

In the absence of research immortality, the real joy of accomplishment has to be in the achievements of one's students and postdoctoral associates and, in turn, their students and postdoctoral associates. There is pride in hearing, "Years ago, when I took organic chemistry in college, I used Roberts and Caserio. It was hard, but I learned a lot." Indeed, that is the point. One does not achieve in a vacuum—people are needed, not only to help, but to *appreciate*. It is especially wonderful to have some degree of international appreciation as well, with the chance to carry the message of one's work to far-off places and to have co-workers from many foreign countries.

Picnic in Boston's Public Garden, 1985. (Left to right) Anne; her future husband, John Arnold; John D. Roberts holding his first grandchild, Emily; Edith; and Don. The picture was taken by Don's wife, pianist Barbara Allen.

I have been unbelievably fortunate, not only in being in the right place at the right time, but also in having help, wonderful help, generous help—from parents, relatives, wife, children, friends, students, postdoctoral associates, colleagues, staff, granting agencies, and Caltech. The external human element that helps to motivate research seems so important that, for years, I have asked my friends what they would do if wholly isolated on a desert island with ample sustenance and a laboratory full of the best equipment and supplies. Would they delve into investigating the wonders of nature with no one, or even the hope of anyone, with whom to share their findings?

The years since I was born have been an extraordinary period in human history, particularly in science and technology. Many of the basic scientific discoveries of the 18th, 19th, and early 20th centuries have been brought to fruition in almost every sphere of human activity, except for interhuman relationships. The process has been greatly facilitated by profligate use of our carbon and hydrocarbon energy resources. The progress of the highly industrialized nations in my lifetime will surely be regarded as unfeeling, probably even greedy, by succeeding generations. These future generations will be faced with depleted energy resources, thinned-out deposits of useful metals,

The members of the organic chemistry professorial staff appointed to Caltech after 1953. One could not ask for, or even hardly imagine, a stronger, more helpful, or more collegial group. (Left to right) Peter B. Dervan, Robert E. Ireland *(now at the University of Virginia),* David A. Evans *(now at Harvard),* Robert H. Grubbs, Robert G. Bergman *(now at Berkeley),* John D. Roberts, Andrew G. Myers, Dennis A. Dougherty, *and* George S. Hammond *(now also at Virginia). Not shown is* John H. Richards, *who defected to the molecular biology group at Caltech. This picture was taken in March 1989.*

despoiled forests and oceans, and monumental trash piles of everything from styrofoam cups to worn-out television sets to spent nuclear fuel rods. Of course, this idea assumes there will be future generations, that sane minds will prevail and dispel the nuclear holocaust that hangs over humanity like the sword of Damocles.

Sane minds are unlikely to be enough; clear minds with perspective are needed, minds free of hoary, ingrown, and probably innate prejudices. Those in a frenzy to produce a space-based nuclear defense, the hierarchies of churches that cannot accept demographic facts, business leaders who will sacrifice their societal reason for existence to immediate personal profits, and the like will surely claim to be acting with great sanity. Whether they act with insight, with perspective, and for the future good of all humanity is doubtful. Grasping and dealing with the complexities of the modern world appear to be almost too much for any human mind. It is perhaps not surprising that, in many quarters, there is a heavy reliance on blind faith, even from astrologers,

Edith and John D. Roberts in far-western China on the Karakoram Highway, Marco Polo's route to India. The small-appearing peak to the right of center is actually the summit of Kongar Shan, 25,600 ft. This trip was made near the end of a six-week trip to China in 1985, sponsored by the World Bank to assess chemical education and to give 36 hours of lectures on NMR at Lanzhou University in central China.

that somehow everything will work out and that individual actions to influence the course of things are fruitless.

Science has many wonderful opportunities, and chemistry, particularly organic chemistry, is right in the middle of it. The greatest challenge, of course, is the chemistry of life. When you look at it closely, the chemistry of life becomes ever more complex and yet ever more rational, as it needs to be for organisms to survive the plethora of onslaughts to which they are actually, or potentially, exposed. The chemistry of the brain, of cognition, is a special challenge. That it will be met is not in doubt—just *when* is another question.

Whether the universe has a real "purpose" for human life is unknowable. But one would hope that the beauty inherent in the creations of such diverse artistic geniuses as Michelangelo, Mozart, or van Gogh would survive for "something" to appreciate, even if not the beauty inherent in the edifice of science and mathematics. For some, faced with these unknowables, the balance between creative activity and the "pursuit of happiness" may seem critical, especially if happiness is not much associated with creative activity. The question of when we get to "rest" from our labors was addressed by H. G. Wells at the end of his script for the motion picture, "Things to Come".[300] The chief protagonist said, ". . . rest, too much of it and too soon, and we call it death. For mankind, there can be no rest. We must go on until we have conquered all of the mysteries of space and all of the mysteries of time, and then we will be but just beginning." I hope to meet you there.

Acknowledgments

I am very appreciative of the decision of the American Chemical Society to undertake the publication of the Profiles, Pathways, and Dreams series. It is a somewhat different kind of publication for the Society to support, but a timely one. The authors involved are hardly spring chickens any longer, and they were each fortunate indeed to be participants in an extraordinary period in the development of organic chemistry, a period in which changes took place fully comparable to those changes during the time that the basic structural theories of organic compounds were formulated and applied. The human side of this remarkable period of expansion in our chemical capabilities should be of interest to future generations of both chemists and historians of science.

Many very helpful people helped to make this particular volume possible. The original draft and its many, many revisions, which now amount to more than 750 pages, complete with the chemical formulas reproduced in the text, was entered with enormous skill, patience, and dedication into a Xerox Viewpoint word processor at Caltech by Rose Meldrum, my secretary and resident graphics artist for 20 years. The task of reducing the draft to a manageable size for the series was undertaken by its editor, Jeffrey I. Seeman. Of course, there is more to tell you about other activities involving the National Science Foundation, the National Academy of Sciences, the L. S. B. Leakey Foundation, river trips, tennis, and Caltech, but these are, at best, peripheral to organic chemistry. Dr. Seeman did an extraordinary, and surprisingly painless, job of culling out what was inappropriate for his series. It has been a pleasure to work with him throughout, despite the many changes in scope, format, and timetable of the project over the several years it has taken to bring it to fruition. Every time Jeff phones, you wonder what *next,* and there always is something next. Nonetheless, what he has requested has always served to make the effort of the authors even more worthwhile. Senior editor, Janet Dodd of ACS Books, has done a wonderful job on the final editing. Her painstaking scholarship, enthusiasm, and encouragement have been extremely helpful. Finally, I want to express my appreciation to Edith Roberts for going over the final draft, helping with getting the photographs together, and showing wonderful forbearance over the many hours of what could have been family time, that I spent hunched over yellow pads penciling in what you have just read.

I trust that it is not necessary for me to be effusive here about my general indebtedness to family, friends, colleagues, and institutions that I have been associated with over the years. If my story does not make clear to you how much they (and many others not mentioned for lack of space) have contributed to my life, I have not achieved what I hoped to achieve in this book.

References

1. Slosson, E. *Creative Chemistry*; New York Century: New York, 1921.
2. Lucas, H. J. *Organic Chemistry*; American Book Company: New York, 1935.
3. Robertson, G. R. *Laboratory Practice of Organic Chemistry*; The Macmillan Company: New York, 1st ed., 1937; 2nd ed., 1943.
4. Crowell, W. R.; Yost, D. M.; Roberts, J. D. *J. Am. Chem. Soc.* **1940,** *62,* 2176–2178.
5. *Opportunities in Chemistry*, National Academy Press: Washington, DC, 1985.
6. Winstein, S.; Young, W. G. *J. Am. Chem. Soc.* **1936,** *58,* 104–107.
7. Pauling, L. *The Nature of the Chemical Bond;* Cornell University Press: Ithaca, NY, 1940; Chapter 4.
8. Young, W. G.; Winstein, S.; Prater, A. *J. Am. Chem. Soc.* **1936,** *58,* 289–291.
9. Winstein, S.; Lucas, H. J. *J. Am. Chem. Soc.* **1939,** *61,* 1576–1581.
10. Young, W. G.; Winstein, S.; Goering, H. L. *J. Am. Chem. Soc.* **1951,** *73,* 1958–1963.
11. Roberts, J. D.; Young, W. G.; Winstein, S. *J. Am. Chem. Soc.* **1942,** *64,* 2157–2164.
12. McMillan, W. G., Jr.; Roberts, J. D.; Coryell, C. D. *J. Am. Chem. Soc.* **1942,** *64,* 398–399.
13. Jacobs, T. L.; Roberts, J. D.; McMillan, W. G. *J. Am. Chem. Soc.* **1944,** *66,* 656.
14. Whitmore, F. C. *J. Am. Chem. Soc.* **1932,** *54,* 3274–3283.
15. Whitmore, F. C.; George, R. S. *J. Am. Chem. Soc.* **1943,** *64,* 1239–1242.
16. Marker, R. *CHOC News* **1987,** *4(2),* 3–5. (Published by the Center for the History of Chemistry, University of Pennsylvania, Philadelphia, PA.)
17. Lane, J. F.; Roberts, J. D.; Young, W. G. *J. Am. Chem. Soc.* **1944,** *66,* 543–545.
18. Siegel, S., M.S. Thesis, University of California at Los Angeles, 1940.

19. Young, W. G.; Roberts, J. D. *J. Am. Chem. Soc.* **1945,** *67,* 319–321.

20. Benkeser, R. A.; Broxterman, W. E. *J. Am. Chem. Soc.* **1969,** *91,* 5162–5163.

21. Roberts, J. D.; Young, W. G. *J. Am. Chem. Soc.* **1945,** *67,* 145–150. Ibid. **1946,** *68,* 649–652.

22. Johnson, J. R. *J. Am. Chem. Soc.* **1933,** *55,* 3029–3021.

23. Johnson, J. R. In *Organic Chemistry,* Gilman, H., Ed.; Wiley: New York, 1943, 2nd ed.; pp 1868–1883.

24. Young, W. G.; Roberts, J. D. *J. Am. Chem. Soc.* **1946,** *68,* 1472–1475.

25. Gilman, H.; Jones, R. G. *J. Am. Chem. Soc.* **1941,** *63,* 1162–1163.

26. Young, W. G.; Roberts, J. D. *J. Am. Chem. Soc.* **1944,** *66,* 2131.

27. Kirkwood, J. G.; Westheimer, F. H. *J. Chem. Phys.* **1938,** *6,* 506–512.

28. Hammett, L. P. *Physical Organic Chemistry;* McGraw–Hill: New York, 1940.

29. Bjerrum, N. *Z. Physik. Chem.* **1923,** *106,* 219–242.

30. Whitmore, F. C. *Organic Chemistry;* D. van Nostrand: New York, 1937.

31. Demjanov, N.; Dojarenko, M. *Ber. Dtsch. Chem. Ges.* **1923,** *56,* 2200–2207; Schlatter, M. J. *J. Am. Chem. Soc.* **1941,** *63,* 1733–1737.

32. Roberts, J. D.; Dirstine, P. H. *J. Am. Chem. Soc.* **1945,** *67,* 1281–1283.

33. Roberts, J. D.; Chambers, V. C. *J. Am. Chem. Soc.* **1951,** *73,* 3176–3180.

34. Margrave, J. K.; Cottle, D. L. *J. Am. Chem. Soc.* **1942,** *64,* 484–487. Stahl, G. W.; Cottle, D. L. Ibid. **1943,** *65,* 1782–1783.

35. (a) Roberts, J. D.; Rogers, M. T. *J. Am. Chem. Soc.* **1946,** *68,* 843–846. (b) Smyth, C. P.; Morgan, S. O. *J. Am. Chem. Soc.* **1928,** *50,* 1547–1560.

36. Djerassi, C. *Chem. Rev.* **1948,** *43,* 271–317.

37. Roberts, J. D.; Bennett, W.; Holroyd, E. W.; Fugitt, C. H. *Anal. Chem.* **1948,** *20,* 904–905.

38. Roberts, J. D.; Sauer, C. W. *J. Am. Chem. Soc.* **1949,** *71,* 3925–3929.

39. Roberts, J. D.; Mazur, R. H. *J. Am. Chem. Soc.* **1951,** *73,* 2509–2520; *Current Contents* **1978,** *18,* 11.

40. Demjanov, N. J. *Ber. Dtsch. Chem. Ges.* **1907,** *40,* 4393–4397, 4961–4963.

41. Nevell, T. P.; de Salas, E.; Wilson, C. L. *J. Chem. Soc.* **1939,** 1188–1199.

42. Roberts, J. D.; Bennett, W.; Armstrong, R. *J. Am. Chem. Soc.* **1950,** *72,* 3329–3333. Ibid. **1954,** *76,* 4623.

43. Winstein, S.; Kosower, E. *J. Am. Chem. Soc.* **1959,** *81,* 4399–4412.

44. Goering, H. L.; Schwene, B. C. *J. Am. Chem. Soc.* **1965,** *87,* 3516–3518.

45. Otvos, J. W.; Stevenson, D. P.; Wagner, C. D.; Beeck, O. *J. Chem. Phys.* **1943,** *16,* 745.

46. Bartlett, P. D.; Condon, F. E.; Schneider, A. *J. Am. Chem. Soc.* **1944,** *66,* 1531–1539.

47. Roberts, J. D.; McMahon, R. E.; Hine, J. S. *J. Am. Chem. Soc.* **1949,** *71,* 1896. Ibid. **1950,** *72,* 4237–4244.

48. Roberts, J. D.; Coraor, G. R. *J. Am. Chem. Soc.* **1952,** *74,* 3586–3588.

49. For example, Schleyer, P. v. R.; Watts, W. E.; Fort, R. C., Jr.; Comisarow, M. B. *J. Am. Chem. Soc.* **1964,** *86,* 5679–5680. Saunders, M.; Schleyer, P. v. R.; Olah, G. Ibid. **1964,** *86,* 5680–5681.

50. Roberts, J. D.; Yancey, J. A. *J. Am. Chem. Soc.* **1955,** *77,* 5558–5562.

51. For examples, *see* Saunders, M.; Kates, M. R. *J. Am. Chem. Soc.* **1978,** *100,* 7082–7083.

52. Roberts, J. D.; Mazur, R. H. *J. Am. Chem. Soc.* **1951,** *73,* 3542.

53. Bartlett, P. D. *Nonclassical Ions;* W. A. Benjamin: New York, 1965; p v.

54. Bergstrom, C. G.; Siegel, S. *J. Am. Chem. Soc.* **1952,** *74,* 145–151.

55. Roberts, J. D.; Caserio, M. C.; Graham, W. H. *Tetrahedron* **1960,** *11,* 171.

56. Dewar, M. J. S. *The Electronic Theory of Organic Chemistry;* Oxford: New York, 1949.

57. Walsh, A. D. *Trans. Faraday Soc.* **1949,** *45,* 179–190.

58. Roberts, J. D.; Lee, C. C. *J. Am. Chem. Soc.* **1951,** *73,* 5009.

59. Roberts, J. D.; Lee, C. C.; Saunders, W. H. *J. Am. Chem. Soc.* **1954,** *76,* 4501–4510.

60. Klopman, G. *J. Am. Chem. Soc.* **1969,** *91,* 89–91.

61. Roberts, J. D.; Johnson, F. O.; Carboni, R. A. *J. Am. Chem. Soc.* **1954,** *76,* 5692–5699.

62. Winstein, S.; Shatavsky, M.; Norton, C.; Woodward, R. B. *J. Am. Chem. Soc.* **1955,** *77,* 4183–4184.

63. Roberts, J. D.; Woods, W. G.; Carboni, R. A. *J. Am. Chem. Soc.* **1956,** *78,* 5653–5657.

64. Roberts, J. D.; McElhill, E. A.; Armstrong, R. *J. Am. Chem. Soc.* **1949,** *71,* 2923–2926.

65. Roberts, J. D.; Regan, C. M. *J. Am. Chem. Soc.* **1953,** *75,* 4102.

66. Roberts, J. D.; McElhill, E. A. *J. Am. Chem. Soc.* **1950,** *72,* 628.

67. Roberts, J. D.; Moreland, W. T., Jr. *J. Am. Chem. Soc.* **1953,** *75,* 2267.

68. Roberts, J. D.; Webb, R. L.; McElhill, E. A. *J. Am. Chem. Soc.* **1950,** *72,* 408–411.

69. Roberts, J. D.; Clement, R. A.; Drysdale, J. J. *J. Am. Chem. Soc.* **1951,** *73,* 2181.

70. Roberts, J. D.; Yancey, J. A. *J. Am. Chem. Soc.* **1951,** *73,* 1011–1013.

71. Roberts, J. D.; Regan, C. M. *J. Am. Chem. Soc.* **1954,** *76,* 939–940.

72. *See,* for example, Streitwieser, A. J., Jr.; Berke, C. M.; Schriver, G. W.; Gries, D.; Collins, J. B. *Tetrahedron* **1981,** *37,* 345–358; Holtz, D. *Chem. Rev.* **1971,** *71,* 139–145; but *see also* Schleyer, P. v. R.; Kos, A. J. *Tetrahedron* **1983,** *39,* 1141–1150.

73. Farnham, W. B.; Smart, B. E.; Middleton, W. J.; Calabrese, J. C.; Dixon, D. A. *J. Am. Chem. Soc.* **1985,** *107,* 4546–4567.

74. Schleyer, P. v. R., lecture at the California Institute of Technology, April 1, 1987.

75. Dewar, M. J. S. *Hyperconjugation;* Ronald Press: New York, 1962; pp 83–85, 157–159.

76. Price, C. C. *Reactions at Carbon–Carbon Double Bonds;* Interscience: New York, 1946; Chapter 1.

77. Brockway, L. O. *J. Phys. Chem.* **1937,** *41,* 185–195; *see also* Ref. 7, pp 235–236.

78. Roberts, J. D.; Sanford, J. K.; Sixma, F. L. J.; Cerfontain, H.; Zagt, R. *J. Am. Chem. Soc.* **1954,** *76,* 4525–4534.

79. Roberts, J. D.; Moreland, W. T., Jr.; Frazer, W. *J. Am. Chem. Soc.* **1953,** *75,* 637–640.

80. Roberts, J. D.; Moreland, W. T., Jr. *J. Am. Chem. Soc.* **1953,** *75,* 2167–2173.

81. Taft, R. W., Jr. In *Steric Effects in Organic Chemistry;* Newman, M. S., Ed.; Wiley: New York, 1956; Chapter 13.

82. Roberts, J. D.; Carboni, R. A. *J. Am. Chem. Soc.* **1955,** *77,* 5554–5558.

83. Norris, J. F.; Strain, W. H. *J. Am. Chem. Soc.* **1935,** *57,* 187–192.

84. Roberts, J. D.; Watanabe, W. *J. Am. Chem. Soc.* **1950,** *72,* 4869–4879.

85. Roberts, J. D.; Watanabe, W.; McMahon, R. E. *J. Am. Chem. Soc.* **1951,** *73,* 760–765. Ibid. **1951,** *73,* 2521–2523.

86. Roberts, J. D.; Regan, C. M.; Allen, I. *J. Am. Chem. Soc.* **1952,** *74,* 3679–3683.

87. Roberts, J. D.; Regan, C. M. *J. Am. Chem. Soc.* **1952,** *74,* 3695.

88. Roberts, J. D. *J. Chem. Educ.* **1975,** *52,* 708; *Rec. Chem. Progr.* **1956,** *17,* 95.

89. Harmon, J., U. S. Patent 2 404 374, 1946; *see also* Henne, A. L.; Ruh, R. P. *J. Am. Chem. Soc.* **1947,** *69,* 279–281.

90. Coffman, D. D.; Barrick, P. L.; Cramer, R. D.; Raasch, M. S. *J. Am. Chem. Soc.* **1949,** *71,* 490–496.

91. Roberts, J. D.; Kline, G. B.; Simmons, H. E., Jr. *J. Am. Chem. Soc.* **1953,** *75,* 4765–4768.

92. Roberts, J. D.; Jenny, E. F. *J. Am. Chem. Soc.* **1956,** *78,* 2005–2009.

93. Roberts, J. D.; Smutny, E. J. *J. Am. Chem. Soc.* **1955,** *77,* 3420.

94. Roberts, J. D.; Smutny, E. J.; Caserio, M. C. *J. Am. Chem. Soc.* **1960,** *82,* 1793–1801.

95. Roberts, J. D.; Sharts, C. M. *J. Am. Chem. Soc.* **1961,** *83,* 871–878.

96. Roberts, J. D.; Nagarajan, K.; Caserio, M. C. *Rev. Chim.* **1962,** *VII,* 1109–1117.

97. Roberts, J. D.; Nagarajan, K.; Caserio, M. C. *J. Am. Chem. Soc.* **1964,** *86,* 449–453.

98. Fritchie, C., Jr.; Hughes, E. W. *J. Am. Chem. Soc.* **1962,** *84,* 2257–2258.

99. Bergstrom, F. W.; Wright, R. E.; Chandler, C.; Gilkey, W. A. *J. Org. Chem.* **1936,** *1,* 170–178.

100. Gilman, H.; Avakian, S. *J. Am. Chem. Soc.* **1945,** *67,* 349–351.

101. Urner, R. S.; Bergstrom, F. W. *J. Am. Chem. Soc.* **1945,** *67,* 2108–2109.

102. Horning, C. H., Ph.D. Thesis, Stanford University (1939). *See* Horning, C. H.; Bergstrom, F. W. *J. Am. Chem. Soc.* **1945,** *67,* 2110–2111; Bergstrom, F. W.; Horning, C. H. *J. Org. Chem.* **1946,** *11,* 334–340.

103. Kosower, E. M., B.S. Thesis, Massachusetts Institute of Technology, 1948.

104. Vaughan, C. W., B.S. Thesis, Massachusetts Institute of Technology, 1951.

105. Benkeser, R. A.; Buting, W. E., paper presented at the Cleveland meeting of the American Chemical Society, April 1951.

106. Wittig, G.; Pieper, G.; Fuhrmann, G. *Ber. Dtsch. Chem. Ges.* **1940,** *73,* 1193–1197.

107. Wittig, G. *Naturwissenschaften* **1942,** *30,* 696–703.

108. Roberts, J. D.; Vaughan, C. W.; Carlsmith, L. A.; Semenow, D. A. *J. Am. Chem. Soc.* **1956,** *78,* 611–614.

109. Benkeser, R. A.; Buting, W. E. *J. Am. Chem. Soc.* **1952,** *74,* 3011–3014.

110. Roberts, J. D.; Carlsmith, L. A.; Simmons, H. E., Jr.; Vaughan, C. W. *J. Am. Chem. Soc.* **1953,** *75,* 3290.

111. Benkeser, R. A.; Schroll, G. *J. Am. Chem. Soc.* **1953,** *75,* 3196–3197.

112. Roberts, J. D.; Semenow, D. A.; Simmons, H. E., Jr.; Carlsmith, L. A. *J. Am. Chem. Soc.* **1956,** *78,* 601–611.

113. Roberts, J. D.; Jenny, E. F. *Helv. Chim. Acta* **1955,** *37,* 1248–1254.

114. Huisgen, R.; Rist, H. R. *Naturwissenschaften* **1954,** *41,* 358–359. Also many later papers.

115. Roberts, J. D.; Panar, M. *J. Am. Chem. Soc.* **1960,** *82,* 3629–3632.

116. Wittig, G.; Pohmer, L. *Angew. Chem.* **1955,** *67,* 348.

117. Roberts, J. D.; Bottini, A. T. *J. Am. Chem. Soc.* **1957,** *79,* 1458–1462.

118. Roberts, J. D.; Scardiglia, F. *Tetrahedron* **1957,** *1,* 343–344.

119. Roberts, J. D.; Montgomery, L. K. *J. Am. Chem. Soc.* **1960,** *82,* 4750.

120. *See,* for example, Hoffman, R. W. *Dehydrobenzene; Cycloalkynes;* Academic Press: New York, 1967.

121. Eyring, H.; Walter, J.; Kimball, G. E. *Quantum Chemistry;* Wiley: New York, 1949.

122. Pauling, L.; Wilson, E. B., Jr. *Introduction to Quantum Mechanics;* McGraw–Hill: New York, 1935.

123. Wheland, G. W. *Theory of Resonance;* Wiley: New York, 1944.

124. Wheland, G. W. *Resonance in Organic Chemistry;* Wiley: New York, 1955; *Advanced Organic Chemistry;* Wiley: New York, 1960.

125. Ref. 123, p 175.

126. Wheland, G. W. *J. Chem. Phys.* **1934,** *2,* 474–481.

127. Coulson, C. A. *Quart. Rev.* **1947,** *1,* 144–178.

128. Ref. 56, pp 9–11.

129. Coulson, C. A.; Longuet-Huggins, H. C. *Proc. R. Soc. London A,* **1947,** *191,* 39–60.

130. Roberts, J. D.; Streitwieser, A. J.; Regan, C. M. *J. Am. Chem. Soc.* **1952,** *74,* 4579–4582.

131. (a) Streitwieser, A. J. *Molecular Orbital Theory for Organic Chemists;* Wiley: New York, 1961. (b) Roberts, J. D. *Notes on Molecular Orbital Calculations;* W. A. Benjamin: New York, 1961.

132. Pariser, R.; Parr, R. *J. Chem. Phys.* **1953,** *21,* 466–471, 767–776.

133. Pauling, L. *J. Chem. Phys.* **1933,** *1,* 280–282.

134. Davenport, D. *Chemtech* **1987,** *17,* 526–531.

135. Ingold, C. K. *Structure and Mechanism in Organic Chemistry;* Cornell University Press: Ithaca, NY, first ed., 1953.

136. Roberts, J. D. *J. Am. Chem. Soc.* **1950,** *72,* 3300–3301.

137. Barton, D. H. R.; Miller, E. *J. Am. Chem. Soc.* **1950,** *72,* 1066–1070.

138. Arnold, J. T.; Dharmatti, S. S.; Packard, M. E.; Bloch, F. *J. Chem. Phys.* **1959,** *19,* 507.

139. Waugh, J. S.; Humphrey, F. B.; Yost, D. M. *J. Phys. Chem.* **1953,** *57,* 490–496.

140. Pake, G. *Am J. Phys.* **1950,** *18,* 438–452; 473–496.

141. Ettlinger, M. G. *J. Am. Chem. Soc.* **1952,** *74,* 5805–5806.

142. Bottini, A. T.; Roberts, J. D. *J. Org. Chem.* **1956,** *B21,* 1169–1170.

143. Roberts, J. D. *J. Am. Chem. Soc.* **1956,** *78,* 4495.

144. Shoolery, J. N.; Roberts, J. D. *Rev. Sci. Instrum.* **1957,** *28,* 61–62.

145. Bottini, A. T.; Roberts, J. D. *J. Am. Chem. Soc.* **1956,** *78,* 5126.

146. Bottini, A. T.; Roberts, J. D. *J. Am. Chem. Soc.* **1958,** *80,* 5203–5208.

147. Sharts, C. M.; Roberts, J. D. *J. Am. Chem. Soc.* **1957,** *79,* 1008.

148. Snyder, E. I.; Roberts, J. D. *J. Am. Chem. Soc.* **1962,** *84,* 1582–1586.

149. Petrakis, L.; Sederholm, C. H. *J. Chem. Phys.* **1961,** *35,* 1243–1248.

150. Servis, K. L.; Roberts, J. D. *J. Am. Chem. Soc.* **1965,** *87,* 1339–1344.

151. Weigert, F. J.; Roberts, J. D. *J. Am. Chem. Soc.* **1968,** *90,* 3577–3578.

152. Mallory, F. B.; Mallory, C. W.; Ricker, W. M. *J. Am. Chem. Soc.* **1975,** *97,* 4770–4771.

153. Drysdale, J. J.; Phillips, W. J. *J. Am. Chem. Soc.* **1957,** *79,* 319–322.

154. Roberts, J. D.; Nair, P. M. *J. Am. Chem. Soc.* **1957,** *79,* 4565–4566.

155. Weigert, F. J.; Winstead, M. B.; Garrels, J. I.; Roberts, J. D. *J. Am. Chem. Soc.* **1970,** *92,* 7359–7368.

156. Griffith, D. L.; Roberts, J. D. *J. Am. Chem. Soc.* **1965,** *87,* 4089–4092.

157. Kaplan, F.; Roberts, J. D. *J. Am. Chem. Soc.* **1961,** *83,* 4666–4667. Shafer, D. R.; Davis, D. R.; Vogel, M.; Nagarajan, K.; Roberts, J. D. *Proc. Natl. Acad. Sci. U.S.A.* **1961,** *47,* 49–51.

158. Binsch, G.; Roberts, J. D. *J. Am. Chem. Soc.* **1965,** *87,* 5157–5162.

159. Cope, A. C.; Pawson, B. A. *J. Am. Chem. Soc.* **1965,** *87,* 3649–3651.

160. (a) Whitesides, G. M.; Holtz, D.; Roberts, J. D. *J. Am. Chem. Soc.* **1964,** *86,* 2628–2634. (b) Whitesides, G. M.; Grocki, J. J.; Holtz, D.; Steinberg, H.; Roberts, J. D. *J. Am. Chem. Soc.* **1965,** *87,* 1058–1064.

161. Brown, R. F. *Organic Chemistry;* Wadsworth: Belmont, CA, 1975; pp 318a–318c.

162. McConnell, H. M.; McLean, A. D.; Reilly, C. A. *J. Chem. Phys.* **1955,** *23,* 1152–1159.

163. Roberts, J. D. *Nuclear Magnetic Resonance;* McGraw–Hill: New York, 1959; p 57.

164. Pople, J. A.; Schneider, W. G.; Bernstein, H. J. *High-Resolution Nuclear Magnetic Resonance;* McGraw–Hill: New York, 1959.

165. Wiberg, K. B. *The Interpretation of NMR Spectra*; W. A. Benjamin: New York, 1962.

166. Bothner-By, A. A. In *Computer Programs in Chemistry*; DeTar, D. F., Ed.; W. A. Benjamin: New York, 1968; Vol. 1, Chapter 3.

167. Swalen, J. D.; Reilly, C. A. *J. Chem. Phys.* **1962**, *37*, 21–29.

168. Roberts, J. D. *An Introduction to the Analysis of Spin–Spin Splitting in High-Resolution Nuclear Magnetic Resonance Spectra*; W. A. Benjamin: New York, 1961.

169. Bax, A.; Lerner, L. *Science* **1986**, *232*, 960–967; Wemmer, D. E.; Reid, B. R. *Ann. Rev. Phys. Chem.* **1985**, *36*, 105–137 for references and discussion.

170. Snyder, E. I.; Altman, L. J.; Roberts, J. D. *J. Am. Chem. Soc.* **1962**, *84*, 2004.

171. Roberts, J. D.; Hawkes, G. E.; Husar, J.; Roberts, A. W.; Roberts, D. W. *Tetrahedron* **1974**, *30*, 1833–1844.

172. Gutowsky, H. S.; Holm, C. H. *J. Chem. Phys.* **1956**, *25*, 1228–1234.

173. Alexander, S. *J. Chem. Phys.* **1962**, *37*, 967–974, 974–980. Ibid. **1963**, *38*, 1787–1788.

174. Kaplan, J. I. *J. Chem. Phys.* **1958**, *28*, 278–282. Ibid. **1958**, *29*, 462.

175. Anders, L. R.; Beauchamp, J. L.; Dunbar, R. C.; Baldeschwieler, J. D. *J. Chem. Phys.* **1966**, *45*, 1062–1063.

176. Bargon, J.; Fischer, H.; Johnsen, U. Z. *Naturforsch.* **1967**, *229*, 1551–1555. Bargon, J.; Fischer, H. Ibid. **1967**, *229*, 1556–1562. Ward, H. R.; Lawler, R. G. *J. Am. Chem. Soc.* **1967**, *89*, 5518. Lawler, R. G. Ibid. **1967**, *89*, 5519–5521.

177. Nordlander, J. E.; Roberts, J. D. *J. Am. Chem. Soc.* **1959**, *81*, 1769.

178. Zieger, H. E.; Roberts, J. D. *J. Org. Chem.* **1969**, *34*, 1976–1977.

179. Nordlander, J. E.; Young, W. G.; Roberts, J. D. *J. Am. Chem. Soc.* **1961**, *83*, 494–495.

180. Whitesides, G. M.; Nordlander, J. E.; Roberts, J. D. *J. Am. Chem. Soc.* **1962**, *84*, 2010–2011.

181. Whitesides, G. M.; Nordlander, J. E.; Roberts, J. D.; *Discuss. Faraday Soc.* **1962**, *34*, 185–190.

182. Whitesides, G. M.; Kaplan, F.; Roberts, J. D. *J. Am. Chem. Soc.* **1963**, *85*, 2167–2168.

183. Witanowski, M. ; Roberts, J. D. *J. Am. Chem. Soc.* **1966,** *88,* 737–741.

184. Whitesides, G. M.; Witanowski, M.; Roberts, J. D. *J. Am. Chem. Soc.* **1965,** *87,* 2854–2862.

185. Whitesides, G. M., Ph.D. Thesis, California Institute of Technology, 1964.

186. Anet, F. A. L. *J. Am. Chem. Soc.* **1962,** *84,* 671–672.

187. Gwynn, D. E.; Whitesides, G. M.; Roberts, J. D. *J. Am. Chem. Soc.* **1965,** *87,* 2862–2864.

188. Hawkins, B. L.; Bremser, W.; Borcic, S.; Roberts, J. D. *J. Am. Chem. Soc.* **1971,** *93,* 4472–4479.

189. Westheimer, F. H.; Mayer, J. *J. Chem. Phys.* **1946,** *14,* 733–738. Westheimer, F. H. Ibid. **1947,** *15,* 252–260. Rieger, M.; Westheimer, F. H. *J. Am. Chem. Soc.* **1950,** *72,* 19–28.

190. Halgren, T. A., Ph.D. Thesis, California Institute of Technology, 1968.

191. Halgren, T. A.; Howden, M. E. H.; Medof, M. E.; Roberts, J. D. *J. Am. Chem. Soc.* **1967,** *89,* 3051–3052.

192. Roberts, J. D. In *Abstracts of the Twentieth National Organic Chemistry Symposium, Burlington, Vermont;* 1st ed., 1967; pp 59–72. (The 2nd edition, distributed after the meeting, has a different abstract.)

193. Anet, F. A. L.; Ahmad, M.; Hall, L. D. *Proc. Chem. Soc.* **1964,** 145–146.

194. Hasek, W. R.; Smith, W. C.; Engelhardt, V. A. *J. Am. Chem. Soc.* **1960,** *82,* 543–551.

195. Lambert, J. B.; Roberts, J. D. *J. Am. Chem. Soc.* **1965,** *87,* 3884–3890. Ibid. **1965,** *87,* 3881–3895.

196. Roberts, J. D. *Angew. Chem. Int. Ed. Engl.* **1963,** *2,* 53–59. Spassov, S. L.; Griffith, D. L.; Glazer, E. S.; Nagarajan, K.; Roberts, J. D. *J. Am. Chem. Soc.* **1967,** *89,* 88–94.

197. Roberts, J. D. *Chem. Br.* **1966,** 2, 529–535.

198. Gerig, J. T.; Roberts, J. D. *J. Am. Chem. Soc.* **1966,** *88,* 2791–2799. Lack, R. E.; Roberts, J. D. Ibid. **1968,** *90,* 6997–7001. Lack, R. E.; Ganter, C.; Roberts, J. D. Ibid. **1968,** *90,* 7001–7007.

199. Glazer, E. S., Ph.D. Thesis, California Institute of Technology, 1965. Knorr, R.; Ganter, C.; Roberts, J. D. *Angew. Chem.* **1967,** *79,* 577–578.

200. Hendrickson, J. B. *J. Am. Chem. Soc.* **1962,** *84,* 3355–3359.

201. Anderson, J. E.; Glazer, E. S.; Griffith, D. L.; Knorr, R.; Roberts, J. D. *J. Am. Chem. Soc.* **1969,** *91,* 1386–1395.

202. Hendrickson, J. B. *J. Am. Chem. Soc.* **1967,** *89,* 7036–7043. Ibid. **1967,** *89,* 7043–7046. Ibid. **1967,** *89,* 7047–7061.

203. Anet, F. A. L.; Basus, V. J. *J. Am. Chem. Soc.* **1973,** *95,* 4424–4426.

204. Noe, E. A.; Roberts, J. D. *J. Am. Chem. Soc.* **1972,** *94,* 2020–2026.

205. *See* Dunitz, J. D.; Prelog, V. *Angew. Chem.* **1960,** *72,* 896–902.

206. Frei, K.; Bernstein, H. J. *J. Chem. Phys.* **1963,** *38,* 1216–1226.

207. Renk, E.; Shafer, P. R.; Graham, W. H.; Mazur, R. H.; Roberts, J. D. *J. Am. Chem. Soc.* **1961,** *83,* 1987–1989.

208. Lauterbur, P. C. *J. Chem. Phys.* **1957,** *26,* 217–218.

209. Weigert, F. J.; Roberts, J. D. *J. Am. Chem. Soc.* **1967,** *89,* 2967–2969.

210. Weigert, F.; Roberts, J. D. *J. Am. Chem. Soc.* **1967,** *89,* 5962.

211. Weigert, F.; Jautelat, M.; Roberts, J. D. *Proc. Natl. Acad. Sci. U.S.A.* **1968,** *60,* 1152–1155.

212. Jautelat, M.; Grutzner, J. B.; Roberts, J. D. *Proc. Natl. Acad. Sci. U.S.A.* **1970,** *65,* 288–292.

213. Reich, H. J.; Jautelat, M.; Messe, M. T.; Weigert, F. J.; Roberts, J. D. *J. Am. Chem. Soc.* **1969,** *91,* 7445–7454.

214. Dorman, D. E.; Roberts, J. D. *J. Am. Chem. Soc.* **1970,** *92,* 1355–1361.

215. Dorman, D. E.; Roberts, J. D. *J. Am. Chem. Soc.* **1971,** *93,* 4463–4472.

216. Nourse, J. G. *Simple Solutions to Rubik's Cube;* Bantam: New York, 1981.

217. Nourse, J. G.; Roberts, J. D. *J. Am. Chem. Soc.* **1975,** *97,* 4584–4594.

218. Zambelli, A.; Gatti, G.; Sacchi, C.; Crain, W. O., Jr.; Roberts, J. D. *Macromolecules* **1971,** *4,* 330–332; 475–477.

219. Johnson, L. F.; Heatley, F.; Bovey, F. A. *Macromolecules,* **1970,** *3,* 175–177.

220. Natta, G.; Mazzanti, G.; Pino, P., U. S. Patent No. 3 112 300, 1963.

221. Flory, P. J.; Baldeschwieler, J. D. *J. Am. Chem. Soc.* **1966,** *88,* 2873–2874. Flory, P. J. *Macromolecules* **1970,** *3,* 613–617.

222. Lambert, J. B.; Binsch, G.; Roberts, J. D. *Proc. Natl. Acad. Sci. U.S.A.* **1964,** *51,* 735–737.

223. Roberts, J. D. *Rice University Studies (Felix Bloch; Twentieth Century Physics)*; Vol. 66, No. 3, Rice University: Houston, TX, 1980; pp 147–178.

224. Pauling, L. *The Nature of the Chemical Bond*; Cornell University Press: Ithaca, NY, 1960; pp 120 ff.

225. Binsch, G.; Lambert, J. B.; Roberts, B. W.; Roberts, J. D. *J. Am. Chem. Soc.* **1964,** *86,* 5564–5570.

226. Lichter, R. L.; Roberts, J. D. *Spectrochim. Acta* **1970,** *26A,* 1813–1814. Lichter, R. L.; Roberts, J. D. *J. Am. Chem. Soc.* **1971,** *93,* 3200–3203.

227. Markowski, V.; Sullivan, G. R.; Roberts, J. D. *J. Am. Chem. Soc.* **1977,** *99,* 714–718.

228. Spielvogel, B. F.; Purser, J. M. *J. Am. Chem. Soc.* **1967,** *89,* 5294–5295.

229. Lichter, R. L.; Roberts, J. D. *J. Am. Chem. Soc.* **1972,** *94,* 2495–2500.

230. Duthaler, R. O.; Roberts, J. D. *J. Am. Chem. Soc.* **1978,** *100,* 3882–3889.

231. Bradley, C. H.; Hawkes, G. E.; Randall, E. W.; Roberts, J. D. *J. Am. Chem. Soc.* **1975,** *97,* 1958–1959.

232. Gust, D.; Moon, R. B.; Roberts, J. D. *Proc. Natl. Acad. Sci. U.S.A.* **1975,** *72,* 4696–4700.

233. Cain, A. H.; Sullivan, G. R.; Roberts, J. D. *J. Am. Chem. Soc.* **1977,** *99,* 6423–6425.

234. Gonnella, N. C.; Roberts, J. D. *J. Am. Chem. Soc.* **1982,** *104,* 3162–3164. Gonnella, N. C.; Nakanishi, H.; Holtwick, J. B.; Horowitz, D. S.; Kanamori, K.; Leonard, N. J.; Roberts, J. D. *J. Am. Chem. Soc.* **1983,** *105,* 2050–2055.

235. Nee, M.; Roberts, J. D. *Biochemistry* **1982,** *21,* 4920–4926.

236. Hunkapillar, M. W.; Smallcombe, S. H.; Whitaker, D. R.; Richards, J. H. *Biochemistry* **1973,** *12,* 4732–4743.

237. Blow, D. M.; Birktoft, J. J.; Hartley, B. S. *Nature (London),* **1969,** *221,* 337–340. Blow, D. M. *Acc. Chem. Res.* **1976,** *9,* 145–152.

238. Bender, M. L.; Kilheffer, J. V. *Crit. Rev. Biochem.* **1973,** *1,* 149–199; references cited therein.

239. Bachovchin, W. W.; Roberts, J. D. *J. Am. Chem. Soc.* **1978,** *100,* 8041–8047.

240. Schuster, I. I.; Roberts, J. D. *J. Org. Chem.* **1979,** *44,* 3864–3867.

241. For example, Reynolds, W. F.; Peat, I. R.; Freedman, M. H.;Lyerla, J. R., Jr. *J. Am. Chem. Soc.* **1973,** *95,* 328–331.

242. Chun, Y.; Flanagan, C.; Birdseye, T. R.; Roberts, J. D. *J. Am. Chem. Soc.* **1982,** *104,* 3945–3949.

243. Bachovchin, W. W.; Wong, W. Y. L.; Farr-Jones, S.; Shenvi, A. B.; Kettner, C. A. *Biochemistry* **1988,** *27,* 7689–7697.

244. Bachovchin, W. W.; Kaiser, R.; Richards, J. H.; Roberts, J. D. *Proc. Natl. Acad. Sci. U.S.A.* **1981,** *78,* 7323–7326.

245. For example, Lipscomb, W. N. *Chem. Soc. Rev.* **1972,** *1,* 319–336.

246. Johansen, J. T.; Vallee, B. L. *Proc. Natl. Acad. Sci. U.S.A.* **1971,** *68,* 2532–2535. Ibid. **1973,** *70,* 2006–2010. *Biochemistry* **1975,** *14,* 649–660.

247. Bachovchin, W. W.; Kanamori, K.; Vallee, B. L.; Roberts, J. D. *Biochemistry* **1982,** *21,* 2885–2892.

248. Gardell, S. J.; Craik, C. S.; Hilvert, D.; Urdea, M. S.; Rutter, W. J. *Nature (London)* **1985,** *317,* 551–555.

249. Becker, N. N.; Roberts, J. D. *Biochemistry* **1984,** *23,* 3336–3340.

250. Kanamori, K.; Roberts, J. D. *Acc. Chem. Res.* **1983,** *16,* 35–41.

251. Irving, C. S.; Lapidot, A. *J. Am. Chem. Soc.* **1975,** *97,* 5945–5946.

252. Legerton, T. L.; Kanamori, K.; Weiss, R. L.; Roberts, J. D. *Proc. Natl. Acad. Sci. U.S.A.* **1981,** *78,* 1495–1498. Kanamori, K.; Legerton, T. L.; Weiss, R. L.; Roberts, J. D. *J. Biol. Chem.* **1982,** *257,* 14168–14172.

253. Kanamori, K.; Weiss, R. L.; Roberts, J. D. *J. Biol. Chem.* **1987,** *262,* 11038–11045.

254. Kanamori, K.; Weiss, R. L.; Roberts, J. D. *J. Bacteriol.* **1987,** *169,* 4692–4695. *J. Biol. Chem.* **1988,** *263,* 2817–2823.

255. Kanamori, K.; Cain, A. H.; Roberts, J. D. *J. Am. Chem. Soc.* **1978,** *100,* 4979–4981. Kanamori, K.; Legerton, T. L.; Weiss, R. L.; Roberts, J. D. *Biochemistry* **1982,** *21,* 4916–4920. Legerton, T. L.; Kanamori, K.; Weiss, R. L.; Roberts, J. D. *Biochemistry* **1983,** *22,* 899–903.

256. Duthaler, R. O.; Fürster, H. G.; Roberts, J. D. *J. Am. Chem. Soc.* **1978,** *100,* 4974–4979.

257. Casewit, C.; Roberts, J. D. *J. Am. Chem. Soc.* **1980,** *102,* 2364–2368.

258. Casewit, C.; Wenninger, J.; Roberts, J. D. *J. Am. Chem. Soc.* **1981,** *103,* 6248–6249.

259. Huisgen, R.; Ugi, I. *Angew. Chem.* **1956,** *68,* 705–706. *Chem. Ber.* **1957,** *90,* 2914–2927.

260. Beringer, F. M.; Brierley, A.; Drexler, M.; Gindler, E. M.; Lumpkin, C. C. *J. Am. Chem. Soc.* **1953,** *75,* 2708–2712. Beringer, F. M.; Gindler, E. M. Ibid. **1955,** *77,* 3203–3207.Beringer, F. M.; Geering, E. J.; Kuntz, I.; Mausner, M. *J. Phys. Chem.* **1954,** *60,* 141–150.

261. Caserio, M. C.; Glusker, D. L.; Roberts, J. D. *J. Am. Chem. Soc.* **1959,** *81,* 336–342.

262. Caserio, M. C.; Glusker, D. L.; Roberts, J. D. In *Theoretical Organic Chemistry;*. Papers presented at the Kekulé Symposium of the Chemical Society, London, September 1958; pp 103–113, Butterworths Scientific Publications: London.

263. Roberts, J. D.; Caserio, M. C. *Basic Principles of Organic Chemistry;* W. A. Benjamin: New York, 1964.

264. Roberts, J. D.; Caserio, M. C. *Basic Principles of Organic Chemistry;* W. A. Benjamin: New York, 2nd ed., 1977.

265. Lambert, J. B.; Roberts, J. D. *Tetrahedron Lett.* **1965,** *20,* 1457–1463.

266. Woodward, R. B.; Katz, T. J. *Tetrahedron* **1959,** *5,* 70–89.

267. Lutz, R. P.; Roberts, J. D. *J. Am. Chem. Soc.* **1961,** *83,* 2198–2199.

268. Berson, J. A.; Reynolds, R. D. *J. Am. Chem. Soc.* **1955,** *77,* 4434. Berson, J. A.; Reynolds, R. D.; Jones, W. M. Ibid. **1956,** *78,* 6049–6053.

269. (a) Scheidegger, U.; Baldwin, J. E.; Roberts, J. D. *J. Am. Chem. Soc.* **1967,** *89,* 894–898. (b) Ganter, C.; Scheidegger, U.; Roberts, J. D. *J. Am. Chem. Soc.* **1965,** *87,* 2771–2772.

270. Bartlett, P. D.; Knox, L. H. *J. Am. Chem. Soc.* **1939,** *61,* 3184–3191.

271. Ref. 224, pp 383–384.

272. Mazur, R. H.; White, W. N.; Semenow, D. A.; Lee, C. C.; Silver, M. S.; Roberts, J. D. *J. Am. Chem. Soc.* **1959,** *81,* 4390–4398.

273. Renk, E.; Roberts, J. D. *J. Am. Chem. Soc.* **1961,** *83,* 878–881.

274. Brown, H. C.; Borkowski, M. *J. Am. Chem. Soc.* **1952,** *74,* 1894–1902. Servis, K. L.; Roberts, J. D. *Tetrahedron Lett.* **1967,** *15,* 1369–1372.

275. Howden, M. E. H.; Roberts, J. D. *Tetrahedron* **1963,** *19,* 403–414.

276. Silver, M. S.; Caserio, M. C.; Rice, H. E.; Roberts, J. D. *J. Am. Chem. Soc.* **1961,** *83,* 3671–3678.

277. Maercker, A.; Roberts, J. D. *J. Am. Chem. Soc.* **1966,** *88,* 1742–1759.

278. Howden, M. E. H.; Maercker, A.; Burdon, J.; Roberts, J. D. *J. Am. Chem. Soc.* **1966,** *88,* 1732–1742.

279. Cox, E. F.; Caserio, M. C.; Silver, M. S.; Roberts, J. D. *J. Am. Chem. Soc.* **1961,** *83,* 2719–2724.

280. (a) Olah, G. A.; Spear, R. J.; Hibberty, P. C.; Hehre, W. J. *J. Am. Chem. Soc.* **1976,** *98,* 7470–7475. Olah, G. A.; Prakash, C. K. S.; Donovan, D. J.; Yavari, I. *J. Am. Chem. Soc.* **1978,** *100,* 7085–7086. (b) Kirchen, R. P.; Sorensen, T. S. *J. Am. Chem. Soc.* **1977,** *99,* 6687–6693.

281. Kover, W. B.; Roberts, J. D. *J. Am. Chem. Soc.* **1969,** *91,* 3687–3688.

282. (a) Brown, H. C. *The Nonclasssical Ion Problem;* Plenum Press: New York, 1977. (b) Brown, H. C. *The Transition State;* Chemical Society: London, *Spec. Publ.* **1962,** *16,* 140–158, 174–178.

283. Davenport, D. *Aldrichimica Acta* **1987,** *20,* 25–27.

284. Olah, G. A.; Kelly, D. P.; Jeuell, C. L.; Porter, R. D. *J. Am. Chem. Soc.* **1970,** *92,* 2244–2546.

285. Pittman, C. U., Jr.; Olah, G. A. *J. Am. Chem. Soc.* **1965,** *87,* 2998–3000, 5123–5132. Kabakoff, D. S.; Namanworth, E., *J. Am. Chem. Soc.* **1970,** *92,* 3234–3235.

286. Staral, J. S.; Roberts, J. D. *J. Am. Chem. Soc.* **1978,** *100,* 8018–8020.

287. For example, Olah, G. A.; White, A. M.; DeMember, J. R.; Commeyras, A.; Lui, C. Y. *J. Am. Chem. Soc.* **1970,** *92,* 4627–4640.

288. Staral, J. S.; Yavari, I.; Prakash, G. K. S.; Donovan, D. J.; Olah, G. A.; Roberts, J. D. *J. Am. Chem. Soc.* **1978,** *100,* 8016,–8018.

289. Saunders, M.; Telkowski, L.; Kates, M. R. *J. Am. Chem. Soc.* **1977,** *99,* 8020–8021; earlier references.

290. Saunders, M.; Siehl, H.-U. *J. Am. Chem. Soc.* **1980,** *102,* 6868–6869.

291. Brittain, W. J., Ph.D. Thesis, California Institute of Technology, 1982.

292. Brittain, W. J.; Squillacote, M. E.; Roberts, J. D. *J. Am. Chem. Soc.* **1984,** *106,* 7280–7282.

293. Ref. 282a, Chapter 5.

294. Hehre, W. J.; Hiberty, P. C. *J. Am. Chem. Soc.* **1974,** *96,* 302–364. Hehre, W. J., *Acc. Chem. Res.* **1975,** *8,* 369–376.

295. Dewar, M. J. S.; Reynolds, C. H. *J. Am. Chem. Soc.* **1984,** *106,* 6388–6392.

296. Schindler, M. *J. Am. Chem. Soc.* **1987,** *109,* 1020–1033.

297. McKee, M. L. *J. Phys. Chem.* **1986,** *90,* 4908–4910; *see also* Koch, W.; Liu, B.; DeFrees, D. J. *J. Am. Chem. Soc.* **1988,** *110,* 7325–7328; and Saunders, M.; Laidig, K. E.; Wiberg, K. B.; Schleyer, P. v. R. *J. Am. Chem. Soc.* **1988,** *110,* 7652–7659.

298. Yannoni, C. S.; Macho, V.; Myhre, P. C. *J. Am. Chem. Soc.* **1982,** *104,* 907–908.

299. Roberts, J. P.; Roberts, J. D.; Skinner, C.; Shires, G. T., III; Illner, H.; Cannizaro, P. C.; Shires, G. T. *Ann. Surg.* **1985,** *102,* 1–8.

300. Wells, H. G. movie script for *Things to Come,* London Films, 1936.

Index

D

T

U

V

W

Production: Paula M. Befard
Indexing: Colleen P. Stamm
Acquisition: Robin Giroux

Printed and bound by Maple Press, York, PA

Paper meets minimum requirements of American National Standard for Information Sciences—Permanence of Paper for Printed Library Materials, ANSI Z39.48–1984 ♾

Profiles, Pathways, and Dreams

Sir Derek H. R. Barton *Some Recollections of Gap Jumping*

Arthur J. Birch *To See the Obvious*

Melvin Calvin *Following the Trail of Light: A Scientific Odyssey*

Donald J. Cram *From Design to Discovery*

Michael J. S. Dewar *A Semiempirical Life*

Carl Djerassi *Organic Chemistry: A View Through Steroid Glasses*

Ernest L. Eliel *From Cologne to Chapel Hill*

Egbert Havinga *Enjoying Organic Chemisty 1927–1987*

Rolf Huisgen *Mechanisms, Novel Reactions, Synthetic Principles*

William S. Johnson *A Fifty-Year Love Affair with Organic Chemistry*

Raymond U. Lemieux *Explorations with Sugars: How Sweet It Was*

Herman Mark *From Small Organic Molecules to Large: A Century of Progress*

Bruce Merrifield *The Concept and Development of Solid-Phase Peptide Synthesis*

Teruaki Mukaiyama *To Catch the Interesting While Running*

Koji Nakanishi *A Wandering Natural Products Chemist*

Tetsuo Nozoe *Seventy Years in Organic Chemistry*

Vladimir Prelog *My 128 Semesters of Studies of Chemistry*

John D. Roberts *The Right Place at the Right Time*

Paul von Rague Schleyer *From the Ivy League into the Honey Pot*

F. G. A. Stone *Organometallic Chemistry*

Andrew Streitwieser, Jr. *A Lifetime of Synergy with Theory and Experiment*

Cheves Walling *Fifty Years of Free Radicals*

$24.95